VANISHING ICE

VIVIEN GORNITZ

VANISHING ICE

Glaciers, Ice Sheets, and Rising Seas

COLUMBIA UNIVERSITY PRESS
NEW YORK

Columbia University Press gratefully acknowledges the generous support for this book provided by Publisher's Circle member David O. Beim.

Columbia University Press
Publishers Since 1893
New York Chichester, West Sussex
cup.columbia.edu

Library of Congress Cataloging-in-Publication Data
Names: Gornitz, Vivien, author.
Title: Vanishing ice : glaciers, ice sheets, and rising seas / Vivien Gornitz.
Description: New York : Columbia University Press, [2019] |
Includes bibliographical references and index.
Identifiers: LCCN 2018041938 | ISBN 9780231168243 (cloth : alk. paper) |
ISBN 9780231548892 (ebk.)
Subjects: LCSH: Climatic changes—Arctic regions. | Glaciers—Climatic factors—Arctic regions. | Ice sheets—Arctic regions. | Sea level—Climatic factors—Arctic regions. | Cryosphere—History.
Classification: LCC QC903.2.A68 G67 2019 | DDC 551.6911/3—dc23
LC record available at https://lccn.loc.gov/2018041938

∞

Columbia University Press books are printed on permanent and durable acid-free paper.

Printed in the United States of America

Cover design: Milenda Nan Ok Lee
Cover photo: Simon Lane / © Alamy

CONTENTS

PREFACE

Why should we care about vanishing ice and snow? *Vanishing Ice: Glaciers, Ice Sheets, and Rising Seas* vividly describes how the atmospheric accumulation of greenhouse gases generated by human activities has already begun to affect the Earth's climate, most visibly near the poles and on high mountains. Arctic temperatures rise twice as fast as in the rest of the world. A once-perilous journey through the fearsome, ice-laden Northwest Passage has lately turned into a leisurely summer cruise through open waters for a growing number of large commercial and tour ships. The giant luxury cruise ship *Crystal Serenity*, carrying over 1,000 passengers with a crew of nearly 600, repeated its record-breaking 2016 voyage through the Northwest Passage the following summer, in August 2017. A sharp decline in Arctic summer sea ice since the 1980s is just one of the most conspicuous manifestations of the changing cryosphere—the world of ice—as our planet warms. Some scientists foresee nearly ice-free summers in the Arctic Ocean within the next several decades! On top of the Greenland Ice Sheet, warmer summers expand the area that thaws, even if only briefly, as occurred in mid-July of 2012, when nearly the entire surface of the Greenland Ice Sheet melted. In spite of the rarity of such widespread events (the previous such incident occurred in 1889), the extent and duration of summer melting has grown significantly in recent years.

Arctic peoples notice changes in polar bear behavior and reindeer foraging habits. The stronger waves of a prolonged ice-free Arctic Ocean erode the shoreline, endangering scores of Native Alaskan fishing villages. The environmental changes impinge on the ability of indigenous Arctic populations to maintain traditional lifestyles. Sheila Watt-Cloutier, an Inuit spokeswoman, has emphasized how important snow and ice are to their traditional hunting culture. She declared: "We are in essence fighting for our right to be cold."

The effects of deicing the permafrost sweep across the Arctic. Trees tilt at crazy angles like drunken sailors, coastal cliffs crumble, wave-battered shorelines wash away, highways buckle, buildings sag and collapse, and lakes in the tundra appear and disappear as the changing climate transforms the regional ecology, hydrology, and landscape. Thawing permafrost and ground instability also affect Arctic construction, transportation, tourism, and development. However, a mixed picture emerges. Shipping will benefit from increasingly ice-free waters and considerably shortened transportation routes, cutting costs substantially. Development of potentially vast Arctic oil and mineral resources will open up many opportunities. At the same time, development of as-yet-untapped mineral wealth exposes fragile tundra and wildlife to new stresses from which recovery may be difficult.

Ice tongues and shelves that ring much of Antarctica, as well as many high Arctic islands, and that hold ice sheets and maritime glaciers in check begin to thin and weaken. At least ten large, fringing ice shelves on the Antarctic Peninsula have crumbled one by one, like a pile of dominoes, since the 1990s. Tributary glaciers feeding the ice shelves have then accelerated seaward. On high mountain slopes around the world, most glaciers are also in retreat. Darkening peaks not only disappoint Alpine tourists, but also lead to more unreliable water supplies. Swollen spring river flows generate more flash floods. Also, as glaciers recede and alpine permafrost thaws, the high mountains become unglued, letting loose more rockfalls and avalanches.

The visible changes may be just the tip of the proverbial iceberg. As permafrost thaws, will greenhouse gases like methane or carbon dioxide bubbling up from soggy, unfrozen Arctic soils and the seabed reinforce the ongoing warming trend? Which of these two greenhouse gases matters more to the permafrost climate feedback? Will Arctic vegetation grow wetter or drier? These questions grip scientists in a heated debate, as described in a later chapter.

The ice-albedo feedback amplifies the unprecedented late twentieth–early twenty-first-century rise in Arctic temperatures and speeds up ice losses. Less summer sea ice means more open ocean water. An open, nearly ice-free broad expanse of open water reflects less sunlight, and instead absorbs more of the sun's heat. As a result, more sea ice melts, which in turn allows additional escape of heat and moisture into the atmosphere. The associated feedbacks and their consequences make the Arctic a major crucible of climate change. But their reverberations echo far beyond the deep north, or other currently frozen realms.

Changing Arctic landscapes and endangered lifestyles are not the only reasons why the vanishing cryosphere matters. What does the thawing cryosphere portend beyond polar regions and high mountain peaks? As succinctly stated in 2012 by former National Oceanic and Atmospheric Administration official Jane Lubchenko, "What happens in the Arctic doesn't stay in the Arctic. What happens there often affects people around the world. Melting ice and glaciers in Greenland can contribute quite significantly to sea level rise, which affects people in coastal areas around the world." Sea level rise represents the other side of the same coin of thawing ice sheets and glaciers. These two sides of the coin have been closely interwoven throughout most of geologic history, as initially outlined in *Rising Seas: Past, Present, Future*, and explored at greater length in *Vanishing Ice: Glaciers, Ice Sheets, and Rising Seas*. This close relationship continues today and will persist into the future.

The worldwide menace of sea level rise looms over the tens of millions of people who live along the world's shorelines and who face an ever-growing vulnerability to coastal storm floods, made worse by rising sea levels. Ice sheet losses, along with shrinking glaciers, already account for over half the observed rising sea level trend of roughly 3 millimeters (0.1 in) per year since 1993, and this share is likely to grow in the future. The vast stores of ice in Greenland and Antarctica would raise sea level by 66 meters (217 feet) if fully melted. The two still-frozen giants hold momentous keys to our future well-being. While a large-scale destabilization of the polar ice sheets appears quite unlikely within the next century or two, such a possibility cannot be ruled out entirely, nor can even greater degrees of melting be discounted still further into the future. Thus it becomes critical to gain a better understanding of the causes and effects of the receding world of ice—a major theme of this volume. *Vanishing Ice* describes the current state of the cryosphere, and offers plausible scenarios for how much and how fast ice may melt in the future, by what means, and with what consequences.

The heavy rains, destructive winds, and high storm surges of three intense hurricanes in close succession that flooded much of southern Texas, devastated Caribbean islands, swept across the Florida Keys, and left most of Florida without power for days during August and September of 2017 starkly illustrate the present dangers faced by low-lying coastal areas. The vulnerability of the world's shorelines to coastal storm floods will grow with rising sea levels. Many of the world's most populous cities line the coast and face serious inundation risks: New York City, Los Angeles, London, Tokyo, Shanghai, Hong Kong, Mumbai (Bombay), Bangkok, and Miami, to name just a few. Some small island states, such as Tuvalu, Kiribati, and the Maldives, are already planning for eventual evacuation and relocation.

The planet's warm-up may bring us to a new climate regime that could ultimately endanger the cryosphere. We are inching toward a climate regime like that of past warm geological periods, such as that of the Last Interglacial, ~125,000 years ago, a time when sea level stood 6 meters (20 feet) or more higher, or even like the early to mid-Pliocene, 3.5–3 million years ago, which had atmospheric carbon dioxide levels similar to today's (i.e., ~400 parts per million) but ~20 meters (66 feet) higher sea level. Climate experts warn against a temperature rise of more than 2°C (3.6°F), calling it a "dangerous anthropogenic interference with the climate system" (UNFCCC, 2010). But the long atmospheric lifetime of today's increasing carbon dioxide emissions virtually ensures a warmer planet, a smaller cryosphere, and higher oceans.

Vanishing Ice looks beyond sensational media coverage or dry technical literature to depict the dramatic changes rapidly transforming the cryosphere by illustrating the mounting evidence of the expanding meltdown. The facts, based on the latest scientific observations, build up a convincing case. Going beyond the findings of an earlier book (*Rising Seas: Past, Present, Future*), this volume meticulously probes the closely interconnected roles of mounting temperatures, a shrinking cryosphere, and higher oceans. This book aims not only to raise public awareness of the meteoric changes already transforming the cryosphere, but also to forcefully substantiate its vital role in planetary processes and in our future well-being. Thus the book highlights the widespread repercussions of ice loss, which will affect countless people far removed from frozen regions. Although *Vanishing Ice* avoids thorny policy issues or mitigation recommendations that are amply covered elsewhere, its straightforward, well-documented presentation opens the eyes of the reader to the serious environmental challenges ahead and enables him or her to make well-informed choices that will help avert disaster. In addition to raising awareness, the book emphasizes that losing ice has global-scale consequences and therefore the big meltdown truly matters to all!

A unique feature of the book is the long-term perspective gained by closely examining changes in the cryosphere and corresponding variations in sea level over the past few million years. The history of expansion and contraction of the Greenland and Antarctic ice sheets holds valuable clues to the future stability of these vast, icy storehouses. Added to geophysical and geochemical records of past climates are new observations and modeling studies probing the dynamic behavior of ice sheets, which will shed light on the critical questions of how much and how soon they will melt and what this means for future generations.

ACKNOWLEDGMENTS

Vanishing Ice: Glaciers, Ice Sheets, and Rising Seas represents the culmination of a long journey into research on global sea level change that initially began years ago at the suggestion of James Hansen, former director of the NASA Goddard Institute for Space Studies (GISS) in New York City. Subsequently, it became clear that the fate of the ice sheets remains a major uncertainty in predicting the future course of sea level rise. This has led to a growing interest in what was happening to the cryosphere, not only the two major ice sheets of Greenland and Antarctica, but also the valley glaciers, permafrost, and floating ice, in particular sea ice and fringing ice shelves, and what the consequences of a meltdown may be for the rest of the world.

I wish to thank Patrick Fitzgerald of Columbia University Press for encouraging me to write this book and for his many beneficial comments along the way. Thanks are also extended to Ryan Groendyk and Brian Smith, assistant editors at Columbia University Press, for their expert editorial support. Appreciation is also expressed to Gavin Schmidt, director of NASA GISS; and Cynthia Rosenzweig of NASA GISS and the Center for Climate Systems Research at Columbia University, for their continued support and patient understanding during this lengthy undertaking. I also thank the reviewers for their constructive comments and suggestions that have contributed to the improvement of this book.

Numerous colleagues have also contributed many perceptive insights and advice. In particular, my sincere thanks are extended to Joerg Schaefer of the Lamont-Doherty Earth Observatory (LDEO) at Columbia University for clarifying the history of Quaternary ice sheet changes in

Greenland; Robin Bell of LDEO for insightful discussion on ice sheet dynamics; to Marco Tedesco of LDEO for proving images of the Greenland Ice Sheet, to Mark Chandler of GISS for helpful conversations on mid-Pliocene climates and sea levels; and Allegra LeGrande of GISS and Anders Carlson of the University of Wisconsin for information on late-Quaternary ice sheet retreat and sea level change. The book has also greatly benefited from stimulating discussions with Michael Oppenheimer of Princeton University and Robert Kopp of Rutgers University on extreme regional sea level–rise scenarios arising from recent models of future hydrofracturing and cliff failure mechanisms in Antarctica. Discussions and papers presented at cryosphere-focused sessions at the American Geophysical Union and Geological Society of America annual meetings, 2012 through 2015, also provided useful information during the early stages of writing.

VANISHING ICE

1

WHITHER THE SNOWS OF YESTERYEAR?

THE WORLD OF ICE

I seemed to vow to myself that some day I would go to the region of ice and snow and go on and on till I came to one of the poles of the earth, the end of the axis upon which this great round ball turns.

—Ernest Shackleton, polar explorer

A space voyager heading toward Earth first notices the deep blues streaked with evanescent, white, swirling clouds. Approaching Earth, the space traveler gradually begins to separate large brown, russet, and dark-green continents from deep-blue oceans. Closer yet, the astronaut discerns differences among ever-changing white clouds, snow-draped mountaintops, and ice sheets. Nearing the polar surface, the traveler now clearly differentiates floating ice floes from icebergs, ice shelves, and the vast ice sheets.

We call our planet Earth, yet water covers three-quarters of its surface. Water is unique in its ability to exist as three distinct phases—vapor, liquid, and solid—at the Earth's surface. The cryosphere—parts of our planet where water exists frozen as ice—occupies a vast territory that extends from the frigid poles to ice-laden polar seas, lofty mountain peaks, and frozen tundra (fig. 1.1). This frozen realm also encompasses seasonal winter snow, lake and river ice, floating ice (sea ice, icebergs,

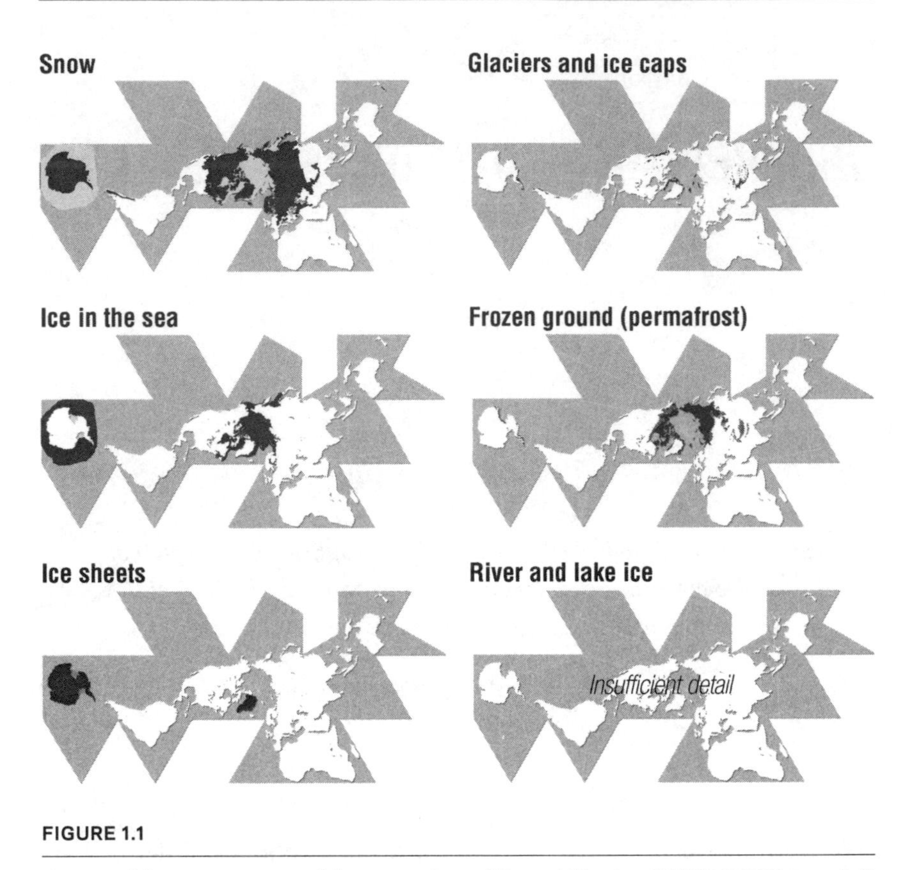

FIGURE 1.1

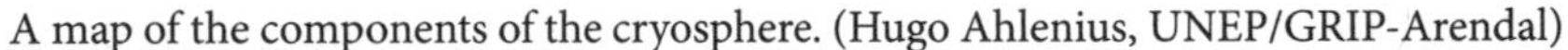

A map of the components of the cryosphere. (Hugo Ahlenius, UNEP/GRIP-Arendal)

and ice shelves), land ice (glaciers, ice sheets, and ice caps), and permafrost (table 1.1). We now meet the individual members of the cryosphere. Succeeding chapters will discuss them in greater depth.

Sea ice, formed by freezing seawater, differs from icebergs—chunks of ice that have broken off from tidewater glaciers that reach the sea, or from ice shelves. Ice shelves are extensions of glaciers or ice sheets that float on a lake or ocean. Glaciers are ice streams that flow downhill under the pull of gravity. Continental-scale ice masses form ice sheets, whereas smaller domes of ice that cover less than 50,000 square kilometers (19,300 square miles) are ice caps. Permafrost, or perennially frozen ground, is soil, sediment, or rock that stays below 0°C (32°F) for two or more years.

TABLE 1.1 Major Components of the Cryosphere

CRYOSPHERIC CONSTITUENT	LAND AREA (10^6 km^2)	ICE VOLUME (10^6 km^3)	SEA LEVEL RISE EQUIVALENT (m)*
Greenland Ice Sheet	1.80	2.85	7.36
Antarctica	12.3	24.7	58.3
Glaciers and ice caps	0.74	0.24	0.41
Permafrost	13.2–18	0.011–0.037	0.02–0.10
Ice shelves (Antarctica)	1.62	0.66	0
Sea ice	6–15 (Arctic), 3–18 (Antarctica)	0.002–0.05	0
Seasonal snow cover (Northern Hemisphere only)	1.3%–30.6%	—	0.001–0.01
Total ice volume		*28.5*	

*Sea level equivalent (in millimeters per year) is the rate of globally averaged sea level rise resulting from a total melting of the indicated ice component. Conversion of ice mass to sea level rise: 362.5 Gt = 1 millimeter per year of sea level rise (SLR) equivalent, where 1 Gt = 1 billion metric tons, 1 trillion kilograms, or 1 cubic kilometer of freshwater (or 1.1 cubic kilometer of ice).

Note: Excluding seasonally frozen ground and Northern Hemisphere winter snow cover.

Source: IPCC (2013a), table 4.1; ice volume from Barry and Gan 2011, table 1.1.

Antarctica houses the largest store of ice by far—a vast continental ice sheet comprising 87 percent of the total by volume—followed by Greenland (10 percent) and ice shelves (2.3 percent), with smaller volumes encased in sea ice, permafrost, and mountain glaciers (table 1.1). However, permafrost and sea ice cover the vastest area, followed by the Antarctic and Greenland ice sheets, and then by ice shelves. Permafrost underlies nearly a quarter of the Northern Hemisphere land surface, while sea ice covers just 2 to 5 percent of the world's oceans. Glaciers and small ice caps, by comparison, occupy a relatively minuscule area.[1]

Some components of the cryosphere grow or shrink dramatically with the seasons (e.g., snow, sea ice), whereas others (e.g., ice sheets, glaciers) change much more slowly. In the Northern Hemisphere, for example, snow blankets ~46 million square kilometers (18 million square miles) in January, shrinking down to 3.8–3.9 million square kilometers (1.4–1.5 square miles) in August.[2] Southern Hemisphere sea ice fluctuates sixfold from winter to summer, in contrast to Northern Hemisphere sea ice, which varies by a factor of 2.5. Glaciers may take decades to advance or retreat. However, glaciers worldwide have generally been retreating since the 1850s, as chapter 4 vividly describes.

FROM SNOW TO ICE

Oh, where are the snows of yesteryear?

—François Villon,
"Ballade des dames du temps jadis" (Ballad of the Ladies of Times Past, 1489)

Snow forms when water vapor condenses and solidifies below 0°C (32°F) onto minute particles of mineral dust, organic matter, soot, or artificial particles (dry ice, silver iodide). The arrangement of atoms within the ice crystal establishes the typical hexagonal snowflake shape. In ice, the bonding between hydrogen and oxygen atoms from individual water molecules produces an array not unlike a three-dimensional chicken coop fence[3] (fig. 1.2a). Cooling water attains its greatest density (1 gram per square centimeter [g/cm^3]) at 4°C. Because the hydrogen bonding creates an open structure, ice, unlike most solids, becomes less dense at 0°C (0.917 g/cm^3) than water (0.999 g/cm^3) as it freezes. Hence, it floats on water.

The growth of snowflakes is extraordinarily sensitive to minor variations in external conditions. A wide variety of crystal shapes appear, ranging from flat hexagonal plates to solid and hollow hexagonal prisms, to six-sided dendritic (branched) plates in endless combinations (fig. 1.2b). Hollow, stepped, or branched forms result from faster growth along crystal edges and side branches, under supersaturated conditions. The growing snowflakes, buffeted by gusts of wind, encounter random fluctuations in air temperature and degrees of water vapor saturation. However, the crystals, in constant motion, experience similar conditions in all directions from moment to moment. Thus they maintain the hexagonal symmetry dictated by their atomic structure. A host of diverse and complex

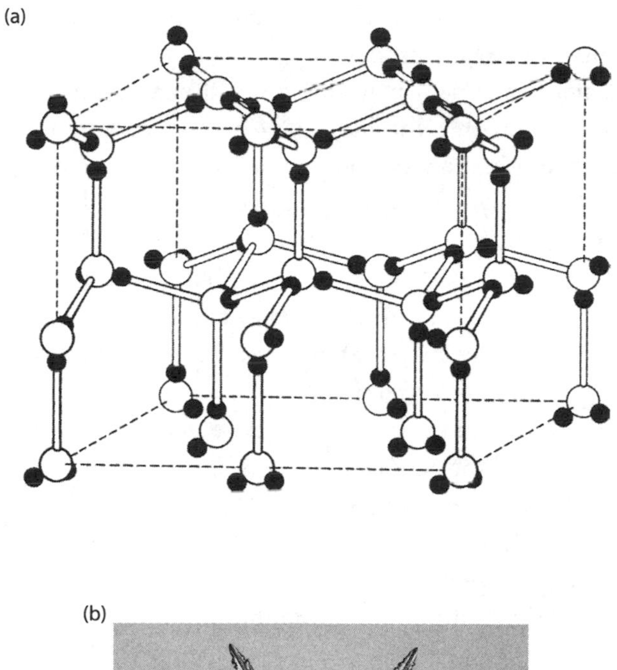

FIGURE 1.2

(a) Crystal structure of ice; (b) snowflake demonstrating hexagonal symmetry. (Kenneth G. Libbrecht and California Institute of Technology, SnowCrystals.com, http://www.its.caltech.edu/~atomic/snowcrystals/primer/primer.htm)

forms arise under these rapidly changing environmental conditions. Thus, no two snowflakes are identical.

Individual snow crystals may clump together and grow large enough for gravity to overcome the buoyancy of air. Or they may partially melt, as sleet. Fresh snow is very porous. As more snow accumulates over time, older snow compresses and recrystallizes. The crystal grains grow larger and rounder. Partially consolidated snow that has survived a summer's melt season is called firn—the first step in snow's transition to ice. Depending on local snowfall and average yearly temperature, firn transforms into pure ice from within several years in temperate mountain glaciers to millennia in Antarctica (chap. 4).

Fresh snow is the brightest natural surface, reflecting 80 to 90 percent of the sunlight striking it. As snow ages, its reflectivity—or albedo—decreases and grain size increases. Minute impurities, such as dust or black carbon (soot), or incipient melting also lower the albedo. Reduced brightness sets up a feedback known as the albedo-snow feedback, which stimulates further melting (see Hastening the Meltdown, later in this chapter).

Water resists changing its state because of the small size of water molecules and strong hydrogen bonds. Water has a high latent heat of fusion, needing 80 calories to melt a gram of ice and even more energy, 677 calories per gram, to go from ice to water vapor, a process known as sublimation. This energy is released when water vapor condenses to ice, or when water freezes. These high energy demands limit the amount of melting or sublimation possible at high elevations, where there is low relative humidity, or in very cold climates, such as in the East Antarctica Plateau.

Because of its high porosity, fresh snow acts as a good insulator, sealing heat beneath sea ice and soil surface. A thick layer of snow preserves underlying sea ice and permafrost through the summer melt season. The thickness and spatial extent of the snow cover also influence the growth and distribution of vegetation.

Snow stores water seasonally, releasing it in spring and early summer. The peak spring stream runoff meets important agricultural, urban, and industrial demands. Therefore, changes in the timing and amount of the maximum spring melt may strongly impact downstream water uses in summer or fall.

VOYAGES TO THE ENDS OF THE EARTH

Despite the danger and hardship of polar exploration, people keep on rising to this extreme challenge. Some who make it are rewarded with a place in history, but many more failed and are forgotten.

—Robert Peary, Arctic explorer

The two poles differ in important ways. The North Pole sits in the Arctic Ocean, surrounded by land, whereas the South Pole lies in Antarctica, a frigid continent entirely encircled by water. Greenland extends to lower latitudes than Antarctica. Its highest point, Gunnbjørnfjeld, is 3,694 meters (12,120 feet) above sea level, while the Vinson Massif in Antarctica soars to an altitude of 4,897 meters (16,066 feet). Antarctica is the continent with the highest average elevation. Its high topography and isolation from warmth by the strong westerly winds and Southern Ocean make it the coldest continent as well.

Owing to their remoteness and extreme climate conditions, the high latitudes were among the last portions of the cryosphere to be scientifically explored and studied. Propelled onward by a mix of curiosity, dreams of riches, and desire for glory, the voyages of courageous early explorers opened the way for subsequent scientific, commercial, and military operations that followed in the Arctic and Antarctica. Over the centuries, these intrepid adventurers and explorers gradually filled in the remaining blank areas on maps. This section briefly highlights some of these pioneering journeys to the ends of the Earth.

The Arctic was by no means uninhabited. Successive waves of paleo-Eskimos[4] had migrated from eastern Siberia into Alaska, northern Canada, and Greenland between four and seven thousand years ago, developing a culture well adapted to the harsh climate and the harvesting of abundant seal, walrus, polar bears, whales, and fish. Norse seafarers had settled Iceland by 874 CE and established colonies on Greenland by the 980s, lured there by Erik the Red's glowing descriptions of a "green land." (In a clever example of good salesmanship, Erik the Red reportedly "named the land Greenland, saying that people would be eager to go there if it had a good name," according to the Norse *Saga of Erik the Red.*) Erik's son, Leif Erikson (c. 970–c. 1020), after having been blown off course on his way to Greenland around 1000 CE, encountered lands farther west—rocky, desolate *Helluland* (likely Baffin Island), forested *Markland* (Labrador), and *Vinland* or Wineland, now identified as Newfoundland, Canada.[5]

Search for the Northwest Passage

The lure of a shortcut to the fabled wealth of the Orient—tea, spices, silks, gems—spurred explorers to seek a sea route either via the northwest, across the top of North America, or the northeast, across the top of Eurasia. Roald Amundsen (1872–1928) was the first to successfully navigate the entire Northwest Passage in 1903–1906, in the *Gjoa*. However, his expedition had been preceded by numerous previous attempts, some ending tragically.[6] John Davis sailed and surveyed Davis Strait and Cumberland Sound near Baffin Island, west of Greenland, between 1585 and 1587. Hoping to cut across the continent, Jacques Cartier journeyed up the St. Lawrence River as far as present-day Montreal in 1535. With a similar goal in mind, Henry Hudson sailed north up the river bearing his name as far as present-day Albany, New York, in 1609. On a subsequent trip to the Arctic and Hudson Bay, a mutinous crew set him adrift in James Bay, leaving him and his companions to die. By the nineteenth century, several segments of the Northwest Passage had already been explored, but the route often proved treacherous. In 1845, Sir John Franklin's attempt to cross the remaining unexplored reaches of the passage ended in disaster when his two ships became icebound and all men perished. Several unsuccessful expeditions to find Franklin and his men followed. Sir Robert McClure (1807–1873), however, was the first non-Native to traverse the Northwest Passage from west to east by boat and sled in 1854, until Amundsen's historic journey.

The mystery of the ill-fated Franklin expedition has now been solved with the discovery of the first of the two missing ships, the HMS *Erebus*, on the ocean floor off King William Island in the Canadian Arctic, on September 7, 2014.[7] A team led by Parks Canada, jointly with the Canadian Coast Guard and several other organizations, used an advanced, remotely operated underwater vehicle equipped with sonar and high-resolution cameras to capture details of the ship's planking and construction (fig. 1.3). The location of the shipwreck corroborated earlier Inuit claims of sighting

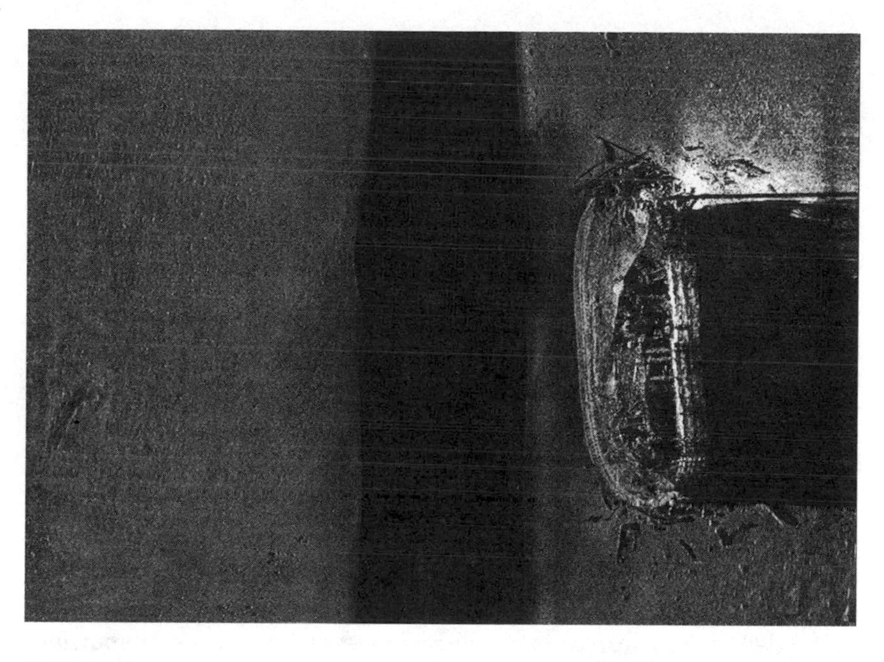

FIGURE 1.3

A sonar image of HMS *Erebus*, one of the ships from the Franklin expedition, resting on the ocean floor. (Parks Canada)

two ships in the late 1840s, one of which sank in deep water west of King William Sound and the other perhaps reaching as far south as Queen Maud Gulf or Wilmot and Crampton Bay. The second ship, the HMS *Terror*, was found in eastern Queen Maud Gulf, nearly two years after the first, on September 3, 2016.[8] Amazingly, the wreck is so well preserved that glass panes still remain in three of four windows of the captain's stern cabin. The recent discovery of the *Terror* was aided by crewman Sammy Kogvik, an Inuit from nearby Gjoa Haven, who, while out fishing with a friend, had noticed a large piece of wood shaped like a ship's mast sticking out of the sea ice on Terror Bay.

Race to the North Pole

Fridtjof Nansen (1861–1930), a Norwegian explorer and adventurer who successfully traversed Greenland over land in 1888–1889, planned to reach the North Pole by means of the natural drift of sea ice. Nansen, Otto Sverdrup, Hjalmar Johansen, and several others set sail in 1893 in the sturdy *Fram* from Bergen, Norway. Growing impatient with the slow drift of the pack ice, he left the ship with Johansen and continued northward by dogsled. They reached as far north as 86°14′N before turning back. Meanwhile, the *Fram* slowly drifted westward, eventually arriving in the North Atlantic by 1896. Conflicting claims by Robert E. Peary (1856–1920) of reaching the North Pole on April 6, 1909, and Dr. Frederick A. Cook (1865–1940) in April 1908, have been subsequently questioned.[9] Richard E. Byrd's claim of being first to fly over the North Pole in 1926 has also been challenged, although that of the Amundsen-Ellsworth-Nobile expedition in 1926 remains undisputed.[10] The Russian SP-1 expedition made the first airplane landing on ice near the North Pole in 1937, while the USS *Nautilus* nuclear submarine was the first to sail beneath the ice pack under the pole in 1958. Cutting through the polar ice, the Russian nuclear-powered icebreaker *Arktika* followed in 1977. Thirty years later, a crew of three descended to the ocean floor beneath the North Pole in the submersible *Arktika 2007*, planting the Russian flag.

The Northeast Passage

Navigators also sought a shorter sea route to the East by heading across the top of Eurasia. The Dutch explorer Willem Barentsz (c. 1550–1597), searching for the Northern Sea Route (or Northeast Passage), discovered Spitsbergen, in the Svalbard archipelago, in 1596, and sailed as far as Novaya Zemlya, an island in the Arctic Ocean northeast of Archangel, Russia. Vitus Bering (1681–1741), a Danish-Russian captain under Peter the Great, reached the Kamchatka Peninsula, in the Russian Far East, in 1728. On a subsequent expedition, he arrived in Alaska in 1741, discovering Kodiak Island and sighting Mt. Saint Elias and part of the southern Alaskan coast. His deputy, Lieutenant Aleksei Chirikof, spotted western Prince of Wales Island, the Kenai Peninsula, and some of the western Aleutian Islands. Years later, another prominent Arctic explorer, Adolf Erik Nordenskiöld (1832–1901), a Swedish-Finnish geologist-mineralogist and naturalist, was the first to successfully navigate the entire Northeast Passage in the *Vega*, as far east as the Bering Strait between Russian and Alaska, in 1878–1879, returning to the south and west around Eurasia.

Terra Australis Incognita: The Elusive Southern Continent

> For scientific leadership, give me Scott; for swift and efficient travel, Amundsen; but when you are in a hopeless situation, when there seems no way out, get on your knees and pray for Shackleton.
>
> —Sir Raymond Edward Priestley, *Antarctic Adventure: Scott's Northern Party*

Stemming from the Greek love of symmetry, Aristotle inferred the existence of a southern landmass to balance that of the Northern Hemisphere—a theme expanded on by Ptolemy and, much later, by some Renaissance mapmakers.[11] Yet the Great South Land remained elusive in spite of extensive European exploration of the Southern Hemisphere during the seventeenth and eighteenth centuries. James Cook (1728–1779) came as far south as 71°10′S on January 30, 1774, but Fabian Gottlieb von Bellingshausen (1778–1852) was first to sight the Antarctic mainland in January, 1820.[12] Sealers and whalers, following in their wake, were quick to exploit the rich waters of the Southern Ocean.

As the twentieth century dawned, Antarctica reluctantly began to yield its secrets. Robert Falcon Scott (1868–1912) led one of the first scientific expeditions to the southern continent, the British National Antarctic Discovery Expedition (1901–1904). Scott explored the area around Ross Island and McMurdo Sound, collecting geological and zoological specimens, and reached 82°17′S. At the same time, the Swedish Antarctic expedition (1901–1904)—led by geologist and polar explorer Nils Otto Gustaf Nordenskjöld (1869–1928, and nephew of Adolf Erik Nordenskjöld) and Carl Anton Larsen (1860–1924)—surveyed the northern tip of the Antarctic Peninsula, also gathering scientific data. Several other British, French, German, and Japanese expeditions closely followed on their heels. By 1910, the South Pole awaited!

Robert F. Scott, now leading the British Antarctic Expedition (Terra Nova Expedition), set sail for Antarctica and the South Pole on June 15, 1910, in the *Terra Nova*, arriving at Ross Island on January 4, 1911, where they built a shelter at Cape Evans. After many months of preparations, Scott and a small team began their arduous trek to the South Pole in late October, 1911. Scott and four men made the final push to the South Pole on January 17, 1912, only to discover that Roald Amundsen had preceded him by about a month. With a harsh wind blowing and a temperature of −20°F (−29°C), Scott declared, "Great God! This is an awful place, and terrible enough for us to have laboured to it without the reward of priority."[13] Even worse, all five men, including Scott, perished on their return trip due to illness, insufficient rations, accidents, and bad weather.

Amundsen had abruptly changed his plans to be first to the North Pole upon learning of the conflicting claims of Peary and Cook.[14] Wanting to beat Scott to the South Pole, in 1910 he secretly headed south instead in the *Fram*, and built a base camp in the Bay of Whales on the Ross Ice Shelf of Antarctica in January, 1911. Preparing for the polar journey during the long Antarctic winter, Amundsen and four others set out across the "Great Ice Barrier" (the Ross Ice Shelf) by dog sledges and skis in October, 1911. They headed toward Axel Heilberg Glacier and the high Antarctic Plateau, reaching the South Pole on December 14, 1911. There, Amundsen planted the Norwegian flag, pitched a tent, and left a letter for Scott to deliver to King Haakon VII of Norway. After they made a rapid descent back to the Bay of Whales, the *Fram* left for Hobart, Tasmania, arriving on March 7, 1912, whereupon Amundsen immediately sent out telegrams announcing his team's achievement to the world. His success, in contrast to Scott's effort, has been attributed to careful planning, skill in handling sledge dogs, proficiency on skis, use of Inuit-style skin clothing, and a clear focus on his goal.

The last major expedition of the "Heroic Age of Antarctic Exploration" was that of Ernest Shackleton (1874–1922) in the *Endurance* in 1914–1917. Its aim was to traverse Antarctica over land "from sea to sea." After arriving in Antarctica in December of 1914, the ship soon became trapped in the ice pack. After drifting for several months, the ship was crushed by ice. The crew removed supplies and equipment to an ice floe along with the ship's three lifeboats. When the ice began to break up, the men boarded their lifeboats and sailed to Elephant Island in the South Shetland Islands, where they made camp. On April 24, 1916, Shackleton and five others sailed 800 nautical miles (1,290 kilometers) through stormy seas to the island of South Georgia (at that time, a whaling station in the

South Atlantic) to obtain help. Arriving in early May of 1916, Shackleton organized a rescue party and was able to rescue the entire crew. Although failing to accomplish its original objectives, the expedition nevertheless remains a remarkable feat of endurance and survival.[15]

SCIENCE AT THE POLES

Early expeditions to the poles collected exotic specimens and sought to exploit valuable minerals and other natural resources for economic gain and national prestige. During the late nineteenth and twentieth centuries, science and technology in the Arctic were often used to further military and strategic objectives in these formerly remote and largely inaccessible lands. However, initial steps toward international scientific cooperation had already begun in the nineteenth century. The Austrian scientist, explorer, and naval officer Karl Weyprecht proposed the program, which later became the first International Polar Year of 1882–1883.[16] Twelve countries, including the United States, participated, focusing on meteorology, terrestrial magnetism, the aurora borealis, and the cryosphere at both poles. Fifty years later, the second International Polar Year was launched in 1932–1933, during which 26 countries participated. Here too, the focus was on meteorology, magnetism, and atmospheric science, including study of the newly discovered "jet stream."

The International Geophysical Year (IGY; 1957–1958), with 67 countries and over 4,000 scientists participating, made major advances in atmospheric science, Antarctic meteorology and topography, magnetism, and the ionosphere. The Van Allen radiation belts were discovered using instruments onboard the U.S. satellite *Explorer 1.* Measurements of the thickness of the Antarctic ice sheet led to the discovery of the Gamburtsev Mountains, an Alps-sized mountain range buried under more than 600 meters (2,000 feet) of ice and snow in East Antarctica. The IGY also led to the establishment of permanent bases on Antarctica and the creation of the Antarctic Treaty in 1959, which was ratified in 1961. The treaty stipulates, among other things, that science in Antarctica is for peaceful and noncommercial uses only, weapons development or testing is prohibited, data should be shared openly among researchers, mineral extraction is banned, and the terrestrial ecosystem is protected.

The growing awareness of the magnified effects of global warming at high latitudes led to the establishment of the International Polar Year (IPY) of 2007–2009 on the 125th anniversary of the original IPY. More than 50,000 polar researchers, educators, students, and Arctic residents from 60 countries participated, spanning a broad range of disciplines. Julie Brigham-Grette of the University of Massachusetts at Amherst, cochair of the National Reseach Council (NRC; 2012) report,[17] noted that "we really confirmed on many levels . . . that the polar regions are changing faster than anyone predicted." Outreach programs at museums, science centers, and schools and on the internet illustrated how "what happens at the poles affects us all." IPY 2007–2009 netted the first high-resolution radar images of the ice-buried Gamburtsev Mountains, initially discovered during IGY 1957–1958. Researchers gained deeper insights into the connections between a shrinking sea ice cover and enhanced global warm-up and why some ice sheets lose mass while elsewhere snowfall increases. The findings raised more questions. Are the record low summer Arctic sea ice extents of 2007, 2012, and 2016 signs of a "tipping point"? Are we approaching an "end of February Arctic sea ice age" in which hardly any multiyear sea ice remains in the Arctic? Why, in contrast, is the summer and autumn sea ice in the Antarctic expanding? In Antarctica, sea ice distribution may be linked to the ozone hole, changes in air pressure, upwelling of deep-ocean water, and possibly even variable climate patterns, such as the El Niño–Southern Oscillation (ENSO). A 3.5 million-year history of the West Antarctic Ice Sheet (WAIS) produced evidence of repeated collapses during past warm periods. Does this imply a potentially unstable WAIS in our future? The IPY 2007–2009 leaves an unfinished legacy and work continues apace to fill knowledge gaps and improve our understanding of a critical part of the global system.

THE CHANGING CRYOSPHERE

We build statues of snow, and weep to see them melt.

—Sir Walter Scott (1771–1832)

The world's average temperature has climbed nearly 1°C (1.8°F) since the late nineteenth century.[18] Arctic annual temperatures have gone up twice as fast as elsewhere since the 1970s.[19] Summer temperatures during the past few decades were the highest in 2,000 years. Most of northern Canada, Alaska, and north-central Asia have warmed year-round. While most Arctic warming has occurred in autumn and winter, spring has also grown milder and come earlier. Further signs of a thawing Arctic come from diminishing snow coverage, particularly in spring. For the last 40 years, annual Arctic snow cover duration has shortened by 2–4 days per decade.[20] Reduced snow cover results in a longer growing season, which can change plant distribution. Snow thickness also impacts animal mobility and foraging activities in winter.

In sharp contrast to early explorers' encounters, the Northwest Passage has recently remained open for weeks, as the summer sea ice cover has plummeted within the last decade.[21] Arctic summer sea ice extent[22] set a record low in mid-September 2012, as compared to satellite observations since 1979 (fig. 1.4). Some scientists foresee a nearly ice-free Arctic Ocean

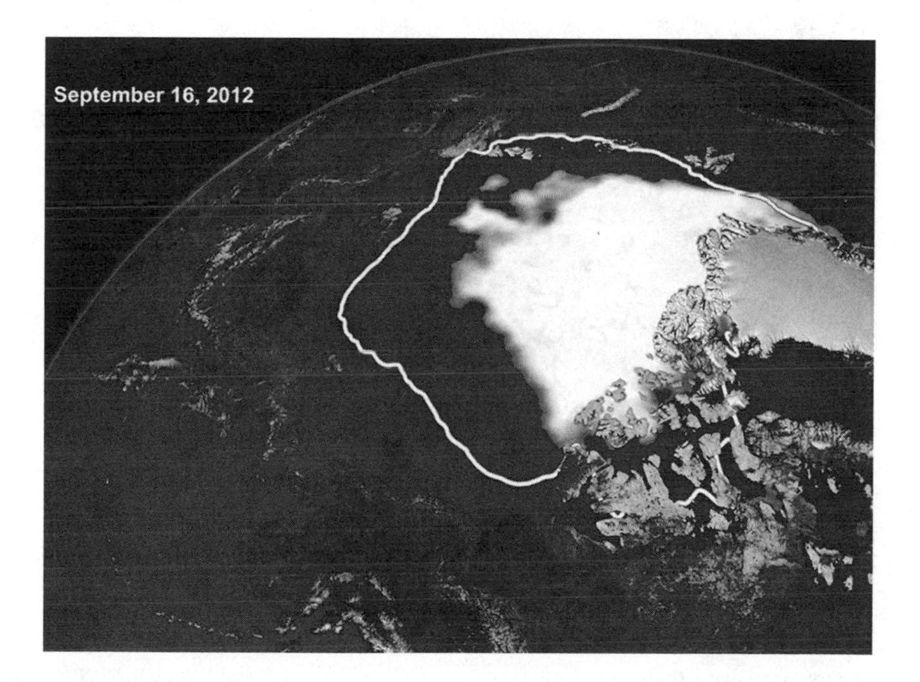

FIGURE 1.4

September 2012 set the record for the lowest Arctic summer sea ice extent since 1979. (NASA)

in summer within a few decades![23] The loss of Arctic summer sea ice is one of the most glaring signs of a changing cryosphere (chap. 2). Its summer minimum extent (in September) dropped by 13.2 percent per decade relative to the 1981–2010 average.[24]

Antarctic sea ice, on the other hand, had been expanding—until very recently. Antarctic *summer minimum* sea ice extent reached a historic low in March 2017, with March 2018 following as second lowest in the satellite era.[25] Meanwhile, October 2017 tied with 2002 for the second-lowest Antarctic *winter maximum* extent.[26] The next chapter explains the wide contrasts between the poles.

A massive iceberg, three times the size of Manhattan, split off the Petermann Glacier in northwest Greenland in July of 2017[27] (fig. 1.5). This and other large calving events on the same glacier in 2012 and 2010 are recent

FIGURE 1.5

Massive ice island that calved off Petermann Glacier, Greenland, August 5, 2010. (NASA Earth Observing-1 satellite)

dramatic episodes in the disintegration of many ice shelves and tongues encircling high Arctic islands and tidewater glaciers. A mere 8 percent of a once near-continuous band of ice shelves ringing the northern shore of Ellesmere Island, Canada, in 1906 survived in 2008.[28]

The breakup of Arctic ice shelves is mirrored by similar phenomena in Antarctica. At least 10 large fringing ice shelves have crumbled one by one since the 1990s on the Antarctic Peninsula, which has warmed by ~2°C (3.6°F) over the past 50 years (chap. 2).

The following chapters highlight the ways in which the rapidly climbing polar temperatures are transforming the cryosphere. The still-frozen poles and high mountaintops are among the first places to feel the heat. The poles, far away and once inaccessible, strongly influence global climate and atmospheric circulation, sending blasts of frigid air toward the equator. Polar realms—the Earth's final frontier—also hold vast biological and mineral resources (chap. 9). The world-encircling, density-driven, deep ocean currents—often called the Great Ocean Conveyor, or sometimes the global ocean conveyor belt (described in the following section)—originate in the cold, dense polar oceans. Yet, in spite of their distance from major population centers, polar regions feel most forcefully many of our civilization's inadvertent side effects, such as air pollution or the ozone "hole" developing over Antarctica since the 1980s (and a weaker one over the Arctic since the late 1990s).

Permanently frozen ground (permafrost) in the Arctic is thawing, and as a consequence the area of wetlands and small lakes is expanding (chap. 3). Thawing soil becomes unstable, undermining building foundations and roads in Alaska, northern Canada, and Siberia. Roads buckle, buildings develop cracks, and black spruce tilts at odd angles ("drunken forests"). Anaerobic bacteria rapidly decompose once-frozen soil organic matter, releasing greenhouse gases. A diminished winter Arctic sea ice cover also reduces protection from pounding waves and coastal erosion, especially of soft, unconsolidated sediments. Similarly, a longer ice-free summer/fall season prolongs exposure to more storm-induced shoreline erosion.

Glaciers, which hold the equivalent of 0.4 meters (1.4 feet) of sea level rise, are also in retreat. They currently lose enough ice to raise the ocean by up to 0.8 millimeters (0.03 inch) per year (see table 4.1). In the early nineteenth century, toward the end of the cold period known as the Little

Ice Age,[29] farmers from the mountain valleys above Brig in Switzerland's Valais canton prayed for the advancing Aletsch Glacier, the largest in the Alps, not to engulf their farms. In 1865, their prayers were answered—the glacier finally began to retreat. Nowadays the local residents, fearful of lost tourist revenue as well as adverse effects on their agriculture, are praying for an end to the pullback of the Aletsch and other Alpine glaciers.[30] Darkening peaks not only disappoint Alpine tourists, but may also decrease the reliability of glacier meltwater runoff as a fraction of total stream runoff (as discussed in the following paragraphs and chap. 9).

The Greenland and Antarctic ice sheets together have shed ice into the ocean over the past decade equivalent to a sea level rise of up to 0.8 to ~1 millimeters (0.03–0.04 inches) per year.[31] Although this amount may seem puny, Greenland and the West Antarctic Ice Sheet (WAIS) hold enough ice to raise global sea levels by ~7 meters (23 feet) and 3 meters (10 feet), respectively, if melted completely.[32] Many glaciers that reach the sea have also sped up as floating ice shelves that otherwise would slow their advance have melted from beneath and broken apart.[33] Thinning of parts of the WAIS bodes ill for the future because of its potential instability, as discussed further in chapters 6 and 8. Claims have been made of an irreversible collapse of the WAIS already under way.[34] Although a total meltdown appears extremely improbable, enough ice could melt to raise sea levels by more than one meter within the next century or two (chaps. 5, 6, and 8).

The world of ice is changing rapidly. The recent sharp decline in Arctic Ocean summer sea ice, crumbling Arctic shorelines, decaying permafrost, shrinking glaciers, reduced and shortened Arctic winter snow cover, mountain avalanches, glacial dam bursts, and flash floods all underscore the big thaw under way.

Hastening the Meltdown

Different elements of the climate system interact in complex ways through feedbacks that either amplify or dampen the forces leading to climate change. The unprecedented rise in Arctic temperatures and ice losses have been reinforced by several positive feedbacks that involve the cryosphere. These include feedbacks associated with changes in snow and ice albedo, water vapor, clouds, and black carbon (fig. 1.6).

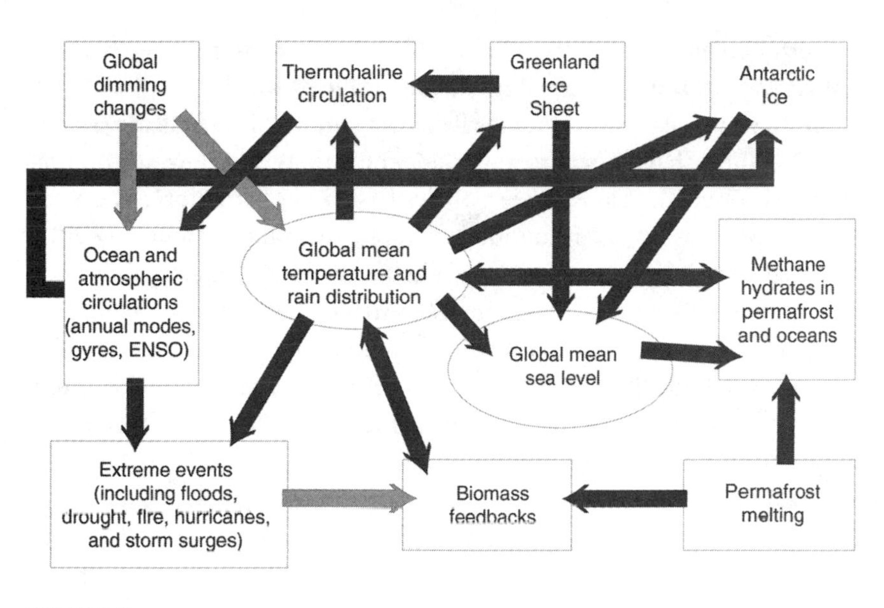

FIGURE 1.6

Climate feedbacks affecting the cryosphere. Heavy black arrows point in the direction of interactions among terrestrial processes that either change or are changed by global warming. Gray arrows show less significant processes. Most of these interactions involve the cryosphere in some manner.

A shrinking snow cover initiates a positive feedback, the snow-albedo feedback. The albedo of forests, tundra, and bare soil is considerably lower than that of snow or ice. Therefore, as more vegetated or bare ground is exposed earlier each spring, more solar heat is absorbed, which in turn melts more snow, and so on.

As with snow, a reduced summer sea ice cover exposes more of the ocean surface to the sun's incoming energy, raising surface water temperatures, which triggers more ice melting (the ice-albedo feedback) (chap. 2). A warmer ocean delays the onset of autumn freeze-up, keeps more ocean open into the winter, and initiates earlier spring melting. Sea ice, a good insulator, seals in the ocean's heat. Stored heat escapes and warms the atmosphere once this lid is removed. More open water therefore fosters evaporation, atmospheric humidity, and cloudiness, especially during months with the least sea ice cover.[35] The added water vapor, also a powerful greenhouse gas (see the Greenhouse Effect later in this chapter), enhances warming and expedites further sea ice loss.

The ice albedo-temperature feedback also indirectly affects the Greenland Ice Sheet. Higher summer air temperatures expand low-albedo surface melt pools on the ice sheet. The enlarged pools reinforce warming that primes the ice sheet for sudden, rapid thawing. In mid-July 2012, nearly the entire surface of the Greenland Ice Sheet melted within days. (No more than half melts in a typical summer.) Early August of 2012—a "goliath year"—set a new record for cumulative melt in Greenland.[36] Although the July 2012 event is still quite rare (the last such incident was in 1889), recent summer surface melting has generally expanded and lasts longer, as compared to the 1981–2010 average.[37]

The surface-elevation feedback also reinforces ice melting. Ice losses lower the elevation of a glacier or ice sheet, bringing the surface into a higher-temperature zone with a longer melt season. Therefore, more ice will melt, especially near the edge of the glacier or ice sheet, further lowering elevation.

Black carbon mixed with sulfates, nitrates, dust, and other pollutants from lower latitudes reach the Arctic in late winter and spring. The resulting "Arctic haze," or smoggy skies, sets up a series of complex feedbacks that have a net warming effect.[38] "Black carbon" refers to airborne particles, or aerosols, of carbon or carbon compounds (i.e., soot) derived from incomplete combustion of vegetation and fossil fuels. It affects climate by absorbing heat in the atmosphere and by lowering the albedo of snow and ice. Because of the very high albedo of snow or ice (~70–80 percent), just 10 to 100 parts per billion of soot can lower the albedo of snow by 1–5 percent.[39] Therefore, dirty snow melts earlier in the spring and exposes darker underlying ground, which induces further melting. Before the 1850s, summer forest fires were a major source of Arctic black carbon.[40] Wintertime fossil fuel combustion (primarily coal) dominated the period between the 1850s and 1950s, but diminished thereafter due to stricter air pollution controls in North America and Europe. (However, black carbon levels may have increased again recently due to air pollution from Asia.)

Loss of sea (and land) ice initiates another feedback loop. When the North Atlantic and Norwegian Currents—the northern extensions of the warm Gulf Stream—reach the Norwegian and Greenland Seas, the now-chilled, salty ocean water sinks as North Atlantic Deep Water (NADW). This dense, sinking current, joined by additional cold, sinking water off southeastern Greenland and the Labrador Sea, forms a deep-water branch of the global ocean circulation—the Atlantic Meridional Overturning Circulation (AMOC), popularly known as the "global ocean conveyor."[41] Computer models suggest that future melting of sea ice, glaciers, and the Greenland Ice Sheet would lower ocean salinity and slow down the sinking of NADW.[42] The cold, fresher ocean water would chill the eastern North Atlantic and northwestern Europe, but otherwise have a minor effect on global temperatures. On the other hand, heat from a more sluggish Gulf Stream would build up along the east coast of North America and raise regional sea levels due to the expansion of warmer ocean water.

Escaping plumes of methane have been spotted in the Arctic Ocean from decomposing methane hydrate (a cage-like form of ice enclosing methane gas), as well as from decaying land and subsea permafrost (see chap. 3). By amplifying greenhouse warming, these gases would help melt more land ice and thus indirectly add to sea level rise.

Figure 1.6 schematically illustrates the complex ways in which these processes interact. For example, a rise in global mean temperature would increase melting of the Greenland and Antarctic ice sheets, which in turn would lead to higher sea levels. This in turn would thaw methane hydrate–rich coastal permafrost drowned by the rising waters. The methane gas released into the atmosphere, whether remaining as methane or rapidly oxidizing to carbon dioxide, would further enhance global warming. Rising temperatures could also alter ocean currents and trigger more frequent and/or more severe floods, droughts, or hurricanes.

THE GREENHOUSE EFFECT

As previously noted, the rapidly metamorphosing cryosphere clearly signals a warming planet. However, because of sharp past geologic climate changes (chap. 7), some argue that this "recent" warm-up merely represents part of a natural cycle after the 1850s—the end of the Little Ice Age, which began ~700 years ago. This ignores the importance of the rate of change, which far exceeds that of most geologic processes. Although natural climate variability accounts for most of the temperature rise in the late nineteenth and early twentieth centuries, the human fingerprint has grown increasingly more pronounced since the 1960s.[43]

Carbon dioxide (CO_2) atop Mauna Loa, a dormant volcano on the island of Hawaii, has increased steadily from 315 parts per million in 1958, when the time series began, to ~406 parts per million by late 2017 (fig. 1.7).[44] Gas bubbles encased in polar ice sheets extend the CO_2 record far beyond the instrumental period (chap. 7). Before the Industrial Revolution, CO_2 levels stood close to 280 parts per million for at least a millennium, but began a slow upward climb in the nineteenth century that has quickened since the 1950s. Other greenhouse gases such as methane (CH_4) and nitrous oxide (N_2O) show similar upswings.[45]

How do these gases affect the Earth's climate? The heat-trapping ability of these gases has been compared to a glass greenhouse, which admits visible and ultraviolet radiation from the sun but traps escaping heat energy in the form of infrared radiation, which warms the interior of the

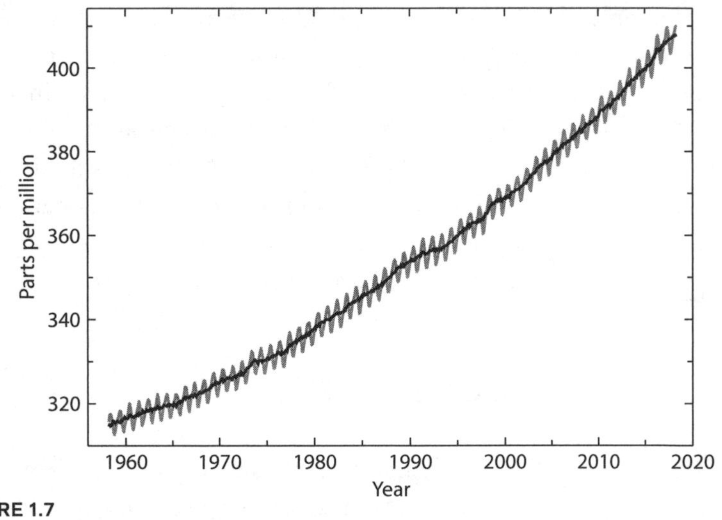

FIGURE 1.7

Rising atmospheric carbon dioxide levels at Mauna Loa Observatory, Hawaii, from 1959 to the present. The wavy pattern in the curve is caused by seasonal uptakes of CO_2 by plants, predominantly in the Northern Hemisphere. (NOAA; Scripps Institution of Oceanography)

greenhouse.[46] In an analogous manner, roughly one-third of incoming solar ultraviolet and visible radiation is reflected directly back into space by clouds or bright, reflective surfaces like snow or ice, while the rest is absorbed either by the atmosphere or at the Earth's surface, raising temperatures (fig. 1.8). The Earth reemits some of this energy at longer wavelengths—in the infrared. Some of this infrared radiation passes through the atmosphere and is lost to space. Atmospheric water vapor, carbon dioxide, and some other trace gases absorb another portion of

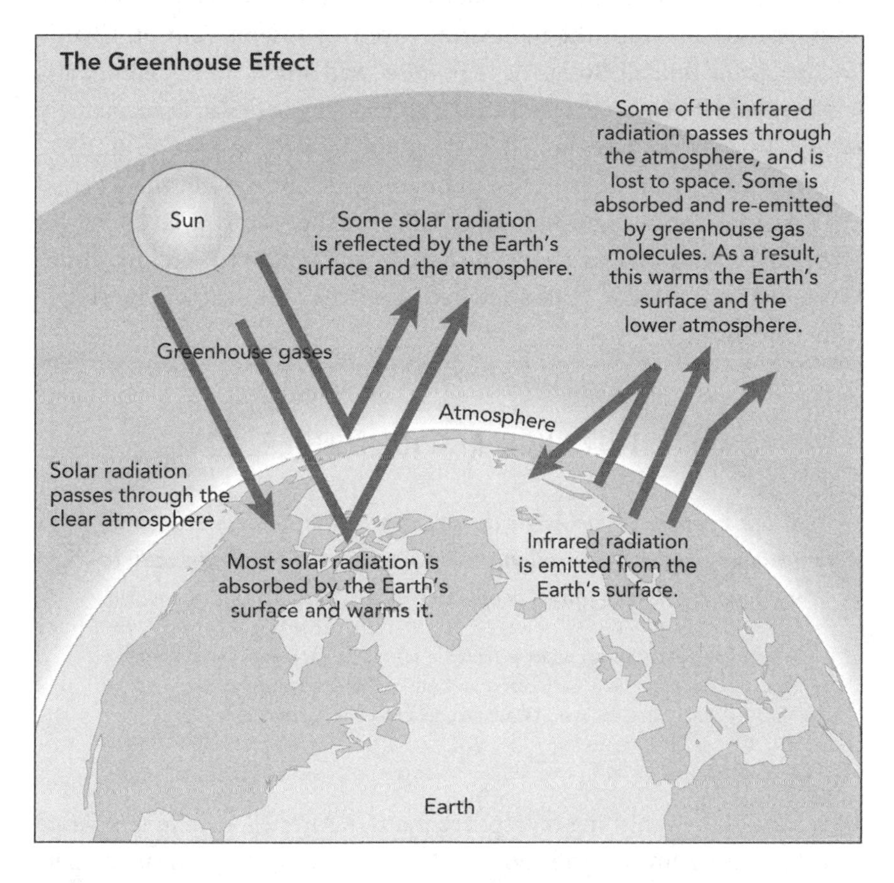

FIGURE 1.8

The greenhouse effect caused by trapping of heat by greenhouse gas molecules, such as CO_2 and CH_4, in the Earth's atmosphere. (Adapted from Mann [2012])

this radiation. Naturally occurring greenhouse gases, primarily water vapor and CO_2 from volcanic eruptions, help warm the Earth by 33°C (59°F). However, the steep twentieth-century rise in greenhouse gases closely parallels increases in fossil fuel combustion and deforestation (major sources of carbon dioxide, CO_2); rice cultivation and livestock (sources of anthropogenic methane, CH_4); and industry (nitrous oxide, N_2O). Natural processes alone fail to explain this sudden, rapid upward surge.

The basic science behind the greenhouse effect that closely relates mounting anthropogenic atmospheric CO_2 levels and temperature has been established since the pioneering work of Joseph Fourier, Claude Pouillet, John Tyndall, Svante A. Arrhenius, and others over a century ago (chaps. 8 and 9). Given the upswing in CO_2 since the 1950s, driven largely by human activities (see fig. 1.7), it should therefore come as no surprise that the recent climate change is an ongoing reality and not the feverish imagination of an ardent environmentalist. As the following chapters will prove, the various components of the cryosphere are heating up, more markedly in the Arctic than elsewhere, and they are reacting by slowly withering away.

WHY SNOW AND ICE MATTER

What happens in the Arctic doesn't stay in the Arctic. . . . Melting ice and glaciers in Greenland can contribute quite significantly to sea level rise, which affects people in coastal areas around the world.

—Jane Lubchenco, former undersecretary of commerce for oceans and atmosphere and director of the National Oceanic and Atmospheric Administration (NOAA), American Geophysical Union Annual Meeting, December 5, 2012

Why should the vanishing cryosphere matter? After all, here in temperate climates, hardly anyone really notices the small rise in global mean temperature. The rapid Arctic warm-up—double that of the rest of the world—is already transforming the far north, with more changes expected. Arctic and mountain dwellers are the first to perceive the changes. The longer ice-free season forces Inuit hunters farther afield to hunt polar bears and other sea mammals for meat and pelts (bearskins are a source of income in the cash-pressed northland). Riskier soft, thin sea ice, coupled with higher gas prices for snowmobiles, imperils their traditional way of life. Polar bears, seals, and walruses lose sea ice habitat, spruce bark beetle outbreaks spread, and grizzly bears encroach northward into polar bear territory. Caribou migration routes are also changing, and herd populations have dropped. As Inuit spokeswoman Sheila Watt-Cloutier so poignantly put it: "As our hunting culture is based on the cold, being frozen with lots of snow and ice, we thrive on it. We are in essence fighting for our right to be cold."[47] In Scandinavia and Russia, ice crusts on snow due to late-winter rainfall hamper foraging reindeer, impacting traditional Saami (Lapp) herders.

Each June, thousands of indigenous pilgrims trudge along an icy path over 4,700 meters (15,000 feet) high in the Peruvian Andes to celebrate Qoyllur Rit'i, a festival dating back to pre-Columbian times. Pieces of the glacier would be carried back to Cusco and surrounding villages to ensure abundant water and crop fertility. Because of the glacier retreat, the Peruvian government now prohibits any ice removal.[48] The loss of Andean glaciers affects much more than local festivals. These glaciers, representing 70 percent of the world's tropical glaciers, have shrunk by 20 percent since 1970. This imperils water, food, and hydroelectric energy resources for a region of 77 million people in South America.[49]

As mountain glaciers dwindle, changes in glacier runoff could affect water availability for hundreds of millions of people throughout Asia, the Andes, and several western U.S. states who depend, to varying degrees, on this critical resource for agriculture, drinking water, and industry. Mountains have been labeled the "water towers of the world." The high mountains of Asia, in particular, are known as the "Third Pole" because of their high elevations and largest concentration of glaciers outside the polar regions. The Himalayas are the source of most major Asian rivers. These mountains provide water to a considerable fraction of South Asia's population. Thus, Himalayan glacier retreat would greatly impact future Asian water resource availability, as shown in chapter 9. Irrigation water for the North Indian "breadbasket" regions comes from rivers draining the southern flanks of the Himalayas, as well as from summer monsoon rainfall. The glacial meltwater helps buffer droughts, especially in the drier watersheds of the western Himalayas and high central Asian plateau, by allowing rivers to maintain a steady flow.[50] But what of the future? Short-term increases in peak river runoff due to increasing meltwater may temporarily offset drought scarcities. But in the long term, continued glacial shrinkage will probably diminish meltwater runoff. Some regions, such as the Karakoram mountain range, can expect more consistent runoff, since local meteorological conditions favor abundant winter snowfall and ice buildup (see "Bucking the Trend," chap. 4). Growing unreliability of summer monsoon rains could add to water insecurities.

In the United States, many western states will also confront growing water scarcity issues. California recently suffered its worst drought in at least 500 years and faced serious water shortages, intensified by a far-below-average Sierra Nevada winter snowpack. Snowmelt from the mountains supplies nearly a third of California's water needs.[51] These climate-induced impacts will be further elaborated on in chapter 9.

As the high summits shed their white snow and ice caps, mountain valleys confront increasing dangers. Swollen spring river flows generate more flash floods.[52] Also, as glaciers recede and alpine permafrost thaws, the high mountains become unglued, letting loose more rockfalls, avalanches, and GLOFs, or glacial lake outburst floods. GLOFs occur when moraine-, ice-, or landslide-formed dams that restrain swollen glacial lakes suddenly crumble and briefly unleash huge torrents of water. GLOFs, particularly in the Nepal and Bhutan Himalayas, have provoked destructive flash floods.[53] Whether in the Alps, the Andes, the Rockies, or the Green Mountains of Vermont, declining snow and ice cover will impact the tourist economy, curtailing winter skiing and endangering summertime mountain climbers as slope instability escalates. Enjoyment of the serene beauty of snow-draped mountain peaks in summer may someday become a distant memory, recalled only in fading postcards and in literature. The fate of the cryosphere no longer concerns the remote Far North, the South Pole, or lofty snow-clad peaks alone; it affects the entire planet (chap. 9).

Melting Ice and Rising Seas

The year 2017 was the second- to third-warmest year since 1880—continuing a long-term trend that began in the late nineteenth century.[54] Sea level has crept upward steadily by 1.2 to 1.9 millimeters per year during most of the twentieth century. Since 1993, satellite altimeters have shown an average sea level rise of around 3 millimeters per year.[55] As air and ocean temperatures rise, so does sea level. Seawater expands as it warms, raising the ocean level by a process known as thermal expansion.[56] Ice shed by glaciers and ice sheets ultimately enters the ocean, also adding to sea level rise. Since the early 2000s, melting ice sheets and glaciers have accounted for over half of the total observed current rise.[57] This fraction is likely to increase with global warming.

The Intergovernmental Panel on Climate Change (IPCC), an international panel of experts in the natural and social sciences, has periodically assessed the state of the Earth's climate since 1990. Its 2013 report anticipated a sea level rise of 26–82 centimeters (10–32 inches) by 2100.[58] On the other hand, semiempirical approaches that assume that the past correlation between global sea level and temperature will continue in the future instead foresee a higher rise of 75–190 centimeters (2.5–6.2 feet) by 2100. But neither type of study accounts for possible speeded motion of glaciers or ice sheets, which will likely become increasingly pronounced during this century.

Recent observations and better understanding of how ice sheets and oceans interact with the atmosphere, outlined further in chapters 5 through 8, raise the prospect of higher sea levels than previously assumed. An accumulating body of evidence points to signs of latent instability of the West Antarctic Ice Sheet, of thinning and weakening Antarctic ice shelves, and of the lurking possibility of ice cliff collapse. This may set off a more dynamic future response of the WAIS to continued climate warming, resulting in higher sea levels than formerly considered.

Ocean Surprises

A dwindling cryosphere and higher ocean levels will ultimately affect hundreds of millions of people far removed from frozen regions. Populations in low-lying coastal areas face higher risks of flooding from storm surges, beach erosion, and saltwater intrusion.

One study anticipates that a 21-centimeter (8.3-inch) sea level rise by 2060 could put up to 411 million people worldwide who live in the

1-in-100-year flood zone at risk for coastal flooding.[59] In the United States, 4.2 million people would be at risk for land inundation from a 90-centimeter (35-inch) sea level rise by 2100.[60] The majority live in the Southeast, along the Atlantic and Gulf Coasts, with the greatest number in Florida. Intense coastal storms, such as Superstorm Sandy that struck New Jersey and New York City in late October of 2012, or Hurricanes Harvey, Irma, and Maria that hit Texas, Florida, and Puerto Rico in August and September of 2017, dramatically underscore the existing vulnerability of low-lying coastal areas to flooding—a hazard that rising seas would only intensify.

Keys to Our Climate Future

Because of the vast stores of ice they hold, the two frozen polar giants—the Greenland and Antarctic ice sheets—hold fateful keys to our future well-being. How stable are the ice sheets in the face of rising temperatures? How much ice will melt, and how fast? And by what means? How much will sea level rise in turn? What lessons can we glean from the past behavior of these two icy behemoths? How would the shrinking cryosphere affect the rest of the world? Chapters 5 through 9 will delve into these important questions in depth.

Changing Arctic landscapes, receding glaciers, disrupted lifestyles, climate feedbacks, and global sea level rise are just a few of the important reasons why vanishing snow and ice matters. Others include water resource availability that could potentially affect millions in Asia and Andean nations, and the emerging threat of geohazards—not only of landslides or flash floods, but also of more unexpected phenomena such as icequakes triggered by accelerating glaciers, increased calving, and land uplift following removal of ice loads.[61] Glacial rebound may occur on two different timescales—a nearly instantaneous, springlike response as the Earth's crust and upper mantle adjust to recent ice mass losses (as in Glacier Bay, Alaska, and in Greenland; chaps. 4 and 5), and a much more sluggish, still-ongoing response to removal of ice sheets after the last ice age, still evident around Hudson Bay, Canada, and the Gulf of Bothnia, Sweden-Finland (chap. 7).

BENEFITS OF ARCTIC WARMING

The picture is not altogether bleak. Increasingly navigable waters will shorten shipping routes, cut costs, and open up the Arctic for development of its vast mineral wealth. As of 2007, 40 billion barrels of oil, 1,136 trillion cubic feet of natural gas, and 8 billion barrels of natural gas had been extracted north of the Arctic Circle.[62] Geologists estimate that 30 percent of the world's undiscovered gas and 13 percent of the undiscovered oil lie in the Arctic, mainly in shallow offshore marine deposits less than 500 meters deep. The Russian Arctic holds most of the undiscovered gas. In addition, Norilsk, Siberia, is a major producer of nickel, palladium, and copper, while the Sakha Republic (Yakutia) accounts for 97 percent of Russian and 25 percent of the world's diamond production. Northwestern Alaska holds one of the world's largest zinc deposits,[63] and an enormous iron ore mining project is planned for Mary River, northern Baffin Island. Nonetheless, this surge of new development will expose fragile Arctic ecosystems to new stresses from which they may not easily recover. Furthermore, large-scale exploitation of new Arctic fossil fuel resources would only augment carbon emissions and exacerbate the very meltdown this book documents.

TREADING ON THIN ICE

A temperature rise of more than 2°C (3.6°F) may constitute a "dangerous anthropogenic interference with the climate system."[64] The planet's feverish direction may ultimately endanger the cryosphere as we know it. The Earth may be inching toward a climate regime like that of the Last Interglacial, ~125,000 years ago, when temperatures were 1° to 2°C (1.8°–3.6°F) warmer and enough ice melted to raise sea level by around 6–9 meters (20–29.5 feet), or perhaps even like that of 3–3.3 million years ago, when atmospheric carbon dioxide levels resembled today's (chap. 7). Sea levels then stood 20 meters (66 feet) above present levels—equivalent to a melting of the Greenland and West Antarctic Ice Sheets, and a bit of the East Antarctic Ice Sheet.[65]

Because of the longevity of carbon dioxide in the atmosphere, today's increasing hydrocarbon emissions virtually ensure a much warmer, less ice-covered, and more aquatic future planet (chap. 8). A prolonged warming could therefore push the Earth into uncharted territory.

The ongoing changes and their associated feedbacks (see "Hastening the Meltdown," earlier in this chapter) make the Arctic, in particular, a crucible of climate change. As noted above and in chapter 9, consequences of the thaw in the Arctic will affect its inhabitants directly. More importantly, the consequences will spill out far beyond the Arctic or Antarctica; hence the urgency in alerting the public to the enormous transformations and potential risks associated with vanishing ice.

The coming chapters describe the rapid metamorphosis of the cryosphere and what this implies for the future well-being of this planet. The journey to the frozen corners of the Earth begins in the next chapter, which zooms in on the different types of floating (marine) ice—sea ice, icebergs, and ice shelves: how they form, how they differ, their current state, how they are changing, and their future.

2

ICE AFLOAT—ICE SHELVES, ICEBERGS, AND SEA ICE

ICE SHELVES

Crumbling Ice Shelves

Antarctic ice shelves are highly impressive. You have to sail up close to them to appreciate their size and apparent solidity. . . . Every so often, a massive slab on the seaward edge of one of these monsters will flex a little too much; perhaps the surface crevasses will start to make their way deeper; tides work the ice up and down, making more crevasses, more cracks until, finally, the slab breaks off and sails away, forming one of those tabular icebergs, flat-topped and square-shouldered, the size of a floating city, or even a county. . . . And it is these shelves, on the [Antarctic] Peninsula, that have started to alarm scientists by falling apart.

—Gabrielle Walker, *Antarctica: An Intimate Portrait of a Mysterious Continent* (2013)

Although overwhelmed by the sheer volume of ice firmly sitting on land, ice afloat occupies a vast yet highly variable expanse. Floating ice occurs in three distinct forms: ice shelves, icebergs, and sea ice. Of these, ice shelves are the most intimately tied to the land-based ice sheet or tidewater glacier from which they protrude seaward. (Floating ice tongues, usually confined to deep, narrow fjords or bays, extend outward from tidewater glaciers.) Icebergs, by contrast, are fully floating pieces of ice that have split off from ice shelves, glaciers, or fast-flowing ice streams that reach the sea. As an extension of glaciers and ice streams, ice shelves develop as the ice slowly creeps seaward and spreads outward over water. In addition to the ice supplied by these sources, ice shelves thicken by snow accumulations and by direct freezing of seawater onto the base. (On northern Ellesmere Island, shelves build up from sea ice over several years.) They shrink by calving icebergs and melting from top and bottom. Like the icebergs they shed, ice shelves retain distinctive features, such as flow structures, layering, or mantles of rocky debris, inherited from their source glaciers.

Ice shelves occur in Antarctica, Greenland, and northern Ellesmere Island, Canada. As Gabrielle Walker notes, they can be quite impressive monsters, towering between 100 and 1,000 meters (328–3,280 feet) thick. Their massive dimensions contrast sharply with free-floating sea ice that freezes and melts annually, or survives a few seasons and is therefore usually much thinner (generally under 3 meters [10 feet] thick). Ice shelves ring roughly three-quarters of the shoreline of Antarctica, occupying an area of more than 1.56 million square kilometers (0.6 million square miles).[1] The two largest include the Ross Ice Shelf, covering nearly 473,000 square kilometers (182,630 square miles), and the Ronne-Filchner Ice Shelf, another 422,000 square kilometers (162,930 square miles).[2] In Greenland, ice tongues cover a much smaller fraction of the coast.

Lately, many ice shelves have been in serious trouble, and are even endangered in some places. Within a matter of weeks in 1995, around 2,000 square kilometers (770 square miles) of the Larsen A Ice Shelf, near the northern tip of the Antarctic Peninsula, shattered into a slivered jumble of icebergs. While ice shelves routinely calve icebergs, the disintegration was then considered unprecedented. In early 2002, an enormous chunk of ice, roughly the size of Rhode Island in area, split off the Larsen B Ice Shelf to the south of Larsen A, and rapidly disintegrated within a month.[3] The size and rapidity of this particular event astounded scientists. Overall, Larsen B lost a total of 9,166 square kilometers (3,540 square miles) between the 1960s and 2008–2009 (table 2.1). Following their disintegration, satellites observed that large glaciers feeding into Larsen A and B had lurched forward, thinned, lowered in elevation, and disgorged myriad small icebergs into the ocean.[4]

TABLE 2.1 Retreat of Ice Shelves on the Antarctic Peninsula

ICE SHELF	PERIOD	CUMULATIVE LOSS (km^2)
Jones	1950s–2008/9	29
Larsen Inlet	1986–1989	407
Larsen A	1950s–2008/9	3,624
Larsen B	1960s–2008/9	9,166
Larsen C	1960s–2008/9	5,295
Larsen C	July 2017	~5,800
Müller	1956–2008/9	38
Northern George VI	1950s–2008/9	1,939
Prince Gustav	1950s–2008/9	1,621
Wilkins	1950s–2008/9	5,434
Wordie	1960s–2008/9	1,281
Total		**34,634**

Sources: Ice shelf data from Cook and Vaughan (2010), table 3; Larsen Inlet data from National Snow & Ice Data Center (http://nsidc.org/cryosphere/sotc/iceshelves.html; last updated June 22, 2018; accessed November 12, 2018); Larsen C data from Viñas (2017).

Glaciers feeding the remnant Larsen B shelf continue to accelerate and thin. Rifts slice across the ice. These features imply that "the final phase of the demise of Larsen B Ice Shelf is most likely in progress" and may now be "nearing its final act" within the decade,[5,6] Ala Khazendar of the NASA Jet Propulsion Laboratory and University of California and his team have been closely monitoring the incipient demise of the shelf with laser altimeters, depth-probing radar, and radar interferometry for measuring glacier velocities. Khazendar says that although "it's fascinating scientifically to have a front-row seat to watch the ice shelf becoming unstable and breaking up, it's bad news for our planet."[6]

The Larsen B Ice Shelf had remained stable for the 11,000-year period since the end of the last ice age—based on information derived from marine sediment cores and seismic reflections.[7] Its sudden collapse was therefore more likely related to recent warming than to surging tributary glaciers or sudden ungrounding (fig. 2.1).

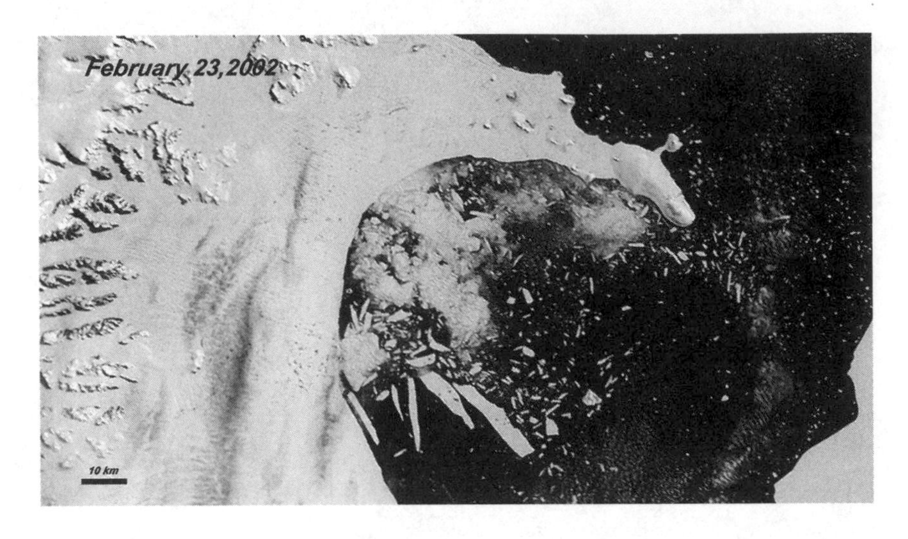

FIGURE 2.1

Satellite view of collapse of the Larsen B Ice Shelf, February 2002. (NASA, "Collapse of the Larsen B Ice Shelf," October 17, 2013, https://svs.gsfc.nasa.gov/30160)

Larsen C, to the south of Larsen A and B and now the largest surviving ice shelf on the Antarctic Peninsula, had remained fairly stable since the 1980s, although the shelf showed signs of thinning and lowered elevation.[8] On July 12, 2017, a giant iceberg the size of Delaware (nearly 5,800 square kilometers [~2,240 square miles]) split off from Larsen C.[9] A fissure in the ice had been slowly growing for decades, but recently began to spread northward until it finally unleashed the berg. What next for Larsen C? While calving is a normal part of an ice shelf's life cycle, icebergs this large are rare. The stability of Larsen C will depend on whether future icebergs break free of at least two "pinning points," or underwater ridges that help anchor the ice shelf. Should the ice shelf lose that support, it would no longer effectively restrain the glaciers feeding into the shelf from speeding up and discharging more ice.

On the southwest side of the Antarctic Peninsula, the Wilkins Ice Shelf, which had changed little during the twentieth century, began to retreat in the late 1990s and broke up further from 2008 through March 2013.[10] South of Wilkins, rifts have developed over a significant portion of the small Verdi Ice Shelf.[11] These could be precursors of instability.

These shelves are not unique. Starting in the 1950s, at least 10 ice shelves on the Antarctic Peninsula shrank by roughly 34,634 square kilometers (13,372 square miles) and fell apart one by one (table 2.1). Nor is ice shelf thinning confined to the Antarctic Peninsula. Since 2003, Antarctic ice shelves have lost 310 cubic kilometers per year (74.4 cubic miles).[12] While Antarctic Peninsula and West Antarctic ice shelves account for over two-thirds of this loss, the colder, more stable East Antarctic shelves also contribute a growing fraction. Unusually warm air temperatures and warm ocean incursions are the likely culprits. If warming persists, disintegration of ice shelves will likely propagate southward to the Antarctic mainland, potentially destabilizing much larger ice shelves. In the words of Theodore Scambos, a glaciologist at the National Snow and Ice Data Center in Boulder, Colorado: "It's the trailer for the movie that's going to unfold over the rest of Antarctica for the next 50 to 100 years."[13]

Analogous Arctic phenomena echo the breakup of Antarctic ice shelves. Ellesmere Island, Canada, lost over 90 percent of a once-continuous band of ice shelves during the twentieth century, with further breakups continuing since 2000. In July 2017, another massive iceberg in northwest Greenland suddenly detached from Petermann Glacier, following two large calving events on the same glacier in 2010 and 2012 (fig. 1.5).

Jakobshavn Isbrae, a large glacier in western Greenland, drains 6.5 percent of the Greenland Ice Sheet.[14] This glacier may have calved the iceberg that sank the *Titanic*.[15] In their 2012 award-winning documentary film *Chasing Ice*, nature photographer James Balog and his Extreme Ice Survey team vividly captured a major calving event in 2008 on this glacier. One team member excitedly recalls that it was like "watching Manhattan break apart in front of your eyes." Before 1997, this large glacier flowed into a 15-kilometer (9.3-mile) floating ice tongue in a deep-ocean fjord. In the early 2000s, its ice tongue began to disintegrate and the glacier sped up—as much as four times faster during the summer of 2012 than in the 1990s.[16] In August of 2015, the glacier shed another huge chunk—some 12.5 square kilometers (nearly 5 square miles), claimed by some to be the biggest calving event on record.[17] Regardless of whether or not this set any record, Eric Rignot, a scientist at NASA's Jet Propulsion Laboratory, was "struck by the sheer size of this calving event, which shows that the glacier continues to retreat at 'galloping speed.'" These dramatic events underscore the recent rapid Arctic warming[18] that triggered the disintegration of ice shelves and tongues encircling many high Arctic islands and tidewater glaciers. The next section probes why and where ice shelves are crumbling.

Undermining Ice Shelves

Some of the fastest-warming places on Earth are located, surprisingly enough, in Antarctica. Over the past 50 years, Byrd Station on the West Antarctic Ice Sheet (WAIS) has warmed by 0.42°C (0.76°F) per decade.[19] Springs have warmed the most. The −9°C (16°F) annual temperature line—the apparent temperature limit for ice shelf stability—has been slowly creeping southward on the Antarctic Peninsula, reducing ice shelf area by nearly 35,000 square kilometers (13,400 square miles) since the 1960s[20] (table 2.1). Referring to the small Sjögren Glacier's fjord on the northern Antarctic Peninsula, which according to Douglas Fox[21] once "held ice 600 meters thick as recently as 1995 . . . but now it holds seawater instead." Regarding the demise of the Larsen A and B Ice Shelves, he wrote: "That more ice shelves will collapse is a foregone conclusion. An average summer temperature of zero degrees Celsius seems to represent the highest temperature at which an ice shelf can exist. And the invisible line where summer averages zero degrees Celsius (32°F) is creeping south along the Antarctic Peninsula tip toward the mainland, along with higher mean annual temperatures. Every ice shelf that the line crosses has collapsed within a decade or so. Next up, south of Larsen B and Scar Inlet, is the Larsen C ice shelf," the largest remaining shelf on the peninsula, which as if on cue unleashed a huge iceberg in July 2017.

Ice shelves disintegrate in several ways. Rising temperatures and changing wind patterns may have hastened ice shelf retreat on the Antarctic Peninsula. Higher air temperatures prolong the melt season and multiply the number of surface meltwater ponds. Pressure from meltwater generates small surface cracks. Meltwater that wedges into fissures widens them upon freezing. Cracks deepen and weaken the ice until it breaks. Other cracks develop as the shelf bends at the ice front and along margins. Rifts appear on a thinning ice shelf as the outlet glacier advances.

Shear margins along the contact area between the grounded ice stream or glacier and its side walls, as zones of weakness, are often riddled with crevasses. These may develop into full-fledged rifts at the point of contact between floating ice and bedrock at the base of the grounding line.

A tipping point is reached when crevasses formed parallel to the ice edge suddenly calve numerous highly fractured, tabular, elongated slabs of ice that topple over like a pile of dominoes, ultimately crumbling into a mush of small icebergs and ice rubble, like Larsen A and B or Wilkins.[22] Numerous rifts had developed prior to their breakup. Although the ice shelves in the Amundsen Sea region of West Antarctica still remain intact, satellite imagery reveals widespread rifting that has preceded ice shelf marginal retreat over the past 40 years.[23]

Ice shelves may be even more vulnerable to basal melting, or thawing from beneath, as the ocean warms. Changes in wind circulation over the past several decades have brought warmer air to the Antarctic Peninsula, increasing surface melting that has helped to destabilize the ice shelves. Changed wind patterns have also facilitated the upwelling of warmer deep water. Comparatively mild, dense, salty Circumpolar Deep Water (CDW) now penetrates into numerous submarine troughs that cut across the continental shelf off West Antarctica beneath the ice shelves.[24] (Submarine troughs—basically drowned fjords—were gouged out by glaciers during past ice ages at times of lower sea level.) While surface water is near the freezing point, CDW can be several degrees warmer. It therefore melts the ice shelf from below. The resulting meltwater cools, freshens, and lightens the CDW. Seaward-outflowing CDW is then replaced by sinking colder, saltier, and denser overlying shelf water. CDW incursions have thinned ice shelves in the Amundsen Sea Embayment sector of West Antarctica within the past two decades. Ice shelves there melted at a rate of around 19 meters (62 feet) per decade between 1994 and 2012.[25] Basal melting is fast outpacing calving as a cause of Antarctic ice loss.[26]

On the Antarctic Peninsula, more so than in West Antarctica, increases in regional air temperature have contributed to the remarkable string of ice shelf breakups (e.g., the Larsen A and B and Wilkins Ice Shelves; fig. 2.1; table 2.1). Less sea ice around the Antarctic Peninsula (although it is expanding elsewhere in Antarctica—see "Why Is Antarctic Sea Ice Expanding?" later in this chapter) may also have facilitated ice shelf collapse. Large calving events tend to occur in summer, when sea ice cover is minimal. (Sea ice inhibits wave action and helps prop up an ice shelf.)

Opening the Dam

Floating ice shelves or ice tongues act like buttresses supporting an old, decaying building. The "backstress" exerted by ice shelves or tongues slows the advance of many Antarctic ice streams, as well as Alaskan and Greenlandic tidewater glaciers. The removal of this support, by processes described in the previous section, is like releasing the floodgates on a dam. Once the ice shelf weakens or collapses, the glacier can surge forward rapidly (fig. 2.2). Following the breakup of the Larsen B Ice Shelf in 2002, its tributary glaciers began to accelerate forward.[27] Around 28 percent of all Antarctic ice losses between 1992 and 2011 came from the Antarctic Peninsula.[28] However, the fairly small glaciers on the Antarctic Peninsula would add at most only half a meter (20 inches) to sea level if they all melted. More importantly, however, surging glaciers following the

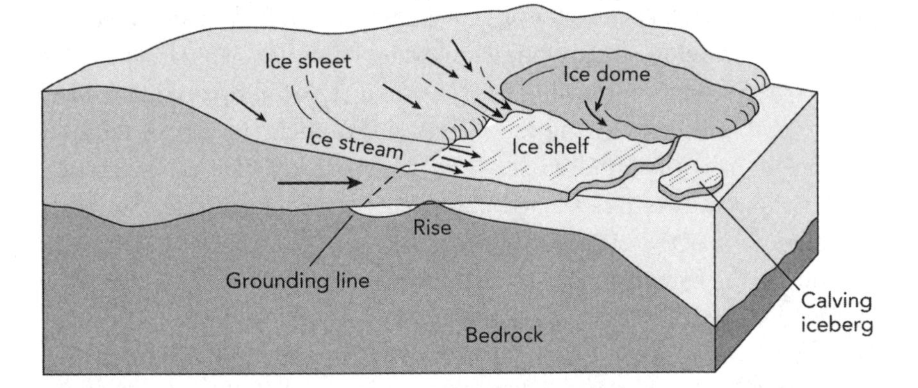

FIGURE 2.2

Sketch of an ice sheet showing the relation of the ice shelf to the grounding line and ice sheet. (Fig. 6.8 from Gornitz [2013])

demise of their fringing shelves may foreshadow future behavior of the much larger West Antarctic Ice Sheet (WAIS).

Although not yet disintegrating, many ice shelves fringing the WAIS are thinning and weakening as a result of the slight increase in nearby ocean temperatures. The Amundsen Sea Embayment region, including the Pine Island, Thwaites, and Smith Glaciers, once described as the "weak underbelly of West Antarctica," is warming at a rate comparable to that of the Antarctic Peninsula. The modest ocean warming may have sufficed to initiate the recent acceleration of the Pine Island and neighboring Thwaites Glaciers by eroding their ice shelf bottoms. Pine Island Glacier alone now discharges 20 percent of the total ice lost from the WAIS.[29]

The WAIS is potentially unstable because large portions of the ice sheet rest on a substrate like a cereal bowl, with edges that slope down steeply landward. A landward-retreating grounding line allows more seawater to penetrate under the ice, setting it afloat (fig. 2.2; see also fig. 6.6). As more ice at the grounding line is eroded and undermined, landward retreat continues, the ice sheet surges forward, and calving increases. This runaway process, originally proposed by J. H. Mercer in 1978, has been called the marine ice sheet instability (MISI).[30] While most experts view a catastrophic collapse of the WAIS as an extremely improbable event this century, chapters 6 through 8 will examine recent evidence for greater potential instability than previously thought, raising the prospects of increased sea level rise.

In Greenland, smaller ice tongues at the mouths of marine glaciers play a role analogous to that of Antarctic ice shelves. Oceanfront changes in Greenland have sped up many of its tidewater glaciers.[31] The Jakobshavn Isbrae glacier accelerated following loss of its ice tongue, as changes in atmospheric circulation beginning in the late 1990s drove warmer ocean water shoreward.[32] Its rapid surging may be related to a MISI already under way, which has led to a 12.5-kilometer (7.8-mile) retreat of the grounding line over a 20-year period.[33] Recent surveys of the glacier bed depth indicate a reverse slope within a deep section of the confining fjord, which persists for at least the next 50 kilometers (31 miles) farther inland (a situation similar to the WAIS MISI, described in the previous paragraph). Fast retreat may persist for many decades until the grounding line reaches a shallow section of the fjord or climbs above sea level.

In contrast to ice shelves attached to land, icebergs float freely in the ocean, detached from their terrestrial origins and subject to the shifting whims of wind and currents. Their transitory existence ends as they melt away and merge with the vast ocean upon which they floated. As tidewater glaciers recede landward and ice shelves fall apart, icebergs grow more numerous. Their chief importance lies in the degree to which a greater quantity of melted icebergs (as well as sea ice) would lower high-latitude sea surface temperatures and dilute ocean salinity.

ICEBERGS

> *And the present sheared asunder from the past, like an iceberg sheared off from its frozen parent cliffs, and went sailing out to sea in lonely pride.*
>
> —Aldous Huxley (1894–1963)

Towering icebergs loom over a barren, brooding Arctic seascape in Frederic Edwin Church's masterpiece painting *The Icebergs* (1861) (fig. 2.3). A broken ship's mast lies trapped in ice in the foreground—a grim reminder of the dangers of the Arctic and the puny scale of human achievements in the face of nature. In 1859, Church (1826–1900), an American artist of the Hudson River School, sailed to Newfoundland and Labrador, where he sketched scenes for his Arctic paintings. The broken mast commemorates Sir John Franklin's ill-fated 1845 voyage to the northern seas in search of the Northwest Passage. Trapped in sea ice, Franklin and his men were forced to abandon their ships, *Erebus* and *Terror*, and set out in lifeboats. One by one, the men succumbed to illness, hypothermia, and lead contamination from food cans or the ships' pipes. After numerous unsuccessful search parties in the following years and retrieval of several relics from the expedition, the remaining mystery was finally solved with the location of the two ships on the ocean floor in 2014 and 2016, respectively (see chap. 1).

Icebergs were "once plentiful," some even "grand and imposing," near Battle Harbor, Labrador, a booming nineteenth-century fishing and

FIGURE 2.3

The Icebergs (1861) by Frederic Edwin Church.

sealing town.[34] Louis Legrand Noble (1813–1882), a writer who accompanied Frederic Church on his journey, described the "wandering alp of the waves," stating that "icebergs, to the imaginative soul, have a kind of individuality and life. They startle, frighten, awe; they astonish, excite, amuse, delight and fascinate; clouds, mountains and structures, angels, demons, animals and men spring to the view of the beholder." Individual icebergs received fanciful names: the "Alpine Berg," the "Great Castle Berg," and the "Rip van Winkle Berg." Other icy giants were compared to England's Windsor Castle, or to Niagara Falls.

Church sketched a mountainous iceberg floating near Twillingate, Newfoundland, on July 4, 1859. In the summer of 2008, another "4th of July Iceberg," reduced to a few exposed masses peeking above water, sat in a sheltered cove, "perspiring" and "melting into rivulets that cascaded down the slope of ice."[35] As ice cracked apart, the sound reverberated "like a cannon." Twillingate is now a prime tourist destination for viewing the passing parade of icebergs.

Icebergs form when chunks of ice break off tidewater glaciers and ice streams, or split off ice shelves. Most Arctic icebergs calve off major tidewater glaciers of Greenland and northern Ellesmere Island, Canada, with lesser inputs from the Svalbard archipelago north of Norway and islands of the Russian Arctic.[36] Greenland glaciers produce ~10,000 to 30,000 icebergs per year, or as many as 40,000 by some estimates. Many pass through the Fram Strait into the East Greenland Current and around southern Greenland. Swept north counterclockwise around Baffin Bay, they head south again along the coast of Baffin Island and Davis Strait toward the Labrador Sea. The Labrador Current carries the bergs farther south to "Iceberg Alley" along the Labrador and Newfoundland coasts. On average, only 1–2 percent make it as far south as St. Johns, Newfoundland (48°N latitude).[37] The peak season for icebergs drifting to 48°N extends from March to June, with two-thirds observed by April. However, actual numbers fluctuate greatly from year to year, reaching a high of ~2,200 in 1984 and none in 1966 and 2006. Many survive as far south as 39°–41°N, until they encounter the warmer waters of the Gulf Stream where they rapidly melt. (The southernmost iceberg ever spotted in the Atlantic reached around 30°N in 1926, some 240 kilometers [150 miles] southeast of Bermuda.)

Unlike the recent decline in Arctic summer sea ice (see "Losing Arctic Sea Ice" later in this chapter), iceberg numbers show no consistent trend, in spite of an apparent decline since the late 1990s. Nevertheless, more icebergs reached 48°N latitude in 2014 than in any year since 1998, underscoring the high degree of annual variability in iceberg production. Yearly iceberg counts may be related to winter sea ice coverage in the Davis Strait and Labrador Sea.[38]

Around Antarctica, icebergs split off ice shelves, such as the Ross and Ronne-Filchner shelves. They drift with the prevailing coastal currents; some run aground in embayments, while others are eventually entrained clockwise by the Antarctic Circumpolar Current and drift as far north as 55°–60°S, with a few occasionally spotted near southern New Zealand.

Tracking Icebergs

Although Inuit hunters, the Vikings, and whalers were quite familiar with icebergs, the tragic sinking of the *Titanic* in April 1912 was probably the most publicized historic encounter with a hulking mass of floating ice.

Long a menace to shipping, icebergs form by breaking off the edge of a glacier that ends in the sea, or off an ice shelf still attached to the shore. Moved by winds and currents, they slowly drift toward warmer waters where they eventually melt. The last tiny bits to melt fizz like seltzer due to the numerous air bubbles trapped in ice, originally snow compressed into firn and ultimately glacial ice.

The International Ice Patrol (IIP), founded in 1913 in response to the sinking of the *Titanic*, monitors the presence and movement of icebergs in the North Atlantic and Arctic Oceans, along major shipping lanes.[39] Operated by the United States Coast Guard, the International Ice Patrol is funded by a consortium of 13 nations from North America, Europe, and Japan involved in transatlantic shipping. The IIP tracks icebergs from aircraft (and increasingly, satellites) between February and August; the balance of the year is covered by the Canadian Ice Service, which cooperates closely with the IIP. Passing ships also provide information on ice conditions.

The United States National Ice Center (NIC), operated jointly by the National Oceanic and Atmospheric Administration, U.S. Coast Guard, and U.S. Navy, provides accurate and timely information on ice and snow conditions. For example, the NIC generates daily and weekly maps of sea ice extent and ice edge around polar regions, maps the sea ice edges around Alaska and the Great Lakes (the latter when ice is present), provides daily maps of Northern Hemisphere snow cover, and provides weekly locations of major Antarctic icebergs. Synthetic aperture radar (SAR) on satellites, such as the Canadian RADARSAT-2 and European CryoSat-2, now provide high-resolution images independent of weather and time of day or night. Other instruments that track icebergs from space include MODIS, Landsat 7 and 8, Thematic Mappers, and ASTER, which acquire multispectral images in the visible, near- tomiddle-infrared regions (appendix B).

ICEBERG CHARACTERISTICS

Icebergs look bright white because of myriad tiny trapped air bubbles that scatter all wavelengths of light. Bubble-free ice, on the other hand, appears a soft aqua to deep neon blue, because ice absorbs more of the

TABLE 2.2 Iceberg Sizes

	HEIGHT		LENGTH	
CATEGORY	METERS	FEET	METERS	FEET
Growler	<1	<3	<5	<16
Bergy bit	1 to 4	3 to 13	5 to 14	15 to 46
Small	5 to 15	14 to 50	15 to 60	47 to 200
Medium	16 to 45	51 to 150	61 to 120	201 to 400
Large	46 to 75	151 to 240	121 to 200	401 to 670
Very large	>75	>240	>200	>670

Sources: Wadhams (2013); Diemand (2001, 2008).

red wavelengths, allowing more blue and blue-green light to be transmitted or reflected.

Icebergs come in a wide variety of sizes and shapes, ranging from the smallest, colorfully named growlers and bergy bits a few meters across to massive giants over hundreds of meters across (table 2.2). Iceberg B-15, which calved off the Ross Ice Shelf, Antarctica, in March of 2000, holds the current world record at 11,000 square kilometers (4,250 square miles).[40] It subsequently broke into several pieces, the largest of which, B-15A, was over half the size of the original. Shapes vary from very large, flat-topped, tabular masses such as B-15 to cube-shaped, steep-sided blocks to alp-like pinnacles or spires to domes, wedges, or quite irregular shapes. Church's paintings vividly capture some of these fantastic forms (fig. 2.3). Spired, domed, and steep, pinnacle-shaped icebergs occur more frequently in the Arctic, whereas flat, tabular masses are more characteristic of Antarctica. (The closest Arctic analogues to Antarctica's tabular icebergs are the flat-topped ice islands calved off the ice shelves of northern Ellesmere Island.) A chaotic mix of ice and icebergs less than 2 meters (6.5 feet) is referred to as brash ice.

Icebergs inherit the physical characteristics of their sources. Arctic icebergs have generally calved off tidewater glaciers, whereas those from Antarctica have broken off flat-topped ice shelves or ice streams. Recent snow atop large, tabular Antarctic icebergs blankets a porous firn layer, with a surface density of 0.45 grams per cubic centimeter, increasing to

0.86 to 0.89 grams per cubic centimeter at depths of 60 meters. In contrast, Arctic icebergs lack this firn layer, since they have split from glaciers whose once-snow-covered surfaces have long since compacted into ice by the time the glaciers reach the sea. Thus, densities of northern icebergs approach that of pure ice.[41] Icebergs with near-vertical sides often preserve annual layering from successive snowfalls. But others display near-parallel or contorted dark stripes. These represent layers of sediment and rock debris that fell on the glacier's surface and were subsequently deformed by the glacier's motion. Reddish-brown streaks signal the presence of iron oxide impurities. Greenish streaks derive from algae growing on the underside of an iceberg or ice floe and later freezing to the base by crystallization of seawater.

The life expectancy of an iceberg varies greatly, depending on size and drift patterns. Large icebergs that travel from Greenland around Baffin Bay may last 3 years or more before exiting Davis Strait into the Labrador Current.[42] During the summer and fall in the Labrador Sea, a small, irregular iceberg may disintegrate within days, while larger ones can survive several weeks.

Tides, storm waves, underwater currents, and collisions with nearby icebergs exert stresses on the floating extension of a marine glacier or ice shelf. Friction along the base and edges of a moving glacier stretches the ice and opens cracks, or crevasses. These crevasses may then become incorporated into the iceberg. Variations in ice velocity set up strains that determine the location and depth to which crevasses will penetrate. Other crevasses develop at the boundary between the floating and grounded portions of the ice (i.e., where ice rests on solid bedrock), or near the shelf's seaward edge (fig. 2.2). Repeated bending and flexing by ocean waves helps break up large, tabular bergs and ice shelves.[43] Major calving events in Antarctica, such as that of B-15, occurred after huge rifts, up to several hundred miles long and a few miles wide, formed in the ice shelf and penetrated the full thickness of the shelf. Large pieces of ice split off more readily once crevasses reach the base.

Icebergs deteriorate further by breaking apart and by melting, especially in warmer water. Seawater, generally several degrees warmer than ice, melts it at or below the waterline.[44] This underwater erosion may unbalance the iceberg and topple it over, posing a danger to passing ships.

In addition, wave action can erode a notch into the ice, eventually undermining the iceberg enough to cause pieces to split off.[45]

Aside from their threat to shipping, exemplified by the tragic sinking of the *Titanic*, icebergs indirectly influence global sea level and ocean circulation. Since icebergs displace their weight in water and float, they do not raise sea level when they melt. However, the number of icebergs shed by glaciers does affect sea level indirectly. All other things staying equal, an increasing quantity of calved icebergs is one measure of diminishing ice mass on glaciers or ice sheets. Unlike ice already afloat, the newly added ice will raise sea level.

Melting of icebergs dilutes the salinity of seawater, which in turn may weaken the sinking of dense, cold North Atlantic Deep Water and slow down the Atlantic Meridional Overturning Circulation (AMOC), as described in chapter 1. Such changes in ocean circulation could boost regional sea level along the east coast of North America. But a different form of floating ice may more strongly influence regional and global climate.

Unlike ice shelves and icebergs, sea ice has no prior connections to land; it freezes directly from seawater. In the winter season, sea ice completely fills the Arctic Ocean and encircles Antarctica. What is sea ice, and in what ways does it differ from an ice shelf or iceberg? How do losses of sea ice affect the vanishing cryosphere? The next section will seek answers to these questions and more.

SEA ICE

Until fairly recently, sea ice has presented a nearly impenetrable obstacle to ships navigating Arctic waters. Most of the historic expeditions to the Arctic in search of a Northwest or Northeast Passage ended in failure or disaster, victims of the harsh climate and treacherous sea ice. This is rapidly changing with the sharp decline in Arctic Ocean sea ice cover since the 1980s—one of the most conspicuous signs of an endangered cryosphere. The September 2012 Arctic sea ice extent set a new record low (see fig. 1.4), while 2016 tied with 2007 as second lowest.[46] A nearly ice-free summer Arctic Ocean could become a reality within decades (see "Losing Sea Ice," later in this chapter).

An open Arctic Ocean would accomplish the fondest dreams and ambitions of the intrepid, early navigators, opening up the Arctic to routine commercial shipping, economic development, and tourism, but could also lead to greater environmental degradation, invasion by foreign species, and new challenges for indigenous populations.

Sea ice strongly influences the climate system. The high albedo of sea ice cools the surroundings, while a thick ice cover insulates the ocean from heat losses and inhibits the exchange of gases, such as water vapor and carbon dioxide, between ocean and atmosphere (see chap. 1). Freezing sea ice expels salt, which modifies the density of seawater, and thereby may alter ocean circulation. Changes in regional temperatures also modulate sea ice extent and thickness, initiating a string of feedbacks that in turn may affect the climate over broader regions. Finally, sea ice constitutes a major habitat for the marine mammal, fish, and plant ecosystems upon which many Arctic indigenous populations depend (chap. 9).

Growth and Decay of Sea Ice

Sea ice, unlike icebergs or ice shelves, forms directly from ocean water. The average ocean water salinity of around 35 parts per thousand (ppt), or 3.5 percent, depresses the freezing point from 0°C (32°F) to around −1.8° to −1.9°C (28.8°–28.6°F). In calm water, sea ice forms tiny, platy ice crystals under 2–3 millimeters (0.08–0.12 inch) in diameter, floating on the surface (box 2.1).

The platy crystals soon spread into a thin layer of frazil, or grease ice (named after its appearance), that coalesces into a thin, translucent sheet of nilas ice (fig. 2.4a). Nilas ice first turns gray, then white, eventually thickening into a more stable, smooth-bottomed sheet called congelation ice, characterized by downward-pointing, elongated crystals (box 2.1). Although the growing ice crystal squeezes out most of the salt in a process called brine rejection, some seawater is trapped in fluid inclusions. When this ice refreezes after the summer melt season, the salinity is reduced even further.

In rougher water, the frazil crystals collide and are compressed into slushy cakes that develop into larger slabs of pancake ice (named after their shape), often with raised rims due to multiple collisions (fig. 2.4b). These coalesce into still larger floes that often override or raft onto one

BOX 2.1

The crystal lattice of ordinary hexagonal ice illustrated in figure 1.2a is characterized by a c-axis with sixfold symmetry (in the vertical direction) and three equal a axes that lie 120° apart in a plane perpendicular to c. Typical ice crystals are flattened hexagonal plates perpendicular to c (like the snowflake in fig. 1.2b) and hexagonal prisms, where c is the long axis. Sea ice initially forms flat, platy hexagonal crystals, oriented with c-axes in the vertical direction, i.e., perpendicular to the ocean surface. Further growth occurs sideways, often as dendritic (branching) arms, like those of snowflakes. These fragile crystals break apart easily, but quickly reassemble and merge into a thin layer of translucent ice. By the time crystal grains are more or less touching, further horizontal expansion is inhibited, and therefore growth continues downward. In a type of "crystalline Darwinism," those plates favorably oriented with their c-axes horizontal (i.e., now parallel to the ocean surface) grow preferentially downward into larger, elongated columnar crystals, characteristic of congelation ice.*

*The preferred growth direction points downward (in the plane of the a-axes); Wadhams (2003).

another, or form ridges, and ultimately expand into a continuous sheet of consolidated pancake ice. Ice ridges that form by collision between thicker floes can rise 2 meters (6.6 feet) or more above water, with a correspondingly thicker submerged portion. In the Arctic, first-year ice that lasts into spring ranges from at least 30 centimeters (1 foot) to 1–2 meters (3.3–6.6 feet) deep.[47] Multiyear ice that has survived two or more summers may grow as thick as 4–6 meters (13–19.71 feet) deep.

By summer, the overlying winter's snow and ice begin to melt, forming surface pools. Meltwater thaws its way or trickles downward through small pores, crevasses, and channels, flushing out much of the remaining brine. Thus, sea ice that survives the summer melt season is purer than first-year ice. The thin, young sea ice admits more of the sun's energy, which enhances bottom melting.

(a)

(b)

FIGURE 2.4

Different types of sea ice: (*a*) nilas ice ("Nilas sea ice 3," September 2005, by Brocken Inaglory/Wikimedia Commons, https://commons.wikimedia.org/wiki/File:Nilas_sea_ice_3.jpg); and (*b*) pancake ice (photo courtesy of Ted Scambos/National Snow and Ice Data Center).

LOSING ARCTIC SEA ICE

November of 1978 ushered in a new era of Arctic sea ice observations with the launch of the Nimbus-7 Scanning Multichannel Microwave Radiometer (1978–1987). It was followed by the Special Sensor Microwave Imagers (SSM/I) aboard U.S. Defense Meteorological Satellite Program (DMSP) satellites beginning in 1987. These and other satellites reveal a steady decline in sea ice covering the Arctic Ocean since 1979, with greater losses since the late 1990s.[48] Recent satellite measurements demonstrate a clear correlation among vanishing sea ice, a darker Arctic Ocean, and greater polar warming.[49] The albedo declined from 0.52 to 0.48 between 1979 and 2011. Arctic air temperatures have been rising faster than those in the rest of the Northern Hemisphere, especially toward the latter part of the twentieth century—a phenomenon known as "polar amplification" or "Arctic amplification." In recent decades, Arctic temperatures have climbed almost twice as fast as those in the rest of the world. The years 2011 to 2015 were warmer than at any time since 1900.[50]

The September (summer) minimum sea ice extent plunged by 13.2 percent per decade between 1979 and 2017 as compared to the average for 1981–2010[51] (fig. 2.5). Across the Arctic, the melt season lengthened by five days per decade between 1979 and 2016.[52] Although more ice melts toward the end of summer than in spring, the earlier melt onset is more significant because it occurs at a time of year with nearly 24-hour daylight and solar energy input to sea ice reaches a peak.

Winter maximum (February/March) sea ice extent has followed a persistent downward trend for the last 39 years and reached record low levels 4 years in a row, from 2015 to 2018. Arctic sea ice hit its lowest winter maximum extent to date on March 7, 2017.[53] A year later, March of 2018 ranked second lowest.[54] Most troubling is the plummeting older, thicker multiyear ice, down from 20 percent of the total Arctic ice extent in the 1980s to less than 5 percent by 2014.[55] Ice older than 4 years constituted only 1.2 percent of the total by 2016; most sea ice is now first-year ice.[56] The marked decline in old sea ice and its replacement by younger, thinner ice may be one of the most important changes affecting the Arctic cryosphere.[57]

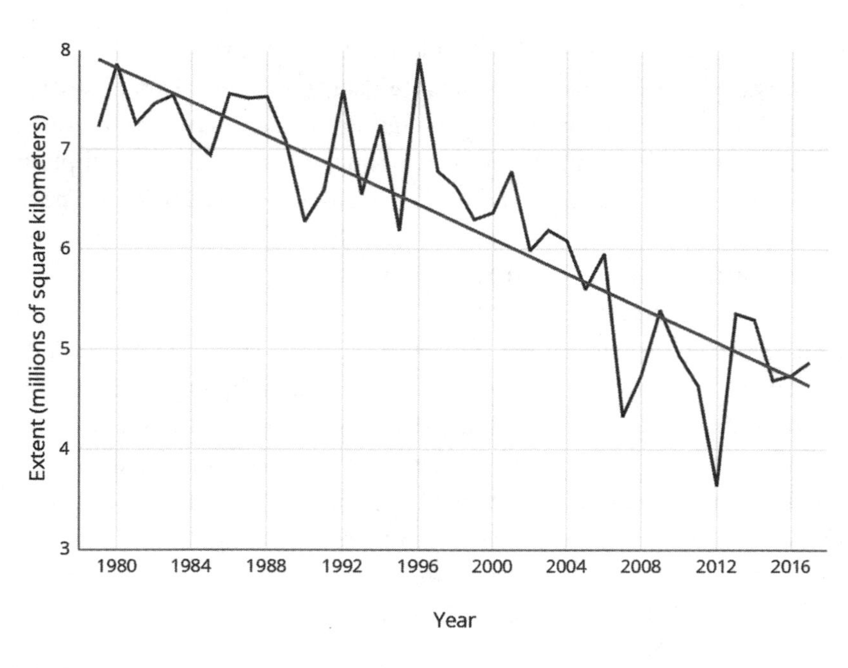

FIGURE 2.5

Declining Arctic sea ice: average September Arctic sea ice extent, 1979–2017. (National Snow and Ice Data Center, "Arctic Sea Ice 2017: Tapping the Brakes in September," October 5, 2017, http://nsidc.org/arcticseaicenews/2017/10/arctic-sea-ice-2017-tapping-the-brakes-in-september/; figure courtesy of the National Snow and Ice Data Center)

Unlike other Arctic regions, the North Pole still possesses a nearly intact sea ice cover at the end of the melt season, ranging from 1.25 to 2.6 meters (4.1–8.5 feet) in thickness.[58] Nevertheless, several recent years have witnessed record summer ice minima, especially 2007, 2012, and 2016. Twice as much ice melted from below between 2008 and 2013 than between 2000 and 2005. Even in this region, sea ice may not remain intact for long.

These trends highlight the growing dominance of thin, first-year ice, which is more fragile and brittle, melts faster in summer, and quickly expands the area of open water. As a consequence, the ocean, which absorbs more heat during the summer, can release its stored heat to the atmosphere in the fall and early winter. The increased autumn warming delays winter freeze-up and lessens sea ice cover. Spring melting now begins about two weeks earlier than in the 1980s.[59] Summer has warmed the least, because it takes considerable energy to melt residual sea ice and heat the upper ocean. During a lengthened melt season, more snow overlying ice thaws, and the number of surface melt ponds multiplies. Darker pond surfaces absorb additional solar energy. The warmer upper ocean also melts sea ice from below. This chain of events reinforces the sea ice-albedo feedback (chap. 1; fig. 2.6).

Have other factors besides Arctic warming amplified recent sea ice losses? How much have natural climate fluctuations, such as shifts in winds or currents, or variations in weather patterns such as the Arctic Oscillation (AO), contributed to the recent negative trend? The AO is a weather pattern characterized by atmospheric pressure differences between the Arctic and midlatitudes. In its positive phase, higher-than-average midlatitude pressures and lower polar pressures strengthen the polar vortex—a west-to-east flow of high-altitude Arctic winds that locks cold air in the Arctic and drives midlatitude storms northward. Precipitation tends to increase in Alaska and Northern Europe. During its negative phase, pressure

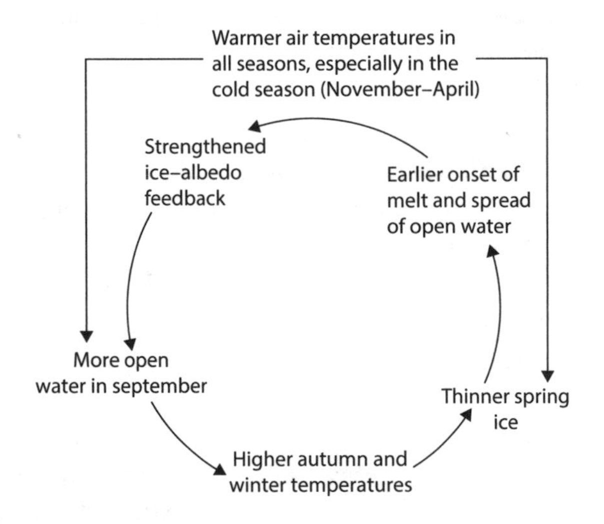

FIGURE 2.6

Is Arctic sea ice caught in a downward spiral? Warming–sea ice–albedo feedbacks that hasten the decline. (After Stroeve et al. [2012], fig. 2)

differences reverse, allowing a weaker polar vortex and wavier jet stream to push midlatitude cold outbreaks farther south. However, AO indices have varied inconsistently since the mid-1990s.

Global warming affects vertical atmospheric temperatures differently near the poles than at the tropics.[60] This too contributes to sea ice loss. As tropical surface temperatures increase, more moist air masses rise aloft and develop into thick storm clouds. Heat released by condensation of water vapor in the clouds then warms the overlying atmosphere more than near the surface. On the other hand, cold, dense Arctic air, even if warmed by a few degrees, stays near the surface and does not mix with air aloft. Instead, any added heat penetrates into the upper ocean, where it melts more sea ice and reinforces Arctic amplification. Abundant late summer–early autumn low clouds at a time of minimal sea ice cover also increase evaporation and relative humidity. The open water and fall-winter release of stored ocean heat help strengthen Arctic warming.[61] The shrinking Arctic sea ice cover therefore results from natural atmospheric variability, together with thinner ice cover and increased summer melting.[62]

Is Arctic sea ice therefore trapped in an irreversible downward spiral? Is summer sea ice approaching a point of no return—a so-called "tipping point"? How soon will we see an ice-free Arctic Ocean in summer? The record-shattering (to date) September of 2012 surprised experts and generated intensive speculation over the fate of Arctic sea ice. Such a speedy retreat took climate modelers by surprise. This led some researchers to speculate that summer Arctic sea ice could disappear within a few decades. But this opinion may be premature because of the large year-to-year variability (see, e.g., fig. 2.5) and capacity for ice recovery under favorable conditions. Negative feedbacks could partially counteract the ice-albedo and cloud feedbacks, at least temporarily. For example, thinner ice grows back faster than thick ice. As thin ice floes converge, they can raft and form thicker and stronger ridges. Therefore, a few cold years could temporarily slow down or reverse the recent negative trend. Such random fluctuations in natural processes could in fact account for as much as 40 percent of the recent sea ice losses.[63]

Even so, at currently high rates of greenhouse gas emissions, most Arctic summer sea ice could disappear by midcentury; lower emissions would postpone this until the late twenty-first century. Some researchers go so far as to suggest that a few more extreme ice-loss events, such as in 2012, or merely continuation of the recent decadal trend could eliminate most summer sea ice as early as the 2020s or 2030s.[64] With sustained warming, recent extreme summers like 2012 could become the new norm.

The Long Reach of Arctic Sea Ice

> *If the jet stream continues to exhibit slower-moving, higher-amplitude waves, these harsh weather conditions will grow even more intense and stay in place longer, multiplying their potential for death and destruction. If the theories . . . are correct, there is no going back to our old climate unless we find a way of growing more Arctic sea ice.*
>
> —Jeff Masters (2014)

The chain of cascading climate changes triggered by diminishing Arctic sea ice may reverberate far to the south. Be prepared for an increasingly "weird" jet stream, according to Jeff Masters, director of meteorology at Weather Underground. "Our recent jet stream behavior could well mark a crossing of a threshold into a new, more threatening, higher-energy climate," he says.[65] Arctic amplification has reduced the temperature contrast between midlatitudes and the North Pole. Some scientists therefore speculate that a weakened north-south temperature gradient makes the jet stream more sluggish and wavier, much like looping meanders in the flat floodplain near a river's mouth. The slowly propagating, sinuous air circulation pushes warmer temperatures north and Arctic cold south and sets the stage for "blocking patterns," or persistent extreme weather in some regions: heat waves or droughts in summer or colder, snowier winters.[66] Low summer/autumn sea ice may also stack the deck for more frequent outbreaks of a "warm Arctic/cold continent" pattern, analogous to the negative phase of the AO, in which a weaker, wavier jet stream brings outbreaks of frigid Arctic temperature and snowstorms to the continental United States, Europe, or eastern Asia, while conversely, balmier conditions prevail in parts of Alaska or Siberia.

The sea ice meltdown, a warming Arctic, and a weaker near-surface north-south temperature gradient could potentially affect the polar vortex, storm tracks, and the jet stream, which in turn govern midlatitude weather patterns. Some evidence points to an association between periods of a wavier midlatitude jet stream and regional weather extremes over the last few decades, such as heat waves over western North America and Central Asia, cold waves over eastern North America, and droughts in Central Asia and Europe.[67] Yet, present-day links among sea ice meltdown, Arctic amplification, and atmospheric waviness remain inconclusive and still poorly understood. Nevertheless, most experts agree that continued Arctic warming will produce far-reaching effects in the future.[68]

Why Is Antarctic Sea Ice Expanding?

While Arctic sea ice extent has been sliding downhill since the 1980s, its Antarctic counterpart has held steady or grown slightly. Antarctic winter sea ice extent reached record maxima in 2012, 2013, and 2014. Why the contrarian Antarctic behavior? The answer, in part, lies in the marked geographic differences between the two polar regions. The Arctic Ocean, unlike the Southern Ocean, is almost entirely surrounded by land, so that most sea ice just stays there. Warmer ocean water can penetrate into the Arctic from the south via the Barents Sea and the Bering Strait, facilitating sea ice melting. Sea ice surrounding Antarctica, on the other hand, unconstrained by landmasses, drifts northward with prevailing winds and currents into warmer waters where it eventually melts. Thus, nearly all Antarctic sea ice that forms in winter melts the following summer and is therefore much thinner than in the Arctic (typically only 1–2 meters [3–6 feet] thick versus 2–3 meters [6–9 feet] thick, respectively). Furthermore, strong westerly winds and currents (e.g., the Antarctic Circumpolar Current) prevent warmer air and water from reaching Antarctica, keeping it cold. The Southern Ocean also delivers more moisture to Antarctica, covering sea ice there with a thicker, better-insulating layer of snow than is found in the drier Arctic.

Paradoxically, global warming may have indirectly led to the expansion of Antarctic sea ice. This apparent contradiction may be understood by closely examining the thread connecting recent thinning of Antarctic ice shelves, ice sheet losses, and sea ice growth.[69] The Southern Ocean surrounding Antarctica has warmed significantly at depths between 100 and 1,000 meters (330–3,300 feet) in recent decades, as have most oceans elsewhere, whereas melting directly beneath ice shelves has cooled and freshened the upper 100 meters.[70] Recall that Antarctic ice shelves can be up to hundreds of meters thick, whereas sea ice is only several meters thick. Antarctic sea ice freezes rapidly in winter and expands into the colder, fresher surface water. On the other hand, the base of the ice shelves is bathed by warmer Circumpolar Deep Water, which thins and weakens them, occasionally to the point of spectacular disintegration, as in the case of the Larsen B and Wilkins Ice Shelves. Glaciers and ice streams abutting thinned ice shelves have often advanced and shed icebergs that cool the upper ocean layer and stimulate sea ice growth.

Changing regional wind patterns may also play an important role.[71] Shifting winds may have physically pushed sea ice to cover wider areas or locally warmed or cooled the ocean surface. Changes in both winds and ocean temperatures may have led to the contrarian growth of Antarctic sea ice, and its paradoxical recent reversal, but further research is needed to resolve this issue.

In spite of any apparent Antarctic gains, sea ice continues its global downward slide.[72] Furthermore, the once-expanding Antarctic sea ice cover has shown signs of contraction ever since September 2016, a period that saw record or near-record lows. The icy continent experienced its second-lowest summer sea ice minimum in late February of 2018, following the record low set on March 3, 2017. Whether this is a fluke or the beginning of a new trend, only time will tell.

While the greatest volume of ice resides on land and sea, a hidden world of ice also lies buried in soil near the poles and within high mountain valleys. The next chapter explores this hidden world and how it too is rapidly changing, with potentially far-reaching consequences that extend beyond its borders.

3

IMPERMANENT PERMAFROST

Some of Phil Camill's trees are drunk. Once, the black spruce trees . . . in northern Manitoba stood as straight and honest as pilgrims. Now an ever-increasing number of them loll about leaning like lager louts. The decline is not in the moral standards of Canadian vegetation, but in the shifting ground beneath their roots.

—Gabrielle Walker, "Climate Change 2007: A World Melting from the Top Down" (2007)

THE ICE BENEATH OUR FEET

The Nature of Permafrost

A hidden world of ice lies concealed in soil beneath our feet near the poles and in high mountain valleys. Permafrost, or perennially frozen ground, underlies nearly a fifth of the world's land area.[1] But this netherworld is also rapidly changing as rising temperatures rapidly defrost the permafrost, sometimes inadvertently uncovering once-buried treasures. Hunters or herders in remote parts of northern Siberia occasionally stumble across long, slender mammoth tusks poking out of the tundra. The newly exposed mammoth ivory has fueled a lucrative trade in recent years, spurred by the combined effects of thawing permafrost, the ban on imports of African elephant ivory, and a growing demand from China, Japan, and Korea, where the material is carved into decorative objects and used for personal seals.[2] The commercialization of mammoth ivory provides income for indigenous people, but could also obstruct scientific analysis of valuable data on ancient environments contained within the fossils. On the other hand, ivory, once exposed to the elements, disintegrates within a few years and will be lost forever unless quickly salvaged. Deicing permafrost not only unlocks buried treasures; its effects sweep across the Arctic. The meltdown tilts trees at crazy angles, topples cliffs, buckles roads, collapses buildings, crumbles shorelines, transforms the landscape and hydrology, and could also alter the regional ecology. But these visible changes may be just the tip of the iceberg. Will we face a carbon apocalypse, as some apprehensive experts have warned? Is decomposing permafrost a ticking carbon time bomb poised to unleash yet more methane into our increasingly heavy load of atmospheric greenhouse gases? Will enough methane bubble up from soggy Arctic soils and the seabed to hasten the ongoing warming trend in a positive feedback loop? What does the thawing permafrost portend for the Arctic and the world beyond? This chapter delves into some of these questions.

Temperature, rather than ice content, defines permafrost: a soil, sediment, or rock in which ground temperatures stay below 0°C (32°F) for at least two consecutive years. Therefore, climate conditions determine the geographic distribution of permafrost. Frozen ground exists worldwide at high latitudes, at high altitudes, and on continental shelves surrounding the Arctic Ocean (fig. 3.1). (Most marine permafrost dates to the last ice age, when a much lower sea level exposed shallow portions of the continental shelves to frigid temperatures. The frozen soils were subsequently drowned when the ice sheets retreated and sea level rose.) Today, the southern limit of most Northern Hemisphere continuous permafrost, where over 90 percent of the area remains frozen, roughly coincides with average yearly air temperatures of −8°C (18°F).[3] Discontinuous permafrost (underlying between 50 and 90 percent of an area) develops where average yearly temperatures are −1°C (30°F).

In the Northern Hemisphere, the permafrost layer thins progressively from north to south and ultimately grades into a patchy, discontinuous zone near its southern boundary. Soil also freezes seasonally wherever ground temperatures remain below freezing for several weeks. Asia boasts the greatest expanse of permafrost—in northeastern Russia,

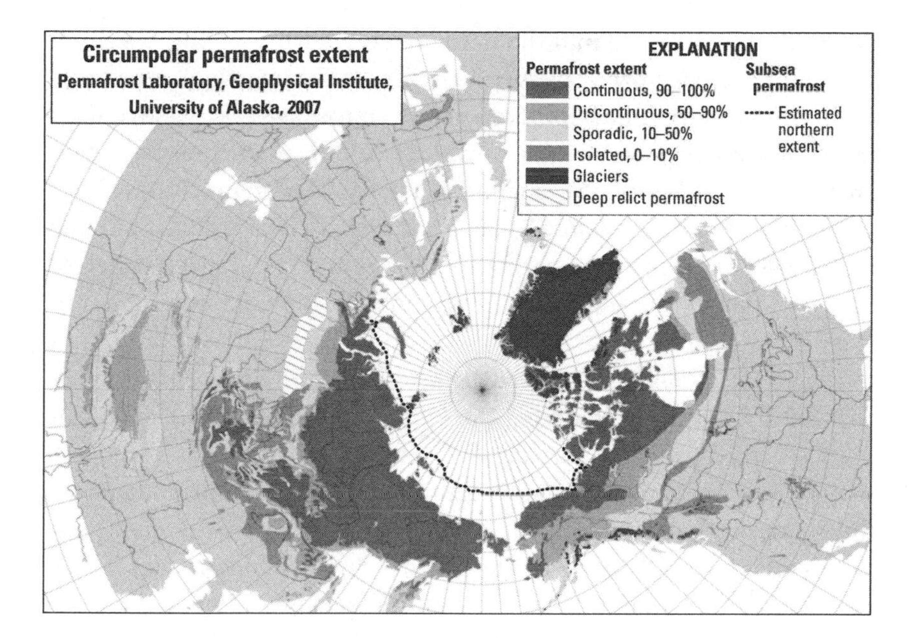

FIGURE 3.1

Distribution of permafrost in the Northern Hemisphere. (Map derived from Brown et al., 1997, "Circum-Arctic Map of Permafrost and Ground-Ice Conditions," http://nsidc.org/fgdc/; http://pubs.usgs.gov/pp/p1386a/images/Cryosphere_Notes/full-res/pp1386a_fig6-1.jpg, accessed October 10, 2017)

including Siberia; the Tibetan Plateau; northeastern China; and parts of Central Asia. Permafrost also extends over most of Alaska and vast areas of northern Canada and Scandinavia. Maximum permafrost thicknesses range from 1,500 meters (5,000 feet) in northern Siberia to around 400 to 700 meters (1,575–2,300 feet) thick in North America.[4] In the Southern Hemisphere, it occupies nearly all of the ice-free regions of Antarctica, islands near Antarctica, and the southern Andes. It also occurs at very high elevations at low latitudes.

In areas of yearlong subfreezing temperatures, some of the frozen ground does not thaw completely in summer. Over time, a permafrost layer develops, gradually thickening each year. The active layer freezes in winter, thaws in summer, and extends about 1 to 10 meters (3.3–32.8 feet) down from the surface into the permafrost. It forms during the short, cool summers. Unfrozen layers or lenses, or taliks, may lie interspersed between permafrost and the base of the active layer. The active layer thins toward the poles or high elevations, and also wherever it is covered by an insulating, thick winter snowpack or thick summertime vegetation. It represents an exchange zone where heat travels back and forth between surface and permafrost, gases and water are transferred between atmosphere and permafrost, and nutrients and water are provided for vegetation.

Permafrost thickness exists delicately balanced between the near-constant upward heat flow from the Earth's interior and heat propagating downward from the surface. The permafrost layer ends at a depth where the temperature rises above freezing (fig. 3.2). Temperatures from the interior rise at a rate known as the geothermal gradient, due to the Earth's internal heat.[5] The flow of heat from below depends on the thermal properties of ice, soil, or rock, and on local geology, being higher near active volcanoes or regions of mountain building than in geologically quiescent areas.

Short-lived and longer temperature perturbations at the surface affect downward-flowing heat. Wide daily and seasonal temperature swings at or near the surface dampen downward to the depth of zero amplitude, around 10–25 meters (32.8–82 feet) below the surface, where temperatures remain constant year-round. Below this level, the geothermal gradient determines any further downward increase in temperature (fig. 3.2). Seasonal temperature variations are generally confined to the upper 20 meters (64 feet) whereas longer (decadal to centennial) temperature changes penetrate to greater depths—100 meters (328 feet) or more.

A thermal disturbance instigated by a climate change propagates downward until it merges at depth with the undisturbed geothermal gradient (fig. 3.2). It also shifts the temperature-depth curve toward higher (or lower) temperatures. As a wave of surface warming gradually moves downward, the top of the permafrost layer begins to thaw. The higher surface temperature lowers both the top of the permafrost layer and the depth of zero temperature amplitude. This shifts the temperature-depth

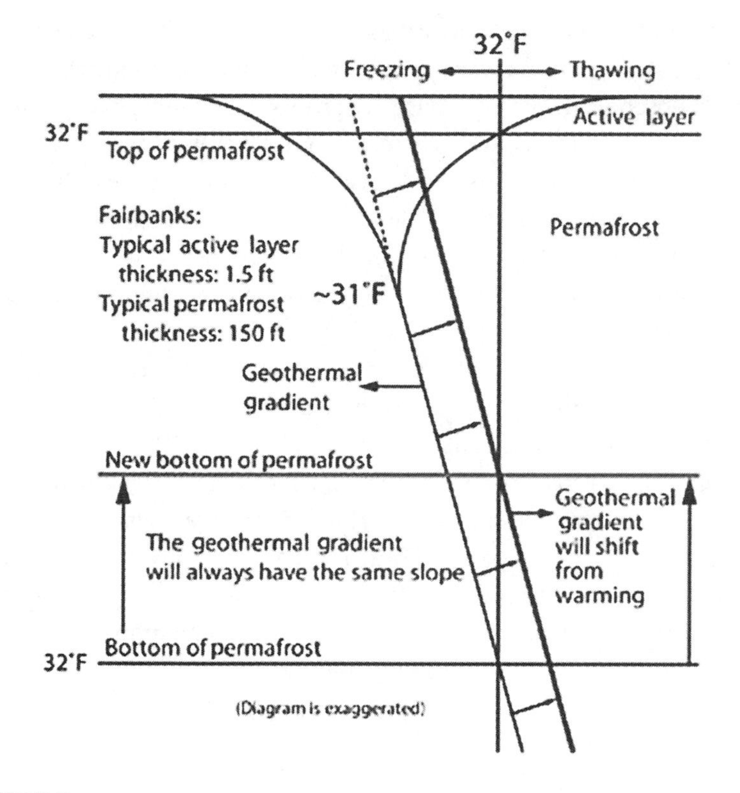

FIGURE 3.2

Schematic temperature profile in permafrost terrain. The active layer at the top is the region that freezes in winter and thaws in summer. The depth of zero amplitude is the point at which annual temperature variations become zero (marked ~31°F [−0.6°C] on diagram). (U.S. Army Corps of Engineers, "Permafrost: Climate Change," https://www.erdc.usace.army.mil/CRREL/Permafrost-Tunnel-Research-Facility/Climate-Change/)

curve toward the right in figure 3.2. The base of the permafrost layer rises because the base intersects the freezing point at a higher elevation under the surface. This causes the permafrost layer to shrink from both top and bottom.

A sensitive thermometer, such as a thermistor or thermocouple, lowered down a narrow-diameter borehole measures soil temperatures at different depths. Temperature profiles vary from place to place because the internal heat flow and temperature history may also vary locally. To reconstruct the thermal history of a given locality, the best match is sought between observed borehole data of downwardly propagating temperature changes and mathematical models of likely temperature histories. Inasmuch as the geothermal gradient remains constant over millennia, any deviation from this theoretical temperature profile at depth represents pulses of recent surface temperature changes traveling downward. The Global Terrestrial Network for Permafrost (GTN-P), organized in the 1990s and managed by the International Permafrost Association, monitors changes in active-layer thickness and permafrost temperatures.[6] A network of over 1,000 boreholes collects ground temperature measurements in North America, the Nordic countries, and Russia.

Some of the earliest boreholes drilled during the 1980s by Arthur H. Lachenbruch and his colleagues from the United States Geological Survey in Alaska revealed temperature anomalies suggesting a twentieth-century regional warming of around 2° to 4°C (3.6–7.2°F).[7] Warming signals from more recent boreholes throughout the Arctic further support an increasingly balmy northland (see the following section).

As Ground Freezes

> *If you want to dig a ditch in the Arctic, you'd better bring more than a shovel.*
>
> —Anonymous[8]

As temperatures plunge below 0°C (32°F), pore ice—water trapped in pores between soil or sediment grains, pebbles, and rocks, and within narrow rock fissures—solidifies. Unlike most liquids, water expands around 9 percent upon freezing, exerting a pressure that pushes soil particles aside.[9] Soil volume expands still further when an external water source builds up ice. The growth of buried ice masses buckles soil upward. Needle ice is one of the earliest types of ice to form in late fall or in winter. Narrow ice needles appear as water in soil pores freezes, drawn upward by capillary action toward the cooling surface. With continued growth, ice needles cluster into vertical columns.[10] The force of crystallization pushes small soil particles aside and upward. This process loosens soil, making it more susceptible to soil creep and erosion on slopes, and creates frost heave.[11]

Frost heave develops as water moves along as extremely thin films between soil particles. Presence of a thin water film is more effective than dry ice expansion alone. Ice initially freezes within tiny soil or sediment pores or in minute rock fractures. Premelted water (a thin molecular film of water adhering to particles) coats surfaces between ice and soil, remaining stable well below the normal freezing point. Water migrates upward along these wet films and feeds a growing ice lens, or layer of buried ice, lying parallel to the surface. The enlarging ice mass in turn dislodges overlying mineral grains and thrusts soil upward forcefully enough to seriously damage roads, bridges, and buildings (see Building on Unsteady Ground, later in this chapter). Often, multiple bands of clear ice develop, sandwiched between ice-free soil layers. Frost heaving occurs optimally in soils of silt-sized particles (0.002–0.05 millimeters in diameter). Sand-sized grains (0.05–2 millimeters) are too coarse (water drains out too fast), whereas fine clays (<0.002 millimeters) are too impermeable. Other favorable conditions include a continuous supply of water, below-freezing soil temperatures, suitable cooling rates (a sudden freeze could halt ice lens accumulation), and a thick, insulating snow or vegetation cover to modulate freezing rates.

Ground ice also forms ice wedges and pingo ice (see the following section). Ice wedges—carrot-shaped ice masses up to 3 meters (9.8 feet) wide at the surface—taper to several centimeters across, some 10 meters (32 feet) below the surface (fig. 3.3a). When winter temperatures dip below −15°C (5°F), soil contracts and becomes brittle, cracking with a sharp noise and tremor. Water enters the contraction cracks during the spring thaw, but refreezes within a narrow, nearly vertical vein the following winter. Ice expansion (9 percent) widens the vein, shoving aside adjacent and overlying soil. The ice-filled vein, weaker than solidly frozen soil, breaks open again the following winter. This recurrent process, year after year, builds up quite large wedges. Vertical banding in the ice wedge marks the scars of repeated breaking and healing cycles. Sand may replace ice in dry or well-drained environments, creating sand wedges instead.

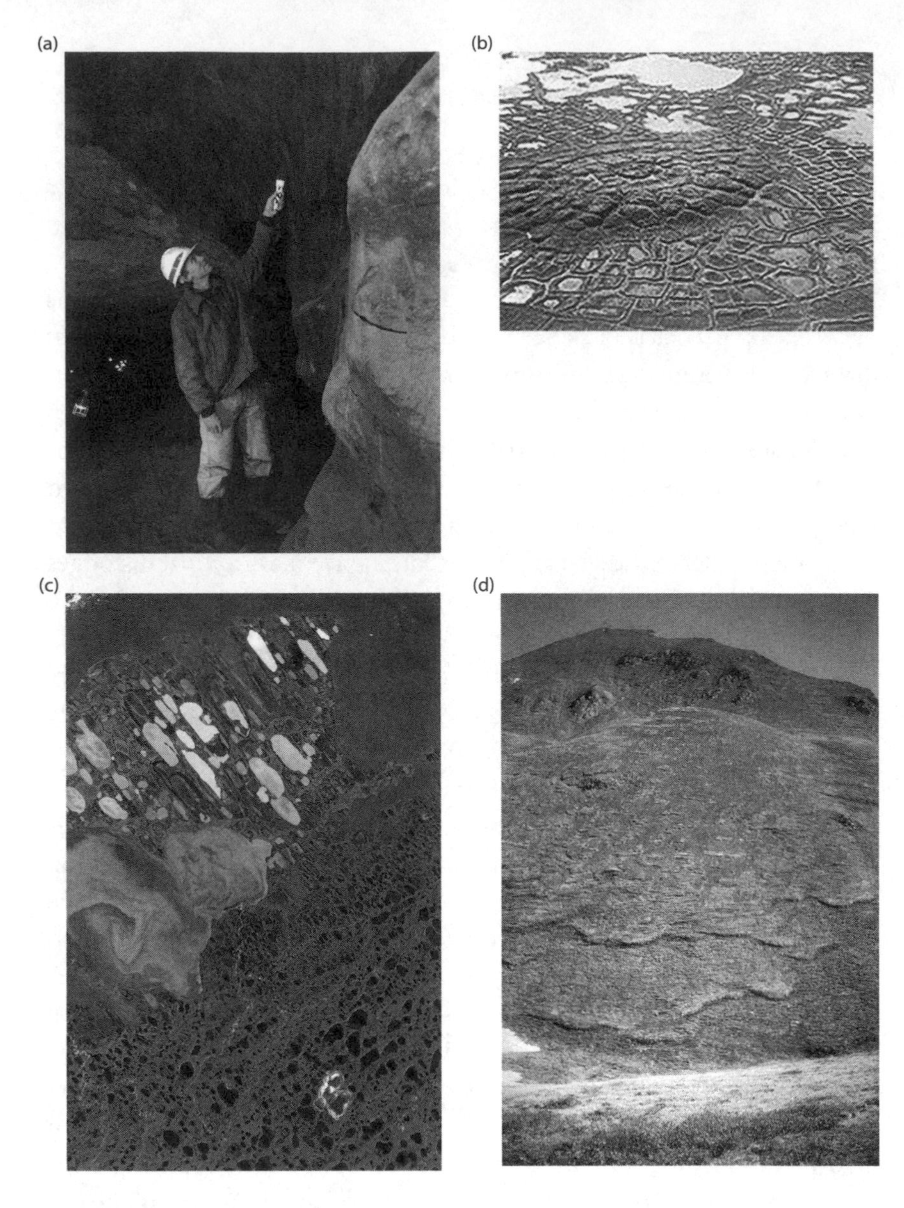

FIGURE 3.3

Characteristic landforms of permafrost regions:
(*a*) ice wedge (man is pointing at the wedge) (U.S. Army Corps of Engineers/CRREL, https://www.erdc.usace.army.mil/Media/Fact-Sheets/Fact-Sheet-Article-View/Article/476646/permafrost-tunnel-research-facility/); (*b*) polygonal networks and pingo ("A melting pino and polygon wedge near Tuktoyakuk, Northwest Territories, Canada," by Emma Pike/Wikimedia Commons, https://commons.wikimedia.org/wiki/File:Melting_pingo_wedge_ice.jpg); (*c*) thermokarst lakes ("Teshekpuk Lake, Alaska's North Slope," August 15, 2000, https://earthobservatory.nasa.gov/images/6391/teshekpuk-lake-alaskas-north-slope; image courtesy of NASA/GSFC/METI/ERSDAC/JAROS, and the U.S./Japan ASTER Science Team); (*d*) solifluction flow in Alaska (B. Bradley/University of Colorado).

IN PERMAFROST TERRAIN

Ice is a powerful sculptor shaping the land, shaving mountaintops, carving valleys, and leaving behind multiple signs of its former passage, as the next chapter will amply illustrate. Although its underground work is subtler, the trained eye quickly discerns the unique features that typify permafrost terrain. Frost action creates distinctive landforms, such as patterned ground, pingos (fig. 3.3b), palsas, thermokarst (fig. 3.3c), and slope movement (fig. 3.3d).

Repeated freeze-thaw cycles modify the surface in characteristic ways. Conspicuous among these are the geometric forms collectively known as patterned ground, which include ice wedge polygons and circles on gentle slopes and steps and stripes on steeper inclines. Ice wedges generally occur in intersecting networks of rectangular or hexagonal polygons 3 to 30 meters (10–98 feet) in diameter. At first glance, they somewhat resemble drying mud cracks or the hexagonal columns in cooling lavas (e.g., the Devil's Causeway in County Antrim, Northern Ireland). However, unlike those, ice-wedge polygons result from contraction cracks in freezing ground (figs. 3.3a and 3.3b). Frost heaving often raises ridges surrounding actively growing ice wedges, producing low-centered polygons that may hold small thaw ponds in summer. By contrast, thawing or inactive wedge networks may form troughs into which surrounding material may slump. In this case, polygon centers may stand above their peripheries. Therefore, subtle height variations among polygonal arrays offer clues to their current status.[12] Drier or better-drained ground may feature sand-wedge polygonal arrays instead.

The adventurous traveler can view some of these features underground, beautifully exposed in the Permafrost Tunnel, near Fairbanks, Alaska (see box 3.1).

Circular, slightly domed features, several meters across and usually bordered by vegetation, often occur on fairly flat terrain. These circles, generally smaller than polygons, typically form in fine-grained soils, occasionally ringed by stones that were emplaced through countless cycles of seasonal frost heaving. The churning action separates fine from coarse material, pushing stones up to the edges of the circles. Irregular clusters of hummocks, with shapes intermediate between circles and polygons, commonly populate the Arctic landscape as well.

BOX 3.1

The Permafrost Tunnel, operated by the Cold Regions Research and Engineering Laboratory (CCREL) of the U.S. Army Corps of Engineers near Fairbanks, Alaska, affords a unique, up-close view of permafrost features.[18] Excavated between 1963 and 1969 in permafrost 15 meters (49 feet) below the surface, the tunnel is approximately 110 meters long (361 feet), 2 to 2.5 meters high (6.6–8.3 feet), and 4 to 5 meters (13.1–16 feet) wide. A walk through the tunnel, the size of a mine or subway tunnel, feels like entering a time machine. An undisturbed cross section of silt, sand, and gravel up to 46,000 years old from the last ice age lines the walls. These sediments preserve abundant mammoth, bison, and horse bones, among other fossils. Still older gravels sit beneath, deposited when the first Northern Hemisphere ice sheets began to grow some 2.5–3 million years ago.

The roof of the tunnel exposes segments of an ice-wedge polygon that tapers downward along the walls of the tunnel into a typical carrot-shaped form (see fig.3.3a). The ice-age ice wedges no longer grow actively. During a past warm period, surface water trickled down into and cut across the ice wedges, eroding channels through which more water penetrated and carved out hollows. The water eventually refroze into thermokarst cave ice.

The successive strata within the tunnel record at least seven distinct episodes of past climate changes during and after the last ice age. Bands of peat indicate the tops of former active layers; interspersed ice layers represent the tops of the permafrost at those times. The tunnel also preserves two separate sets of ice wedges, ranging in age from 25,000 to 33,000 years ago at the peak of the last ice age, and 10,000 to 14,000 years ago near its termination. A thaw during the intervening missing period presumably separated the two sets.

Bench-like steps, or terraces, often run parallel to slope contours on steeper hills. Series of stripe-like vertical bands may appear on hillsides. These develop as moist soil slowly slides down over permafrost on slopes steeper than those harboring circles or polygons.

Large, circular to elliptical mounds often punctuate the permafrost landscape. Pingos typically stand 3–70 meters (9.8–230 feet) high and 15–450 meters (49–1480 feet) across. Water injected under pressure from below creates these landforms. Upon freezing, the ice pushes up overlying soil, forming a mound similar to an intrusion of igneous magma that pushes up overlying rocks into a dome.[13] Two different kinds exist. Closed-system pingos—more common in continuous permafrost, such as along Alaska's North Slope and in northwestern Canada, develop as underground water under pressure forces its way upward and eventually freezes. Increasing ice accumulations from below project the ground upward into a small hill. Most of these occur in or near former lake beds.[14] Open-system pingos, on the other hand, occur mostly in warmer, discontinuous permafrost—usually in valleys or on slopes where artesian water can flow downhill within unfrozen soil layers in the permafrost. Hydraulic pressure ultimately forces the water to the surface via natural cracks or channels. Near the surface, the water freezes into a subsurface lens and lifts up the overlying permafrost.

Palsas, like pingos, are small, ice-cored mounds that protrude above their environs. Smaller than pingos and with more variable shapes, palsas stand up to 10 meters (32.8 feet) high and 15–150 meters (49–492 feet) long. They generally occur in discontinuous permafrost. However, unlike pingos, palsas usually grow in wet, boggy, peaty soils. Growth begins where freezing takes place faster than in surrounding areas, due either to weaker insulation (e.g., a thinner snow cover) or greater heat loss. The bog supplies enough water to the upward-expanding lens, which bulges the surface up into a palsa. Eventually, the mound surface is disrupted, then melts and collapses, until the cycle is renewed nearby.

The ground, rock solid in winter, becomes mushy and unstable during late spring–summer thaw. Fine-grained soil and sediment, loosened by repeated heaving and thawing of frost, creeps slowly downslope by gravity, even on gentle slopes. Solifluction, the slow downhill movement of soil and sediment, creates festoons of tonguelike lobes and sheets (fig. 3.3d). Gelifluction is a form of solifluction in which the moving mass glides over a slick permafrost layer. Sometimes the ground gives way suddenly in a rapid, ribbonlike mud or earth flow.

Repeated rounds of extreme seasonal temperature fluctuations can crack bare rocks on steep slopes. Frost wedging and weathering enlarge the fractures until the rock shatters and collapses in massive rockfalls. Rocks thus eroded from mountain slopes accumulate at the base and in the valley. A rock glacier develops when the frost-shattered, angular debris blankets an ice glacier, or when permafrost ice seals empty spaces between rocky debris, or some combination of both. Regardless of the exact origin, rock glaciers, or ice/debris mixtures, creep slowly downhill within the permafrost domain.

Caving In

Because ground ice lies so near the surface, Arctic tundra is extremely sensitive to any surface perturbations that initiate thawing. These include regional warming, destruction of the vegetation cover by fire or human activities, or river incision, causing localized melting and ground subsidence. This creates a distinctive thermokarst landscape. Telltale signs include small, marshy hollows; hummocks; beaded streams; irregular drainage; and multiple-thaw (or thermokarst) lakes, often elliptical or elongated in shape.[15] Tens of thousands of thaw lakes and depressions dot northern lowland permafrost regions, many of which are relicts of a once–widely distributed network of thaw lakes formed after the last ice age (fig. 3.3c). In fact, roughly a quarter of the Earth's lakes lie in permafrost regions.[16]

The active layer grows quite soggy in summer because permafrost presents an impermeable barrier to drainage. Seasonal wetlands thrive thanks to the ample supply of near-surface liquid water. As rising ground temperatures deepen the active layer, the ground compacts and settles. Small, thawed depressions form within low-centered polygons or coalesced troughs in melting ice-wedge networks and collect standing water. The lower albedo of thaw ponds and higher heat absorption in summer localize further melting and degradation of the underlying permafrost. (The albedo of water is much lower than that of most Arctic vegetation—another aspect of the albedo feedback.) Ponds enlarge as sides cave in and bottoms thaw. Growing ponds merge with adjacent lakes, forming larger water bodies. Some small lakes connect to others via short channels in beaded streams, like beads strung on a necklace; others remain isolated, without inlet or outlet streams. As ice-rich permafrost continues to thaw, water may eventually cut through to lower ground until an outlet channel forms. Slumps and earth flows frequently develop where rivers intersect permafrost. Along low-lying shorelines, pounding waves undercut and erode exposed ice wedges, lenses, and ponds, leading to shoreline retreat (fig. 3.4).

Clusters of variably shaped thermokarst lakes often align along a common direction. Elongated lakes may form when prevailing summer winds set up wave patterns that focus erosion toward both ends of the lake, in a direction at right angles to the wind.[17] Thaw lakes eventually drain and empty due to changes in climate, overflow of excess water into drainage outlets, or intersection with streams, other lakes, or an encroaching shoreline. Drainage may occur suddenly. Mining, construction, or road traffic on sensitive tundra, particularly in the discontinuous permafrost zone, may also initiate thawing, ground subsidence, or thermokarst, and alter drainage patterns. Even though thaw lakes are geologically short-lived features, today's warming permafrost and new land developments are rapidly transforming thermokarst features.

DEFROSTING THE PERMAFROST

Stinky Bluffs, a steel-gray, sheer ice cliff 40 meters (131 feet) high in northeastern Alaska, provides another journey back in time some 45,000 years. The frozen ice cliff is beginning to thaw, releasing large quantities of methane and carbon dioxide—two potent greenhouse gases. The ancient decaying organic matter, recently exposed to the atmosphere, smells like rotten cabbage; hence its odoriferous name. The thawing permafrost oozes mud along the sides of the cliff and at its summit. Traversing the mucky ground feels like wading in quicksand. "We are unplugging the refrigerator in the far north. Everything that is preserved there is going to start to rot," says Philip Camill, an ecologist from Carlton College in Northfield, Minnesota.[19] Stinky Bluffs is just one smelly indication of the defrosting permafrost.[20]

Trees, careening like drunken sailors, are another indication. They have been likened to a visual "canary in the coal mine," portending global warming. The trees, most commonly black spruce (*Picea mariana*) and birch (*Betula* sp.) in Alaska and larch (*Larix* sp.) in Siberia, grow in northern subarctic boreal forests, or taiga, over discontinuous permafrost or over ice wedges that have melted beneath. Other active permafrost processes such as frost heaving, palsa formation, earth flows, or development of thermokarst also make trees slant at precarious angles. Thermokarst undermines the shallow root systems of the trees, withdrawing the necessary support to maintain trees upright. Drunken trees often ring thermokarst lakes. Landslides or earthquakes may also make trees tilt. The seemingly intoxicated trees are not newcomers to the northern forests. Permafrost reached its southernmost limit most recently during the Little Ice Age, a cold period that peaked during the fifteenth through late eighteenth centuries. It has been in retreat since the mid-ninetreenth century. However, more and more trees are tilting crazily as northern regions are thawing.

Ground temperatures in the discontinuous permafrost zone, especially in southern Alaska and interior Canada, are between −2° and 0°C (28°–32°F) and are thus particularly vulnerable to further warming. Over the past three decades, permafrost temperatures have increased by several degrees in most regions, along with air temperatures.[21] The greatest permafrost warming, around 2°–3°C (36°–37°F), has occurred in the treeless tundra regions of northern Alaska and the Canadian High Arctic. Elsewhere, permafrost temperatures have climbed by more than 0.5°C (0.9°F) since 2007–2009, with consequent deepening of the active layer. The southern borders of permafrost terrain are also shifting northward. For example, the southern boundary of permafrost in northern European Russia now lies 80 kilometers (50 miles) farther north than in the mid-1970s; that of continuous permafrost is now 15–50 kilometers (9.3–31 miles) farther north.[22] In Quebec, the southern boundary of the permafrost has migrated 130 kilometers (81 miles) north during the past half century. Significant permafrost degradation has occurred in other parts of Canada and Alaska as well.

Defrosting of the permafrost manifests in diverse ways. The summer thaw penetrates deeper; more permafrost melts from top and bottom; new thermokarst terrain appears; thaw lakes expand; thaw slumps, slope failures, and rock falls multiply; and southern boundaries of permafrost shift northward.[23] In the Alps, rock glaciers are on the move, as warming softens the underlying permafrost. Diminishing ground stability may also generate more rockfalls. Central Asian mountain permafrost holds considerable volumes of water as ice. Meltwater from thawing mountain permafrost could compensate for losses from receding glaciers.

Curiously, disappearing lakes may be another portent of Arctic warming. Comparison of satellite images of thousands of Siberian lakes between the 1970s and 2004 reveals a widespread drop in the number of lakes, particularly in the southern, discontinuous permafrost regions.[24] More connections develop between surface and subsurface water in thinning permafrost, which allows greater infiltration to the water table and more efficient subsurface drainage. Conversely, the number of lakes in continuous permafrost has increased or expanded as hollows have filled with water and more ground has slumped.

Can permafrost, contrary to expectations, resist the coming thaw? Complex feedbacks cloud our ability to forecast the fate of permafrost. For example, more moss grows on the surface over a wider active layer. Moss is an excellent insulator that protects remaining frozen soil from rising surface temperatures. It also cools the air in the warm season by evaporation. A thick snow cover, also a superior insulator, shields the soil from further cooling in winter. However, as the snow cover thins (chap. 1), this protection weakens. In another feedback, water draining downward from the active-layer pools above the impermeable permafrost layer chills, impeding further deepening of the active layer.[25] While these feedbacks could temporarily restrain the big thaw, increased defrosting (and eventual drying) of Arctic soils is likely to prevail in the long run. Over 40 percent of the permafrost could be lost by 2100, assuming high greenhouse gas emissions.[26] At first, Arctic wetlands would prosper due to greater soil moisture from melting permafrost and more rainfall. But the growing warmth would also severely degrade permafrost and dry the landscape, leading to significant wetland losses. The losses, initially concentrated in the south, would advance northward as temperatures rise. Melting the permafrost not only modifies the Arctic landscape and hydrology, but also destabilizes the ground, alters the shape of the coastline, changes plant and animal communities, and could even impact global climate (see "Carbon Aloft" later in this chapter). The changes sweeping across the Arctic are just one manifestation of a larger-scale transformation extending far beyond the Arctic, as the next series of chapters will show.

PERILOUS PERMAFROST

> *Question*: Is permafrost good?
> *Answer*: It is not so much that permafrost is good, as losing it is bad.
>
> —Answers.com, http://wiki.answers.com/Q/Is_permafrost_good/

Building on Unsteady Ground

Trees are not the only things that tilt at crazy angles as the ground gives way underfoot. Buildings and roads often develop large cracks that are the first warning signs of an impending collapse. Heat leaked from buildings or absorbed by road asphalt thaws underlying permafrost, leading to ground settlement that buckles the structures. Heat from these man-made structures alters the normal flow of heat into and out of the ground. Uneven thawing, such as by heat from boilers or from construction over ice wedges or atop pingos, has therefore created problems. These examples tell why Arctic construction requires special techniques to cope with permafrost.

"The best approach to building on permafrost is to avoid it."[27] But if avoidance is not an option, basic principles involve keeping the permafrost solid in order to maintain rigidity and strength.[28] A building should be located on bedrock or on "thaw-stable" permafrost, such as gravels, which remain strong even when thawed. Other strategies include providing insulation and ventilation, for example, by placing a well-compacted gravel pad a few meters thick over the frozen ground and raising the structure above the pad. This provides a cold air space for heat to dissipate from the overlying structure, ventilating the space and leaving the soil frozen. Alternatively, a smaller building can be raised on concrete blocks. Insulating the structure prevents excess heat from reaching the ground. Pilings of wood, concrete, or steel driven into bedrock or a solid, stable ground layer also offer a secure building foundation. Arctic towns often employ utilidors to carry utilities such as water or sewage.

Road construction employs variants of these principles, such as leaving a layer of insulation between road and ground or, in extreme cases, digging out the permafrost (a solution practical only for fairly short segments). Another method involves using a porous rock matrix (as in an air-cooled embankment) to enable cold air to enter pores and cool the ground in winter, while trapping colder air in summer.

Construction of the Trans-Alaska oil pipeline between 1975 and 1977 required elevating much of the pipeline over continuous permafrost. The metal pipes were well insulated to prevent them from becoming brittle in the extreme cold of winter, in spite of the hot oil flowing inside. To allow caribou crossings, segments of the raised pipeline were buried. Chilled brine refrigerated adjacent ground, to dissipate heat generated by the pipeline. The pipeline in these segments was placed in gravel-covered, Styrofoam-lined trenches for insulation. Special engineering designs enabled the pipeline to cross the active Denali earthquake zone safely.[29]

Much of the high-altitude Tibetan Plateau also lies on permafrost. Chinese engineers building the Qinghai-Xizang railway between China and Lhasa, Tibet (completed in 2007), used crushed-rock embankments to insulate the ground and elevated the train tracks like bridges in the most fragile areas to keep them off the permafrost. As with the Trans-Alaska oil pipeline, portions of the track are cooled with ammonia-based heat exchangers.

The changing climate in the far north can impair the stability of existing buildings and new construction. Regions most sensitive to future warming lie within permafrost that is already close to thawing. Both existing and new structures will likely need to be retrofitted to survive in the transformed Arctic environment. Problems affecting infrastructure in coastal areas impacted by both melting permafrost and rising seas will require special, creative engineering solutions.

Crumbling Arctic Coasts

Arctic permafrost is at the mercy of sea ice.

—Barnhart et al. (2014)

More than 70 percent of the world's beaches are eroding, according to Australian geomorphologist Eric Bird.[30] He lists as many as 21 processes, both natural and anthropogenic, that nibble away at the shoreline. These include sea level rise; increased storminess and higher waves; cliffs undercut by waves; lack of sediments from cliffs, dunes, or sea floor, as well as sediments trapped behind artificial reservoirs; engineering structures that block the littoral flow of sand; and beach "mining" (excessive excavation of beach sand or gravel). Because of these myriad interacting processes, most worldwide coastal erosion cannot be blamed directly on sea level rise or climate change. Nevertheless, Arctic coastlines are crumbling—victims of the great thaw: regional warming, loss of sea ice (chap. 2), and rising seas.

Arctic coasts are especially erosion prone because permafrost-bonded sediments occupy over 65 percent of the shoreline.[31] These sediments that now line the coast were deposited during the multiple ice ages of the late Quaternary Period over the last 800,000 years, at times of much lower sea level than present. Coastal permafrost possesses the mechanical strength of hard rock, but upon thawing acquires the consistency of mushy mire—easily suspended and carried away by waves—just as inland, fine-silt-sized coastal permafrost soils cemented by a high percentage of ice are the most vulnerable to ground subsidence upon thawing.[32]

A number of processes help drive up Arctic coastal erosion rates. These include higher Arctic Ocean summer sea surface temperatures, a longer open-water season, less summer sea ice, and generally increasing wave heights superimposed on rising sea level.[33] Regional warming quickens the spring thaw and delays fall sea ice freeze-up, prolonging the open-water season. Lack of protective sea ice gives the more frequent, stronger fall storms and higher waves greater erosional opportunities. Coastal permafrost also suffers "thermal abrasion"—the impact of thaw combined with mechanical wave action—when exposed to warmer seawater. Thus, Arctic coastal cliffs and bluffs suffer a dual-front assault by air and sea.

FIGURE 3.4

Collapse of coastal permafrost bluff, Alaska. Note the ice lens (white) overlain by peat layer. (Benjamin Jones/USGS, "Changing Arctic Ecosystems," December 2012, https://pubs.usgs.gov/fs/2012/3144/pdf/fs20123144.pdf)

With sea ice thin or absent, waves cut notches into the base of icy bluffs, breaking off large, ice-rich blocks that rapidly disintegrate upon falling into shallow water (fig. 3.4). Erosion increases when the sea intersects drained thermokarst (thaw lakes), ice-rich soil in ground ice, or exposed ice wedges along coastal cliffs. Keels of ice floes that gouge shallow grooves or "bulldoze" into deeper seafloor sediments also amplify erosion.[34] On the other hand, mounds or ridges of sand and gravel left on the beach by melted piles of rafted ice slabs temporarily protect the shore from erosion. Because of the complex relationships that exist among the various causes of Arctic coastal erosion, their relative importance varies from place to place.

On average, Arctic coasts recede by 0.5 meters (20 inches) per year, with large regional variations.[35] The Alaskan and Canadian Beaufort Sea coasts, as well as the East Siberian Arctic, experience rates greater than 2 meters per year. Alaska endures among the highest erosion rates in the world.[36] For example, along a 60-kilometer (37-mile) stretch of the Alaskan Beaufort Sea, erosion rates increased from 6.8 meters (22.3 feet) per year between 1955 and 1979 to 8.7 meters (28.5 feet) annually between 1979 and 2002, and up to 13.6 meters (44.6 feet) annually between 2002 and 2007.[37] A larger percentage of the coastline overall also eroded over this period. Crumbling Arctic shores not only alter the geography, but also exact a toll on those whose livelihoods depend on the land-sea interface. One such community is Shishmaref, Alaska.

Relentlessly pounding waves that undermine the shoreline could soon wash the Iñupiaq Eskimo village of Shishmaref, Alaska, into the Chukchi Sea.[38] Decreasing sea ice cover and protracted wave damage magnify historic beach erosion on an open coastline with a high tidal range. The sea has already claimed a number of homes. Thawing permafrost also undermines village building foundations. Attempts to strengthen or build new seawalls have proven unsuccessful thus far. The 600 villagers, traditional hunters and fishers, seriously consider relocating to a safer spot on the nearby mainland, but face prohibitive moving costs and lack of suitable nearby land. The low elevation precludes moving houses to "safer" parts of the island. Relocation to larger towns such as Nome, Kotzebue, or Anchorage may threaten loss of community or cultural identity. Several dozen other coastal native Alaskan villages share Shishmaref's dilemma.

A significant store of Arctic permafrost lies underwater. Submerged permafrost in relatively warmer seawater can degrade from above and also below due to the heat from the Earth's interior. (The geothermal gradient determines the depth to the base of subsea permafrost, as on land.) Destabilization of underwater permafrost could release significant quantities of methane, in addition to that liberated by land-based boreal wetlands. The next section investigates claims of a ticking Arctic methane time bomb.

CARBON ALOFT

Northern soils hold a massive store of organic carbon—an estimated 1,700 billion metric tons—formed by remains of plants and animals that have accumulated over millennia.[39] Of this store, around 1,030 billion tons lie within the top 3 meters (10 feet), with the remainder buried at depth. Microbial activity quickly degrades the freshly defrosted, carbon-rich soil, releasing methane, CH_4, and carbon dioxide (CO_2) into the atmosphere. These newly released greenhouse gases could reinforce the ongoing warming trend and thereby magnify the big thaw.

A Carbon Menace?

Researchers have long speculated that warming could unleash vast stores of the greenhouse gas [methane] from where it lies frozen beneath the sea floor and locked up in Arctic soils. If those deposits were to melt, it would almost certainly trigger abrupt climate change.

—Amanda Leigh Mascarelli (2009)

Will thawing permafrost awaken a "sleeping giant,"[40] release an "economic time bomb,"[41] and pose a "methane apocalypse," as some worried researchers have asserted?[42] Or instead, will it fizz gently? Methane bubbling from thaw lakes, melting permafrost, and offshore continental shelves in the Arctic is entering the atmosphere at higher rates than previously estimated.[43] Methane also migrates through permeable sedimentary strata and reaches the surface along rock fissures and faults, unless confined beneath impermeable strata or favorable geologic structures. Until thawed, the impermeable permafrost cover in the Arctic generally traps and pools migrating natural gas. However, numerous new gas seeps have been discovered in Alaska and Greenland escaping along the boundaries of recently retreating glaciers and deeply buried sedimentary basins, adding to the methane bubbling out of thawing thermokarst lakes.[44]

Arctic permafrost soils, especially the type known as yedoma, hold 2 to 5 percent carbon by weight. Following the end of the last ice age, microbes at the bottom of oxygen-deficient thaw lakes left by retreating ice sheets generated methane as they decomposed organic matter in yedoma.[45] The thawing lakes expanded, drained, accumulated sediments, and sequestered carbon in organic matter, such that by the late Holocene around 5,000 years ago, the region had transformed from a source of atmospheric carbon into a sink.[46] Will this carbon sink revert back to a carbon source once again as regional warming continues? Do recent reports of rising methane emissions from land and sea (see the following paragraphs) signal the onset of a significant greenhouse gas feedback? How much of the carbon gases will be released, which ones, and how fast?

Northern wetlands and lakes constitute a potentially major methane source. But future greenhouse gas emissions will depend on whether wetlands and lakes will expand or shrink as the Arctic heats up. Satellite images show declining numbers and extent of Siberian lakes from the 1970s to the late 1990s, because of thawing permafrost. While the lakes expanded in the zone of continuous permafrost, they declined sharply in regions of discontinuous or patchy permafrost farther south, where soils had drained or dried out.[47] What of the future? Wetlands abound in permafrost terrain because of poor drainage. A warming climate could initially augment wetland abundance because of increased thawing, moister soils, and greater rainfall.[48] Whereas such conditions deepen the active layer, they also improve drainage to underlying soil layers, which in turn dries out the surface. Thus, an initial expansion of wetlands would eventually give way to a degraded and reduced permafrost cover. The decomposing permafrost releases methane and carbon dioxide, initiating a positive greenhouse gas feedback.

Will Arctic ecosystems dry out or grow wetter? What becomes of the carbon released from thawing permafrost? Which of the two greenhouse gases will dominate? How much additional warming will the permafrost-carbon feedback cause? Many processes compete and send contradictory messages. Streams carry some carbon from thawed permafrost to the sea, where it is buried in sediments before it can transform into greenhouse gases. To complicate matters further, extensive coastal erosion along the East Siberian Arctic Shelf relinquishes old carbon from coastal-cliff sediments.[49] Two-thirds of this carbon could escape into the atmosphere as carbon dioxide, while the rest is reburied in shelf sediments.

Land-based thermokarst-sensitive landscapes concentrate soil organic carbon.[50] Thermokarst terrain is likely to expand, not only due to warming, but also because of more frequent wildfires and increased human development. Areas most vulnerable to sudden thaw are clustered along Arctic Ocean coastal regions and in the tundra thermokarst terrains of Eurasia. Expansion of thaw lakes has increased CH_4 and CO_2 emissions during the last 60 years.[51] The extent to which microbes act on thawing permafrost and release methane and/or carbon dioxide may depend on soil moisture content, among other factors. In lab experiments simulating waterlogged soils, such as in peat bogs or wetlands, carbon emissions remain low, whereas well-drained soils release more carbon.[52] The added moisture from thawing permafrost also spurs vegetation growth that could consume carbon emissions, but simultaneously activates methane-generating microbes. Thus, soil moisture, as well as temperature, exerts a strong control over permafrost carbon emissions.

A heated debate also swirls around the relative importance of methane versus carbon dioxide. Microbes from well-oxygenated soils (generally, drier uplands) generate mostly CO_2, whereas in anaerobic (oxygen-poor) soils, such as in wetlands or peat bogs, they release a mix of both CH_4 and CO_2. Furthermore, aerobic (well-oxygenated) soils produce higher carbon emissions overall compared to anaerobic soils. Samples from around the world show CO_2 as the dominant gas released by weight, regardless of soil environment.[53] A panel of experts therefore concludes that by 2100, between 5 and 15 percent of permafrost carbon is likely to be released into the atmosphere, largely as carbon dioxide.[54] However, further experiments indicate that wet, poorly oxygenated soils can generate a higher proportion of CH_4 relative to CO_2 than do drier, oxygenated soils.[55] Inasmuch as CH_4 has a much higher global warming potential than does CO_2, even a small CH_4 increase could thus affect climate disproportionately. On the other hand, Xiang Gao of the Massachusetts Institute of Technology in Cambridge and colleagues strongly discount the climate impact of permafrost-derived methane, which they estimate would raise temperatures by a mere 0.1°C (0.2°F) by century's end.[56] Even a 25-fold emissions increase would raise temperature by just 0.2°C, and a more unrealistically high 100-fold increase by only 0.8°C.

Regardless of which greenhouse gas dominates, the permafrost-carbon feedback would likely intensify the Arctic thaw. But questions still remain. For example, to what extent would plant regrowth counterbalance the permafrost-released gases? The projected prolonged growing seasons, hotter summers, added plant nutrients from decaying soils, and elevated greenhouse gas levels would all stimulate vegetation growth, which would remove some atmospheric CO_2, but how much? Furthermore, how much of the added gases would microbes consume, particularly those in thawing submerged permafrost?[57]

In summary, the climatic effects of greenhouse gases released by thawing permafrost is a highly complex issue, depending on diverse competing processes that can operate in opposing directions. Among these are (1) the increase in atmospheric temperature, (2) the extent of future thermokarst expansion, (3) the moisture content of the soil, (4) the soil oxidation state, and (5) the proportion of CO_2 to CH_4 emitted, to name the main ones.

Is another kind of carbon menace lurking offshore beneath the Arctic Ocean? Offshore marine sediments around the Arctic Ocean harbor another vast storehouse of buried carbon, the fate of which is also in question, as discussed in the following section.

The 120-meter rise in sea level after the last ice age drowned an immense reservoir of organic matter trapped in permafrost. Arctic shelf seawater is typically at least −2°C (28.4°F), as compared to permafrost temperatures as low as −17°C (1.4°F).[58] This marked temperature contrast potentially exposes submerged permafrost to higher future ocean temperatures. This has raised fears that recent warming Arctic Ocean water could destabilize vast quantities of methane locked in marine sediments. Are plumes of methane spewing out of the ocean a sign of recent Arctic warming or relicts from the deglacial marine incursion? The concluding section of this chapter examines the large hidden store of underwater methane in an unusual form of ice.

The Ice That Burns

Gas hydrates (also known as clathrates) are a form of ice with an open, cage-like crystal structure that can enclose gases such as methane, carbon dioxide, or hydrogen sulfide. Methane hydrate is the most common example in nature, with an approximate formula of $8CH_4{\cdot}46H_2O$ (fig. 3.5a). This form of ice exists only under pressure at very low temperatures (fig. 3.5b). When methane hydrate decomposes, methane gas expands some 160 times by volume. Enormous quantities of methane hydrates are believed to be buried in Arctic permafrost and in marine sediments on the continental slope and rise (conservatively estimated at 1,800 billion metric tons globally[59]).

In the Arctic, the marine hydrate stability zone ranges from depths of at least 300 meters to more than 1,000 meters (980–3280 feet) on continental shelf and slope sediments, and several hundred meters deep within and below land permafrost. The thickness of the marine hydrate stability zone is determined by the temperature and seafloor depth and the intersection of the methane hydrate stability curve with the geothermal gradient in

(a)

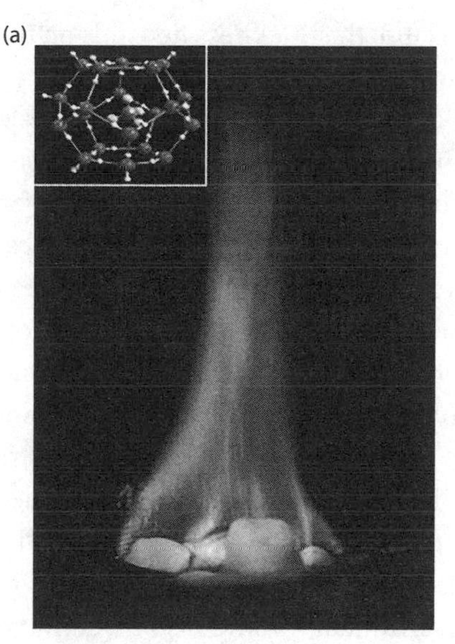

(b)

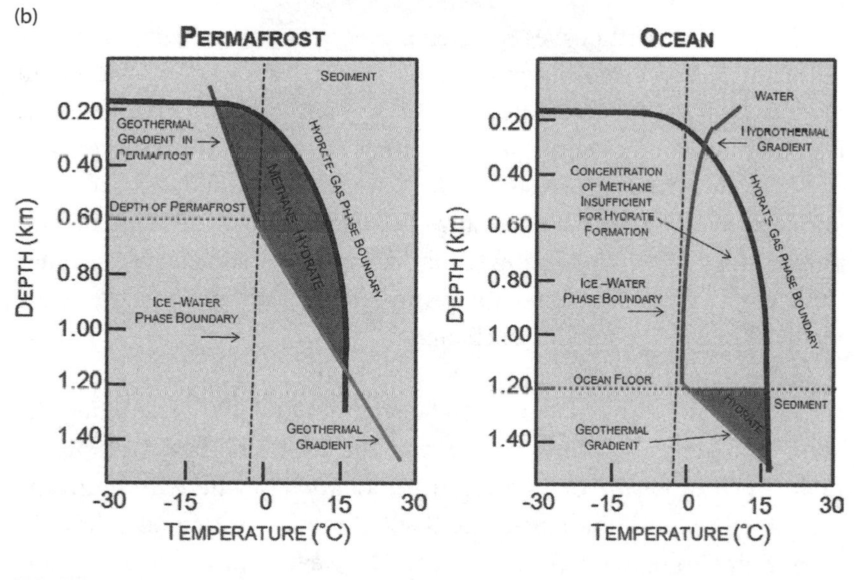

FIGURE 3.5

(*a*) Methane hydrate—the ice that burns and gas hydrate structure (*inset upper left*). (U.S. Geological Survey, https://en.wikipedia.org/wiki/Methane_clathrate); (*b*) methane hydrate stability curve in permafrost and marine sediments (Sara E. Harrison, "Natural Gas Hydrates," October 24, 2010, http://large.stanford.edu/courses/2010/ph240/harrison1/)

ocean sediment. On land, the thickness depends on the intersection of the hydrate stability curve with the geothermal gradient in both permafrost and sediment (fig. 3.5b). Methane hydrate dissociates violently into methane and water when either temperatures rise or pressures are relieved by lowered sea level. The base of the hydrate stability zone defines an impenetrable surface that traps any free methane gas escaping from depth.[60] The gas hydrate layer may begin to decompose after a slight increase in shallow bottom-water temperatures or a sea level drop. If enough free gas accumulates beneath the marine hydrate stability zone on the continental slope, pressures could exceed that of overlying sediment or water. A slight perturbation could then potentially trigger a catastrophic slope failure. Like uncorking a champagne bottle, the sudden decrease in confining pressure would cause slumping of the destabilized hydrate layer, triggering submarine landslides, or even tsunamis.[61] In a snowballing effect, underlying gas-rich sediments would also be entrained into the growing landslide (fig. 3.6).

Alternatively, earthquakes could shake loose unstable overlying marine sediments, also generating massive underwater avalanches that would rupture hydrate-bearing strata and release any confined methane gas. However, marine landslides also occur in the absence of past or present gas hydrates. Other processes, such as weak, unconsolidated sediment horizons or high fluid pore pressures in rapidly deposited sediment, also induce slope instability and eventual collapse. No conclusive evidence links the timing of major submarine landslide events to large-scale methane releases or to climate change.[62]

Methane gas is bubbling up from more than 250 active plumes on the seabed at depths of 150 to 400 meters (490–1,300 feet) along the West Spitsbergen continental margin, the East Siberian Arctic Shelf (a drowned extension of Siberian tundra), and elsewhere in the Arctic Ocean.[63] The East Siberian Arctic Shelf alone is emitting an estimated 17 million tons of methane gas per year. However, most of this methane may not even reach the atmosphere. Some dissolves in seawater; microbes also quickly convert it into biomass and carbon dioxide. Are observed methane emissions even caused by the recent ocean warming, or are they simply ongoing processes dating back thousands of years? Not all marine methane may derive from decomposing gas hydrates;

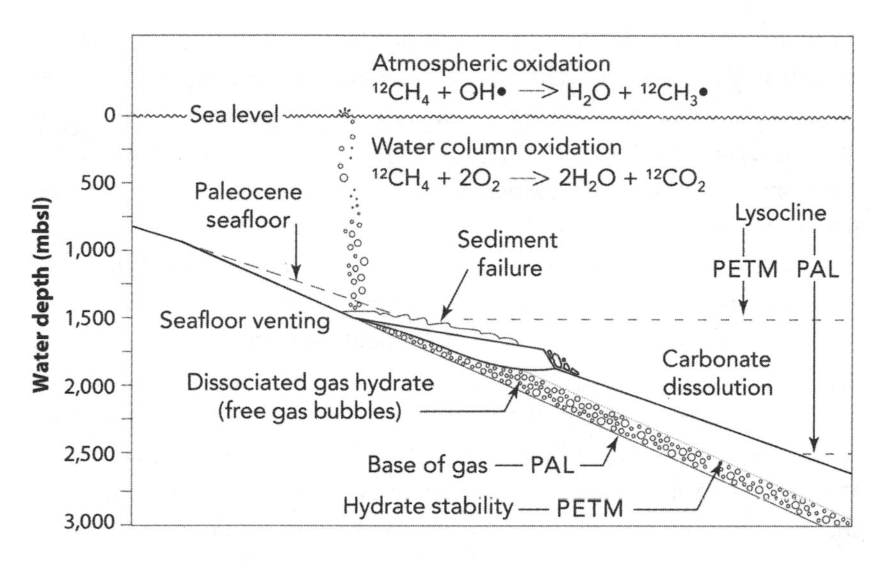

FIGURE 3.6

Submarine landslides triggered by decomposing hydrates in marine sediments. The liberated methane gas quickly travels to the ocean surface and atmosphere, where it rapidly oxidizes to carbon dioxide. (G. R. Dickens and C. Forswall, "Methane Hydrates, Carbon Cycling, and Environmental Change," in *Encyclopedia of Paleoclimatology and Ancient Environments*, ed. V. Gornitz [Dordrecht: Springer, 2009], 563. Reprinted by permission from Springer Nature)

instead they may derive from now-thawing, drowned, carbon-rich permafrost; from eroding shore cliffs; or from gas escaping from deep-seated ocean reservoirs. Microbes living in oxygen-deficient seawater feed on methane in the presence of dissolved sulfate, quickly oxidizing it to carbon dioxide.[64] Stores of subsea hydrates may have also been overestimated, and much of what exists may be buried deep enough to remain undisturbed by ocean warming for centuries to come.[65] Can the escaping methane therefore awaken a sleeping giant?

As Anil Ananthaswamy, a reporter for the *New Scientist*, assures us: "The one thing we don't need to worry about is the so-called methane time bomb. . . . An imminent release massive enough to accelerate warming can be ruled out."[66] Continued climate warming over the next century or two is less likely to unleash an explosive time bomb than to create a gentle fizz like that from an opened can of soda. Nevertheless, hydrate deposits on the shallow continental shelves and slopes surrounding the Arctic Ocean will remain vulnerable to dissociation under ongoing and future warming. Although their present-day contribution is still fairly small, methane and carbon dioxide emissions from both thawing land permafrost and Arctic Ocean continental shelf/slope sediments will likely augment the growing anthropogenic greenhouse gas burden.

As this and the previous chapters have shown, climate change is already under way throughout most of the Arctic, faster there than elsewhere on the planet. Losses of permafrost and floating ice will amplify climate change in the Arctic, creating distant ripple effects. However, the real bull in the china shop is the ultimate fate of glaciers and ice sheets and their roles in sea level rise and in water resources. These impacts could ultimately affect hundreds of millions of people worldwide. A closer look into these processes guides the remaining chapters of this book.

4

DARKENING MOUNTAINS—DISAPPEARING GLACIERS

SHRINKING RIVERS OF ICE

Many of these Alaska glaciers are rapidly melting and are now but the fragments of their former selves.

—John Burroughs (1899)

Glacier Bay National Park and Preserve, in southeastern Alaska, hosts 50 named glaciers and some of the most spectacular fjords in the world—a premiere showcase of glacial action amidst breathtaking scenery. When the British explorer George Vancouver (1757–1798) and his ship's master Joseph Whidbey sailed into Icy Strait in southern Alaska in 1794, a towering wall of ice blocked any further advance. Glacier Bay was a "compact sheet of ice as far as the eye could distinguish." It formed a mere five-mile dent in a massive glacier as much as 1,220 meters (4,000 feet) thick in places and extending more than 160 kilometers (100 miles) northwest toward the St. Elias Mountains. In 1879 naturalist John Muir noted that the ice had retreated around 78 kilometers (44 miles), most of the way up the bay. Twenty years later, in 1899, he wrote:

> Here, then, we have the work of glacial earth-sculpture going on before our eyes. . . . Evidently, all the glaciers hereabouts were no great time ago united, and with the multitude of glaciers which loaded the mountains to the south, once formed a grand continuous ice-sheet that flowed over all the island region of the coast and extended at least as far south as the Strait of Juan de Fuca. . . . And as we have seen, this action is still going on

> and new islands and channels are being added to the famous archipelago. The steamer trip to the fronts of the glaciers of Glacier Bay is now from 2 to 8 miles longer than it was only 20 years ago.[1]

By 1916 the Grand Pacific Glacier had retreated 100 kilometers (60 miles) from Glacier Bay's entrance to Tarr Inlet. During the twentieth century, as the ice continued to recede, other side inlets opened up. For example, the Muir Glacier, on an eastern branch, melted back from the main Glacier Bay ice mass starting in 1860 and continues receding to this very day (figs. 4.1a and 4.1b). Reid Glacier also shrank by several kilometers

FIGURE 4.1

Muir Inlet in the Glacier Bay National Park and Preserve, Alaska, in (*a*) August 1941 and (*b*) August 2004. (U.S. Geological Society, "Repeat Photography of Alaskan Glaciers," https://www2.usgs.gov/climate_landuse/glaciers/repeat_photography.asp)

between 1899 and 2004. Johns Hopkins Glacier separated from the Grand Pacific Glacier around 1890. Since then, both glaciers have mainly retreated, although Johns Hopkins has recovered somewhat since the late 1920s. It is the only actively advancing tidewater glacier in the park that terminates at the coast.

Each glacier has a unique history, and the behavior of regional glaciers varies considerably. Nevertheless, most of the park's glaciers continue to shed ice. Furthermore, glaciers that extend below around 1,500 meters (4,920 feet) of elevation are currently thinning and/or retreating in most areas of Alaska,[2] largely in response to the average ~2°C (3.6°F) temperature increase since the mid-twentieth century.

Meanwhile, the highest rates of land uplift in the world occur in southeast Alaska, particularly the area around Glacier Bay National Park and nearby Yakutat—a striking consequence of the recent glacial recession. Geodetic and precision GPS surveys register the rapid uplift. Regional tide gauges register a falling sea level, in contrast to the globally rising sea levels of the twentieth and early twenty-first centuries (chap. 1). Why, then, should southern Alaskan sea levels contrarily fall by −2.3 millimeters (0.09 inch) per year from 1924 to 2016 in Sitka, −13.1 millimeters (0.52 inch) per year (1936–2016) in Juneau, −14.1 millimeters (0.56 inch) per year (1988–2016) in Yakutat, and an astounding −17.6 millimeters (0.7 inch) per year (1944–2016) in Skagway?[3]

Local sea levels can vary from place to place due to land uplift or subsidence. Land motions from sudden earthquakes, slow cumulative creep along active faults, or subsidence due to overpumping of subsurface groundwater or hydrocarbons result in elevation changes. Furthermore, because the Earth's subsurface is not perfectly rigid, it flexes under the burden of thick piles of sediment, lava flows, mountain massifs, or ice masses in a process known as isostasy. Glacial isostasy, in particular, refers to the way land sinks or rises under the weight of growing or shrinking glaciers and ice sheets.

Glacial isostasy therefore affects the relative elevation between land and sea in coastal areas. Areas weighed down under thick ice sheets rebounded once the ice began to melt around 20,000 years ago, toward the end of the last ice age. Rebound still continues around the northern Baltic Sea and Hudson Bay, Canada, where sea level falls at rates of up to ~−10 to −12 millimeters (0.4–0.5 inch) per year. Land once along the edges of ice sheets was squeezed upward, but began to sink after ice melted. Sea level is rising faster than the global average in these still-subsiding regions, such as the U.S. mid-Atlantic coast and southern England, because the subsidence adds to climate-related sea level rise.

In southeastern Alaska, the strongest land uplift closely corresponds to areas of greatest ice loss and lowered sea level.[4] Furthermore, dating of the oldest trees just below raised shorelines indicates that tree growth began when land emerged from the sea, over 200 years ago. The onset of uplift and its affected area correlates closely with the history of recent glacial retreat in and around Glacier Bay. The regional uplift is therefore tied mainly to fairly recent glacial rebound, rather than to ongoing tectonic activity or to lingering glacial isostatic adjustments from the last ice age.

Glaciers began to recede earlier in southeastern Alaska than elsewhere.[5] By the mid-nineteenth century, glaciers in many other parts of the world had joined in the retreat marking the end of the Little Ice Age, a relatively cold period in the Northern Hemisphere that lasted from approximately 1400 to 1850. The retreat has accelerated, particularly from the second half of the twentieth century to the present. Between 2003 and 2009, Alaska alone delivered almost 20 percent of the world's total glacier ice losses![6]

Not only are glaciers sensitive barometers of climate change; their attrition will significantly impact sea level, mountain-fed freshwater resources, ocean salinity, mountain hazards, and even the shape and rotation of the Earth.[7] Before we investigate the worldwide decline of glaciers, we briefly review how they form and transform the landscape, and how to assess their state of health.

GLACIER BASICS

In a nutshell, glaciers build up wherever more snow falls than melts. The piles of snow eventually compact and recrystallize into solid ice that gradually flows downslope under the pull of gravity (fig. 4.2). Around 197,650 glaciers and ice caps cover an area of 0.73 million square kilometers (0.28 million square miles) worldwide, occupying a volume of 0.15 million cubic kilometers (0.04 million cubic miles). A total meltdown of all glacial ice with the meltwater spread evenly across the sea would raise the ocean level, on average, by 0.35 to 0.52 meters (1.15–1.71 feet).[8]

FIGURE 4.2

Aletsch Glacier in Switzerland, the largest glacier in the Alps. The curved, dark bands are medial moraines. Clouds in the background partly cover sharp crests (arêtes) and ice-filled cirques, or bowl-shaped depressions. A lateral moraine is partially visible at lower left. ("Aletschgletscher mit Pinus cembra," July 22, 2007, by Jo Simon/Wikimedia Commons, https://commons.wikimedia.org/wiki/File:Aletschgletscher_mit_Pinus_cembra2.jpg)

Over three-quarters of the world's glaciers lie along the fringes of Antarctica and Greenland and in the Canadian Arctic, Alaska, and the high Asian peaks. But glaciers blanket high mountain summits on all continents, including in the tropics. Their global distribution depends on a number of factors, including topography, temperature, and moisture supply. Glaciers in the cold climate near the poles extend over broad areas at lower elevations. In the tropics, small, high-elevation glaciers still whiten peaks such as on Mt. Kilimanjaro in Tanzania, Mt. Kenya in Kenya, and Puncak Jaya in New Guinea. However, these glaciers are among the planet's most endangered, as we shall see in the following discussion.

Glaciers exist in many environments. Mountain glaciers confined to narrow valleys flow downhill like rivers of ice. Streams of meltwater often emerge from the mouth of a mountain glacier and pool into small alpine lakes. Glaciers may build up and merge into a massive broad dome that spreads out in all directions, ultimately becoming a continental-scale ice sheet that completely buries underlying terrain. Outlet or piedmont glaciers, on the other hand, spread outward from glaciated terrain onto the lowlands. Tidewater glaciers calve off numerous small icebergs. Along mountainous or cliffed coasts, they often end in fjords—former glacial valleys drowned by rising seas at the end of the last ice age.

Near the poles, glaciers remain frozen to their beds year-round (cold polar glaciers, or cold-based ice). The absence of a bottom meltwater layer and firm attachment to the substrate inhibits forward motion. Such glaciers occur largely in the Arctic and Antarctica. By contrast, temperate glaciers (or wet-based ice) lie close to their melting point under pressure, except when freezing near the surface in winter. Therefore, summer meltwater reaches the base readily, generally affording temperate glaciers greater mobility (see the next section). Wet-based temperate glaciers are common in moist, temperate maritime climates, such as the northwest coasts of Canada and Alaska, southern Chile, New Zealand, the Alps, and elsewhere. Nearly all ice in polythermal glaciers of subpolar regions remains below freezing, except for some summertime surface melting. However, polythermal glaciers often display more complex behavior. They may exhibit characteristics of either cold, polar glaciers or temperate ones on different segments along their length, depending on the internal temperature regime.[9]

Glacier ice often glows a deep aqua blue. Ice preferentially absorbs the longer wavelengths (i.e., yellow and red), mainly scattering blue light. The aquamarine glow shows up best in densely packed, well-crystallized ice with few included air bubbles, as seen in freshly exposed crevasses or newly calved icebergs.

The birth of glaciers begins with gently falling flakes of snow. Once on the ground, the initial intricate six-branched shapes gradually become rounder and more compact. Fine-grained (~1 millimeter) near the surface, the ice crystals slowly grow coarser with depth (although smaller crystals form in the presence of impurities that inhibit growth). As snow is compressed into ice, crystals assume an increasingly nonrandom arrangement, or crystal fabric. The crystals align such that their glide planes tend to lie in the direction of flow, roughly parallel to the bed.[10] Ice with this preferred crystal orientation slides many times faster than for randomly oriented crystals. The arrangement of crystals in ice therefore underpins a deeper understanding of ice flow in glaciers and ice sheets (see "Glaciers on the Move" later in this chapter) that can help in developing more realistic models of their dynamic motion.

The growing pile of successive years' loosely packed snow gradually transforms into ice. Snow that has survived a summer's melt season becomes firn, with an initial surface density of ~0.35 grams per cubic centimeter (0.013 lb/in^3) and a high degree of porosity (i.e., 60–70 percent air by volume).[11] The increasing weight of overlying snow slowly compresses older, deeper deposits. Under pressure, the ice grains pack tightly and slowly recrystallize. The sealed pores trap residual air in tiny, isolated bubbles, and firn slowly transitions to glacier ice with densities approaching 0.83 grams per cubic centimeter (0.030 lb/in^3). Once the density reaches 0.917 grams per cubic centimeter (0.033 lb/in^3) after multiple annual freeze-thaw cycles and continued compression, firn completes its transformation into ice. Depending on climate, this changeover takes from 3 to 5 years at fairly shallow depths on temperate glaciers, over a century in Greenland, and several thousand years in drier, colder Antarctica, on average.

Glaciers grow in the accumulation zone where gains in mass from seasonal snowfall, freezing rain, avalanches, and wind drift exceed losses. Rain or meltwater that penetrates into firn and freezes also builds up more ice. Mountainous midlatitude maritime climates, such as those of Norway, New Zealand, Iceland, and southern Alaska, enjoy heavy winter snowfall and high accumulation rates. The pileup of snow and ice at high altitudes is balanced by the downward flow of ice toward lower elevations, where losses increase in the ablation zone. There, due to higher temperatures, losses through surface and basal melting, runoff, sublimation (see glossary), wind scour, avalanching, and calving exceed mass gains. Sublimation becomes important in cold, dry climates such as Antarctica as well as on high-altitude tropical glaciers. A thick, insulating snow or debris cover on snow or ice curtails ablation. Conversely, a darker, scattered snow cover or thin debris cover enhances melting.

The equilibrium line altitude (ELA) marks the boundary between zones of accumulation and ablation. It delineates the elevation at which annual snow accumulation exactly matches losses. The ELA responds quickly to local climate fluctuations, such as summer air temperatures and the form of winter precipitation (snow versus rain). ELA shifts to higher elevations with decreasing snowfall and/or rising temperature, and vice versa; thus it is closely tied to the annual mass balance—a measure of the glacier's status. It is also a sensitive barometer of regional climate change, inasmuch as changes in glacier mass balance are well correlated over distances of up to 500 kilometers (311 miles).[12]

The glacier mass balance comprises the sum of all processes that add or remove mass from a glacier. More specifically, it represents the difference between accumulation and ablation over the glacier in a specified time interval, usually 1 year. A cooling trend, when annual snowfall or ice buildup exceeds losses by ablation, leads to a positive mass balance. Glaciers grow, the accumulation area expands, the melt season shortens, and/or snowfall thickens. The reverse occurs during warming: glacier mass balance becomes negative, the ablation zone encroaches into the accumulation zone, the melt season lengthens, and a greater percentage of precipitation falls as rain rather than snow. Mass balance becomes zero when accumulation matches ablation. The mass balance thus shows the high sensitivity of glaciers to minor climate variations.

How Glaciers Shape the Land

> *Glaciers are one of nature's most efficient landscape architects, constantly reshaping their environments through the processes of erosion, entrainment, transportation, and deposition.*
>
> —R. D. Karpilo Jr. (2009)

Ice is a powerful landscape architect, sculpting rugged alpine mountain scenery and excavating valleys and fjord basins. As a thick mass of ice slowly descends down the mountainside into the valley and beyond, it abrades exposed rocks, smoothing and rounding their surfaces; plucks boulders and transports them; and quarries shattered rock fragments from valley walls. Laden with broken rocks, pebbles, soil, and meltwater, the relentlessly advancing ice scours the underlying surface, scraping and wearing down bedrock and loose sediment. It leaves telltale signs of its passage in distinctively shaped landforms, rock outcroppings, and debris that accumulates in characteristic deposits. Continual grinding by an advancing glacier grinds entrained rocks into a fine flour that, like jewelers' rouge,[13] polishes bare surfaces to a high luster, but also gouges rocks with scratches, grooves, and linear striations marking its flow direction.

Perched high on the mountainside, the heads of most glacial valleys originate in cirques, or bowl-shaped hollows (figs. 4.2 and 4.3). Sharp

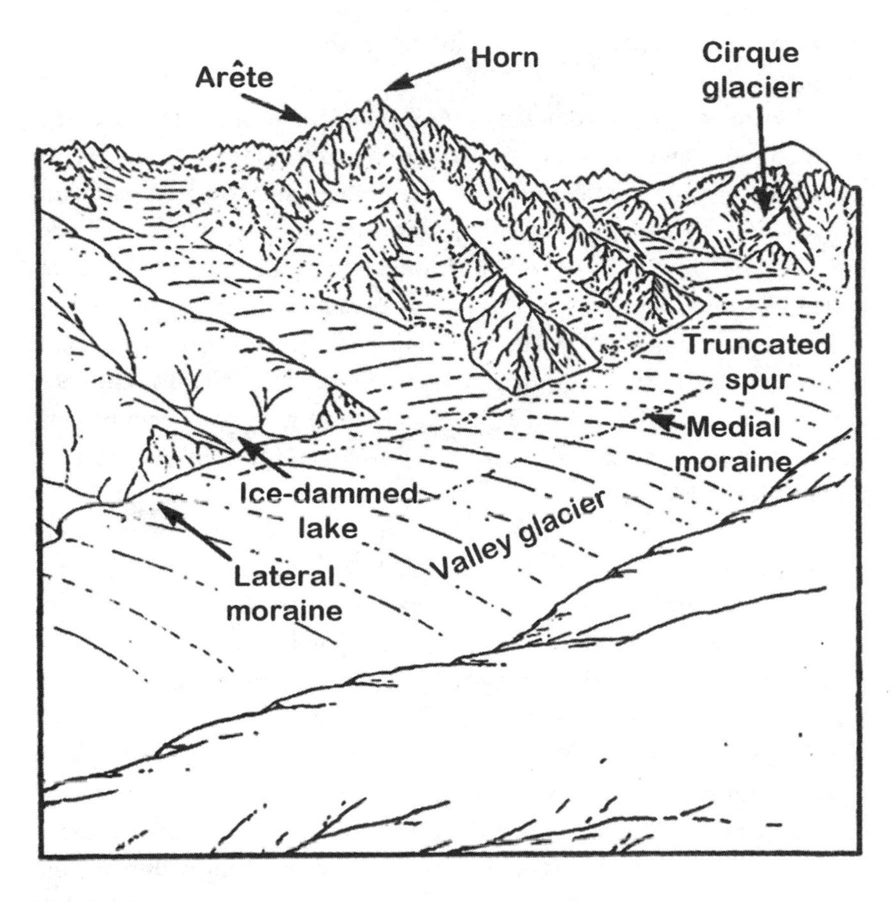

FIGURE 4.3

Mountain glacier features. (NASA, after Davis [1909])

ridges, or arêtes, separate cirques between mountain slopes. Jagged pyramidal peaks, such as the Matterhorn in Switzerland, form where several cirques intersect. A glacier bulldozing its way downhill carves distinctive, U-shaped valleys as ice fills the entire valley to the enclosing valley walls. The glacier's forward motion abrades, smooths, and widens the valley. In marked contrast, rivers flowing in steep-sloped, ice-free valleys etch out narrow channels along the lowest part of the narrow valley, forming the characteristic V profile. Hanging glaciers, from which waterfalls often cascade in summer, sit stranded high above the main glacier, which eroded a much deeper valley at a time when the ice was more extensive (fig. 4.3).

Meltwater emerges from a mountain glacier's terminus (or snout), forms streams, and collects into small, milky, greenish-blue glacial lakes colored by the high number of finely suspended ice-pulverized particles, or "rock flour." These glacial lakes are often confined by end moraines (see more in the following paragraphs). However, moraine-dammed lakes are potentially unstable when overtopped or breached by excess meltwater. As increasing numbers of glaciers retreat, flood outbursts from such lakes pose a growing threat to downstream communities (see "GLOFs," chap. 9). On a much larger scale, former ice sheets have gouged out numerous lakes, such as the Great Lakes in the Midwest and the Finger Lakes in upstate New York.

Ice moving over the landscape streamlines exposed rocky outcrops into rôches moutonnées (Fr., sheep-like rocks) that are smooth on the side facing the oncoming glacier but shattered and jagged on the lee flank. Found throughout glaciated country, good examples can be seen even in Central Park in New York City (fig. 4.4). Ice overriding rock or sediments carves drumlins— elongated hills streamlined by glacial abrasion, with a steeper and wider slope on the side facing the oncoming glacier side and a gentler, more tapered slope on the lee side. Drumlins may occur alone or in swarms numbering hundreds to thousands. Their origin remains uncertain: they may be erosional remnants of preexisting sediments or, alternatively, sediments left beneath glaciers that stripped away more mobile material.

The work of glaciers does not end with glaciers scouring bare rocks or streamlining sediments. Mountain glaciers pile mounds of unsorted sediments against valley walls (lateral moraines) and leave ridges that delineate their most extreme advance (terminal moraines). Recessional moraines denote the successive stages of a glacier's retreat. A medial moraine builds up as the lateral moraines from two converging glaciers join (figs. 4.2 and 4.3). Tributary glaciers that join the main glacier dump alternating dark and light ribbonlike stripes along its length. Bands of relatively pure ice separate dark, rocky, debris-strewn strips.

As glaciers and large ice sheets sweep across the land, they grind up rocks and transport large boulders, cobbles, gravel, silt, and fine-grained

FIGURE 4.4

Roche moutonnée in Central Park, New York City. (Photo by V. Gornitz)

clay. The receding ice dumps huge piles of rocky rubble hundreds of kilometers from source areas. Left behind in its wake are various types of deposits, collectively known as glacial drift, including till. Unsorted deposits, or glacial till, originated as rockfalls, loose sediment, pebbles, and rock fragments on the glacier's surface; were dragged along at its base; or were scattered within the ice. Exotic rocks and boulders, or glacial erratics, some weighing several tons, populate glaciated terrains.

Formerly glaciated terrain is often traversed by long, sinuous ridges, or eskers. Eskers represent a fossil drainage system of channels or tunnels under the ice that later became clogged with sand and gravel. These accumulations were subsequently exposed when the ice melted and retreated. In the true-life adventure story *Barren Lands: An Epic Search for Diamonds in the North American Arctic* by Kevin Krajick, prospectors Chuck Fipke and his longtime partner Stew Blusson hunted for diamonds in Canada's far north by tracking particular "pathfinder" minerals commonly associated with the precious gemstones scattered in the eskers. These telltale "indicator" minerals, such as garnet, chrome diopside, chromite, and the occasional rare diamond sprinkled in esker deposits, along with specific geochemical data, provided important clues in their search. Fipke and Blusson assumed (correctly, as it turned out) that the concentration of the indicator minerals would increase as one approached the source, much in the way gold prospectors sample streambeds for increasing gold content upstream to locate the mother lode. Many unproductive years of exploration in difficult, remote terrain were finally rewarded when, in 1991, they finally stumbled upon their mother lode—diamond-bearing kimberlite pipes near Lac de Gras, the first major diamond discovery in Canada. The two men became rich and Canada soon ranked among the world's top diamond producers.

Stagnant or deteriorating glaciers and ice sheets leave behind other distinctive calling cards as well. Outwash plains form along the margins of ice sheets or piedmont glaciers where braided meltwater streams deposit unsorted, ground-up glacial sediments. Generally, the farther from the glacier's edge, the finer the sediment. Much of Long Island, New York, consists of outwash plains south of the terminal moraine—a relic of the last ice age—that roughly divides the island in half along its length from east to west. In Iceland, extensive outwash plains, known locally as *sandur*, occur along parts of the southern coast. Unlike the cream-colored sands of Long Island (mostly quartz grains, with a sprinkling of dark, heavy minerals such as garnet or magnetite), those of the Icelandic sandur are nearly black—essentially glacially pulverized basalt—testimony to the island's still-active volcanic heritage. Other extensive outwash plains occur in southwestern Alaska and western Greenland. Streams and ponds on top of a stagnant glacier often collect layered sediments that fill hollows or depressions. After the ice melts, small mounds, knobs, or hummocks called kames remain. When an ice block buried in glacial drift melts, it creates a depression known as a kettle. These are common on outwash plains.

Meltwater also finds many ways of traveling through or across active glaciers in what at first glance may appear as a solid mass of ice. But the interiors of glaciers are interlaced with a complex, ever-evolving array of conduits through which water can pass. The next section peers inside glaciers and exposes some of these indirect or hidden routes. The myriad ways in which water moves through glaciers and ice sheets ultimately affects their speed. The stability of glaciers and the ice sheets may thus depend to a large extent on their internal plumbing, as discussed here and further in chapters 5 and 6.

Glacial Plumbing

Water travels through a web of circuitous routes on, within, and beneath a glacier (fig. 4.5). Drainage pathways adjust rapidly to changing conditions as ice melts, refreezes, or deforms. During summer, water from rainfall and melting of snow or firn accumulates on the glacier surface. Because of the generally impermeable nature of ice, meltwater collects in surface pools, or runs off over the surface in sheets or in channels, wherever it incises faster than ablation can lower the surrounding ice level.[14] Like rivers, these ice channels meander and tributaries join into an incipient dendritic (treelike) drainage pattern. Meltwater either moves swiftly through the glacier system, exiting at the snout, or remains within the glacier for varying periods depending on flow path or obstructions encountered along the way. Moulins, or vertical shafts, offer a rapid means of descent to the glacier bed. These rounded tubes often develop near crevasses, which capture upstream surface water flow. Crevasses, or cracks in the ice, that riddle the glacier surface also convey surface meltwater into the glacier interior. Within the glacier, branching passageways may close or freeze over in winter and reopen in summer; these act as further conduits for water circulation.

In temperate glaciers, a thin layer of water generally separates the glacier from its bed, which may also be littered with sediment, cobbles, boulders, or solid rock. At the glacier bed, sections of conduits or tunnels can be large enough to admit a tall man or small vehicle comfortably, although the ride may be rocky. The interface between a glacier or ice sheet and its bed also greatly influences the speed at which the ice can flow.

Surface water penetrates into the glacier by quickly filling crevasses. The water-filled cracks exert added pressure that forces the cracks to deepen further, like a knife slicing butter. These fractures propagate downward until pressure from the sides and overburden exceeds the tensional stresses pulling the ice apart, and the cracks are sealed shut. Water-filled crevasses can attain great depths, often reaching the base, because of the added pressure of water wedging aside the walls in a process analogous to the "hydrofracking" used by the oil and gas industry

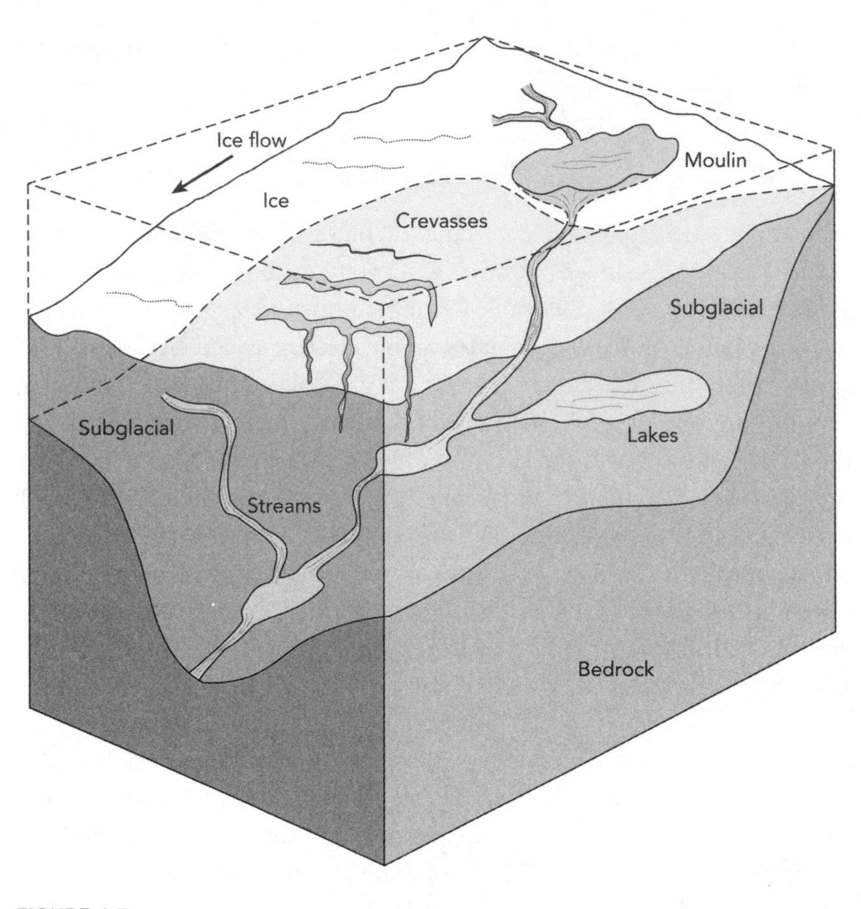

FIGURE 4.5

Schematic diagram of the internal plumbing of a glacier. (From fig. 7.4 in Gornitz [2013])

to extract tightly bound hydrocarbons from rocks. The depth to which a crevasse penetrates represents a delicate balancing act between the stretching of ice that pulls the crevasse apart at the surface and the overlying weight of ice that forces the crevasse to close farther down. The twin requirements for bottom-reaching crevasses are tensional stresses to create fissures in the first place and ample water to fill them.[15]

Entire channel networks lie concealed in ice. The ease with which meltwater courses through the glacier depends to a large extent on the resistance to flow that it encounters. This resistance, in turn, is governed by the size and connectivity of openings within the ice. Intrepid glaciologists, combining the skills of speleologists and mountaineers, have descended into moulins and explored glacial tunnels. By dumping dyes down clusters of moulins and recording exit points, researchers could outline distinct drainage patterns. They found that streams often develop along the base of crevasses and become isolated from the surface when upper levels close because of overhead ice creep. Streams also become deeply entrenched as they incise their way into ice faster than surrounding ice melts.[16] Snowdrifts, large blocks of ice, refrozen meltwater, and ice creep may later create interior passageways by roofing over these channels. Refreezing of water and ice creep may in turn block these tunnels, pushing water into alternate pathways.

Meltwater that enters crevasses and moulins may encounter a series of small cavities within the glacier linked by narrow passageways, or a network of wider interconnected channels and large tunnels that speedily convey water through the ice.[17] In the former case of narrowly constricted channels or poorly developed connections between cavities, water is forced to move slowly or take more roundabout routes. Constricted passageways or circuitously linked networks hinder efficient water drainage.

Daily and seasonal variations in meltwater production alter water pressure in channels within and at the base of glaciers, in turn controlling runoff and ice motion. In winter, the interior drainage system recharges by basal melting and redistribution of accumulated water.[18] In spring, fresh meltwater collects at the glacier bed and emerges at the glacier snout in intermittent waves. An initial high pulse of meltwater may fill cavities and raise water pressures sufficiently to temporarily lift the glacier off its bed, allowing the ice to slide forward faster.[19] Cavities initially linked to narrow openings eventually merge into a branched network. The nature of the interface between a glacier or ice sheet and its bed, as well as the bed's degree of roughness or water content, also affects its ability to flow.

As melting progresses in summer, high volumes of meltwater continue to reach the bed and enlarge internal channels, which develop additional branches. Larger channels expand at the expense of smaller ones. This establishes a more efficient channelized drainage system that can transport greater volumes of water in a steady, even flow. However, sudden summer surface warming episodes may temporarily overwhelm the subglacial drainage system, resulting in short-lived, rapid spurts of ice movement. At the beginning of autumn, meltwater gradually slows to a trickle, many of the once-open passageways begin to refreeze, and a linked cavity drainage system reemerges until the next melt season. Glacier movement and runoff therefore fluctuate greatly over time in response to the volume of injected water and to the ever-changing nature and complexity of these internal pathways.

For the past few years, torrential volumes of water have abruptly burst forth beneath Mendenhall Glacier near Juneau, Alaska, in July.[20] Fed by snowmelt and thawing ice, large volumes of water built up sufficient pressure at the base to temporarily lift the ice and forge a narrow passageway from which water gushed forward with explosive force. (In Iceland, the process is known as a *jökuhlaup*, or "glacier leap," usually triggered there by subglacial volcanic eruptions.) Such sudden glacial outbursts could potentially threaten homes and properties along the Mendenhall River. Unlike less accessible glaciers where such events go unnoticed, Mendenhall Glacier is an urban "drive-up" glacier, attracting 400,000 tourists each summer (including the writer). Thus, such events make vividly apparent one hazardous aspect of glacial plumbing, one that local residents could experience more frequently if spring runoff increases (chap. 9).

Meltwater thus takes many direct and indirect paths toward its ultimate destination—the sea. Having explored some of these pathways, we now turn to the modes by which the entire solid mass of ice advances down mountains or across the land.

GLACIERS ON THE MOVE

Glacier Flow

Despite their proverbial sluggishness, glaciers do slowly flow downhill under the pull of gravity. While ice is as solid as a brick, under prolonged pressure it behaves like a very thick, viscous fluid, like Silly Putty. The combined effects of the weight of compacted ice and gravity set glaciers into motion. Their speed depends on factors such as ice thickness, surface slope, amount of meltwater reaching the bottom, and ease of flow over the substrate (whether smooth or lumpy, or with frozen or thawed surface). A thick, broad accumulation zone and/or thin ablation zone steepens the slope and accelerates the glacier. Conversely, a thinner accumulation zone and/or thicker ablation zone reduces the slope and slows down the glacier's motion. Accumulation and ablation rates vary in response to seasonal or longer-term climatic changes. Therefore, glacier flow is also intimately related to the glacier's surface mass balance.

The stresses and strains that act on a glacier determine how fast it can flow. Put simply, stress refers to a force acting upon an object over an area that causes it to move or change shape, while strain represents any consequent deformation. For example, squeezing a tube of toothpaste (the stress) results in extrusion of the paste (the strain).

The weight of the overlying ice block and gravitational acceleration impose various stresses on the glacier. A *normal* stress acts in the vertical direction. The normal stress at the glacier's bed derives chiefly from the weight of overlying ice. The thicker an ice mass, the greater the stress at its base. Therefore, ice thickness strongly influences consequent glacier flow. A *longitudinal* stress along the direction of flow either presses ice together (*compressive* stress) or pulls it apart (*extensional* stress). Extensional stresses initiate cracks or crevasses where the glacier speeds up over a steeper slope, or at the terminus of a tidewater glacier. Conversely, a slowdown over a flatter slope or overriding a buried ridge induces compressional stresses. *Shear stress*, in contrast, acts in a direction parallel to the surface. Differential motion at different depths or locations on the glacier introduces shear stresses.

An object subjected to a stress that alters its shape or size suffers a strain. In *elastic* strain, the object withstands a temporary distortion that reverses upon release of the stress, like release of a taut rubber band or spring. Application of a high enough stress results in *plastic* strain, where the distortion remains permanent, like taffy when pulled or salt and ice that deform under pressure. The object may further deform either in a ductile manner (i.e., where it undergoes flow or creep, as in metals at high temperatures) or as in brittle failure (e.g., breaking along a crack in cement, a fissure or fault in rocks, or a crevasse in a glacier). These two styles of deformation play important roles in glacier flow.

The interaction between the initiating force (i.e., gravity) and resistance to this force (e.g., drag, friction) sets a glacier in motion. How fast it moves depends on the stresses acting on it, the way it deforms (e.g., by creep), and how easily it slides on its bed.[21] Glaciers move in three main ways:

1. *Internal deformation*, or *creep*. Creep in ice closely resembles the recrystallization and deformation of metals near their melting point (annealing) or in metamorphic rocks, such as schists and gneisses, subjected to intense tectonic forces. In metals, prolonged exposure to heating initiates creep, whereas in ice it results from stresses within the glacier, applied slowly and over time. Creep or slip in ice under stress begins with deformation within and between individual ice crystals. Creep motion is facilitated by crystal glide planes that line up roughly parallel to the direction of motion.[22] In addition, ice slides preferentially along weakened crystal planes densely populated by internal defects, or by recrystallization at grain boundaries. Creep occurs more readily at higher temperatures near the pressure-melting point, as grain boundaries begin to melt.[23] Conversely, ice stiffens at low temperatures, inhibiting internal deformation and creep rates and slowing its motion. Impurities such as gas bubbles, dissolved ions, or solid particles, as well as temperature, also affect creep.

All glaciers advance by internal deformation to some extent. However, the presence of water greatly speeds up motion. Two additional mechanisms describe how a thin water layer or a soft, deformable substrate enhances glacier flow.

2. *Basal sliding*. A thin layer of water or a perfectly smooth surface greatly reduces friction between the ice and substrate, facilitating glacial sliding, in the absence of obstructions. Ice flows past large obstacles on its bed by *enhanced creep*. On irregular surfaces, higher stresses on the upstream side of larger obstacles amplify the amount of creep around and over the obstructions, carrying the ice along. The increased localized pressure on the upstream side of smaller protuberances also lowers the melting point of ice, a process known as *regelation*. The glacier is slowly propelled forward as ice refreezes on the downstream side. Furthermore, very high water pressures pluck cavities or voids on the downstream side that, when filled with enough water, temporarily lift the glacier partly

off its bed. Ice then advances rapidly because basal friction has been significantly reduced. An ample water supply and networks of interlinked passages within the glacier therefore help drive it forward (see also the previous section, "Glacial Plumbing").

3. *Motion along a soft, deformable substrate.* The low cohesiveness of unconsolidated, poorly sorted, waterlogged basal sediments (e.g., fine-grained sands and clays) makes them deform readily under the shearing stresses of the overriding glacier. This deformation carries the ice mass along.

Ice moves at different speeds within the glacier. Drag along the sides of the glacier reduces its speed relative to the center. Similarly, friction at the glacier bed slows it down relative to the surface. Glaciers dominated by internal deformation flow fastest near the surface, but slow to near zero as they approach the base.[24] On the other hand, deformation proceeds more rapidly in ice near its pressure-melting point, where creep becomes more efficient. Glacier bed characteristics, as well as topography, also strongly affect the speed of ice. Glaciers accelerate over soft, fine-grained, unconsolidated, deformable, and waterlogged sediments. They also speed up and stretch out over steeper slopes where extensional forces open up crevasses. Conversely, glaciers slow down over a rough, bumpy substrate riddled with many large boulders, or under extremely cold conditions where very low bottom temperatures effectively cement ice to the bed, as in polar glaciers. They also decelerate and thicken locally on a gentler slope.

The bane of mountain climbers and glaciologists alike, crevasses pose dangers to limb and life, but at the same time illustrate patterns of strain scoring the surface of a moving glacier. Crevasses tell of a glacier's history—its motion, strain, and deformation locked in ice. Crevasses form when forces tearing a glacier apart exceed the mechanical strength of ice. Transverse crevasses that cut across the glacier occur where tensional stresses pull ice apart in the direction of flow down the valley. These "stretch marks" appear in the upper part of the glacier and also where the valley slope steepens. Other crevasses develop at an angle to the valley walls, due to drag that forces the ice to move slower along the sides of the glacier than near its center. Crevasses move along with the glacier and evolve in response to changing stresses along its downhill journey.

New ones form as old ones close, although scars of earlier cracks may still persist.

Although crevasses scar glacier surfaces, they play an important role in calving of icebergs in tidewater glaciers and in ice shelf weakening and breakup, as shown in the "Calving Glaciers" section. But before turning to calving, we probe an unusual form of glacier behavior—one in which glaciers suddenly sprint rapidly down-valley.

Racing Glaciers

Surging glaciers are known from the Austrian Alps, Iceland, the Caucasus, the Karakoram, Alaska, and the Canadian Arctic.[25] Some glaciers may crawl at a snail's pace for many years (10–200 meters per year; 33–656 feet per year) and then suddenly spurt forward by leaps and bounds, some 10 to 1,000 times faster. For example, the Bering Glacier in Alaska surged forward at rates of up to 100 meters (328 feet) per day[26] in 1993. Large ice masses may advance kilometers within a few months. This surging behavior may respond to some internal ice flow instability triggered by outside factors such as seasonal or short-term climate variations or, more rarely, earthquakes. Other triggers may exist within the glacier itself, for example, higher water temperatures within the ice or large amounts of basal meltwater.[27]

In general, glaciers transport a large volume of ice from the accumulation zone to the ablation zone to maintain a steady-state profile. But a small percentage of glaciers surge quasiperiodically, in between long periods of intervening quiescence interrupted by short, rapid outbursts. A surging glacier produces noticeable changes in morphology and surface appearance that can be used later to recognize dormant surge-prone glaciers. Although an uncommon form of glacier behavior, surging is important in understanding the nature of ice flow and anticipating future glacial activity.

A typical surge cycle consists of three stages: (1) a *buildup phase* in the upper zone, where ice builds up faster than it can exit and the glacier thickens and bulges, steepening its profile; (2) a *surge phase* during which the overthickened glacier abruptly speeds up and delivers ice to thinner segments farther downstream, thereby lowering the profile. During this phase, extensional stresses tear apart numerous crevasses in the rapidly moving ice, while the slower-moving ice farther down-valley suffers compression. At the terminus, the glacier front advances markedly, ultimately resulting in (3) a *stagnated* or *depleted phase* following the surge, where the glacier stagnates, leaving pitted and irregular, debris-covered ice patches. The thinned ice undergoes extensive ablation and a short-lived frontal retreat that ends as the glacier gradually returns to the buildup phase.

Calving Glaciers

The calving of a massive glacier believed to have produced the ice that sank the Titanic is like watching a city break apart.

—Duane Dudek (2012)

On May 28, 2008, the largest calving event ever captured on film revealed to astonished observers the sudden disintegration of the margin of the Ilulissat Glacier in western Greenland (also known as Jakobshavn Glacier, or Jakobshavn Isbrae), which retreated a full mile across a three-mile-wide (4.8-kilometer) front over 75 minutes (chap. 2; see also fig. 1.5 for another massive Greenland calving event).

Jakobshavn Glacier continues to retreat landward. Between August 14 and 16, 2015, another huge mass of ice measuring around 12.5 square kilometers (nearly 5 square miles) broke off. The estimated volume would "cover the whole of Manhattan Island by a layer of ice about 300 meters thick."[28] Experts argue whether this represents the largest calving event ever. Nevertheless, it was big enough to impress a NASA glaciologist by its "sheer size" and by the glacier's continuing retreat at "galloping speed."

In spite of such occasional dramatic iceberg breakouts, glaciers that end at a lake or ocean normally shed ice by calving. A steep cliff often defines the edge of tidewater glaciers. Once touching water, ice begins to float. No longer constrained by drag along the base or sides, the glacier speeds up.[29] This sets up tensional stresses that stretch the glacier lengthwise, causing multiple crevasses to develop near the edge. Strong bending forces that act on the buoyant ice at the grounding line boundary generate additional fractures. Others appear as a result of sudden or large changes in water level caused by tides, wave action, or storm surges. Still more crevasses develop when ice at or below the waterline melts, undercuts, and weakens the terminus.[30] As crevasses propagate from top to bottom, large blocks of ice break off with a loud splash.

The rate of calving depends on glacier speed at the terminus, ice thickness, and water depth, the latter two of which also influence glacier buoyancy. The factors that govern the extent of calving also exert a strong control on the forward motion of the glacier.

Abrupt glacier advances and thinning follow major calving events, often succeeded by retreat. Calving plays a major role in ice loss from Greenland, Alaska, and other Arctic tidewater glaciers, which have accelerated and receded in recent years.[31] Calving also contributes to ice loss on Antarctica, where low temperatures prevent surface melting. Undercutting of Antarctic ice shelves by warmer ocean water, especially as has happened in recent years, thins and weakens the ice shelves to a point where calving can occur (see chaps. 2, 5, and 6).

More than a decade of warming preceded the catastrophic breakup of the Larsen A and B Ice Shelves on the Antarctic Peninsula in 1995 and 2002, respectively (see fig. 2.1). Higher air and sea temperatures had increased melting and thinning of the shelves from top and bottom. Several unusually warm summers expanded the number of surface melt pools, which fed numerous surface cracks. Filled with meltwater, the fissures deepened and propagated downward through the ice, setting the stage for sudden fragmentation.[32] The last straw occurred when multiple crevasses opened parallel to the ice edge, unleashing a chaotic mess of highly fractured, tabular ice blocks that tipped over like a stack of dominoes. In the final stage, the former ice shelf was reduced to a morass of flat, tabular ice shelf remnants and rubble. Bending stresses initiated by the buoyancy of ice and extensive crevasse hydrofracturing led to the rapid destruction of these two (and possibly other) ice shelves.

While a casual observer might read a massive calving event as an imminent sign of a tidewater glacier's demise, calving tells only part of the tale. The full story comes from the glacier mass balance, as discussed in the next section.

ASSESSING GLACIER WELL-BEING

The fairly quick response of mountain glaciers to varying climatic conditions makes them important barometers of climate change. A vital sign is whether they are growing, shrinking, or remaining stable. Changes in various physical characteristics—length, width, terminus position, total area, relative areas of accumulation versus ablation zones, precipitation, or meltwater runoff—paint an overall picture of a glacier's health. A healthy, growing glacier accumulates more snow and thickens, advancing toward lower elevations as it reestablishes a new equilibrium profile. A deteriorating glacier shows symptoms of thinning along its entire length, a shrinking accumulation zone, increasing mass losses at lower elevations, and retreat at the terminus. Changes in the equilibrium line altitude (ELA) or in the areas of accumulation versus ablation serve as useful regional climate proxies because of their close ties to climate. The snow-firn boundary, which closely approximates the ELA, is recognized on aerial photographs or satellite images by the brighter and smoother appearance of snow, as compared with firn or ice when viewed at the end of the ablation season (see next section).

Glacier mass balance, however, depicts a glacier's true health status more accurately. Traditionally, the glaciologist digs multiple small snow pits and measures annual layer thickness in the snowpack to determine accumulation rate, generally by the end of summer. Signs of melting and refreezing before next winter's snowfall, or by a thin dust layer marking the end of the ablation season, demarcate annual snow layers.[33] In addition, crevasses expose a lengthy two-dimensional picture of annual snowpack layer thicknesses with depth. Ablation is gauged by setting a grid of stakes vertically into the glacier at the start of the melt (ablation) season. The glaciologist subsequently records the lowering of the ice surface at the end of the melt season. The change in mass at each site is obtained by multiplying the height difference of the exposed stake (or snow thickness changes in pits) by the appropriate density of snow or ice. Glacier mass balance is then estimated by interpolating between individual grid points and extrapolating results over the entire glacier. (For calving glaciers, mass balance calculations also involve monitoring changes in terminus locations and glacier velocity.)

More current approaches rely on climatological and glacier flow data, with selected field measurements for calibration and verification. Climatological models exploit theoretical relationships among temperature, precipitation, air moisture, and winds, and their variations with elevation. These are frequently coupled with ice flow models to obtain mass balance of tidewater glaciers.

Geodetic surveys measure changes in glacier volume by measuring changes in ice areas derived from ground surveys, aerial photographs, or satellite imagery on two or more separate dates, combined with elevation data from topographic maps or digital elevation models (DEMs). Global Positioning System (GPS) satellite instruments detect surface elevation changes with high precision. Volume changes can be converted to mass change if the ice density at different points along the glacier is known. However, density variations of snow and ice within the firn zone may introduce errors into volume calculations. Furthermore, rugged terrain in the high-altitude accumulation zone makes accurate elevation measurements difficult. Nevertheless, the geodetic approach complements classic field methods, particularly for larger areas over longer time intervals.

Glacier area can also be scaled to volume by means of simple empirical power laws. For example, the volume of a cube increases relative to the area as $V = A^{3/2}$. If its area doubled, the volume would increase by a factor of 2.83. The actual exponent for a glacier is less than 3/2, because of its elongated shape and irregularly curved surface. Glacier area/volume scaling laws, therefore, take the general form:

$$V = kA^{\gamma}$$

where V = glacier volume, A = surface area, and k and γ are constants.[34] Some representative values for k and γ fall between 0.04 and 0.07 and 1.20 and 1.37, respectively. For example, if $k = 0.055$ and $\gamma = 1.3$, the volume would increase by a factor of 2.46 for a doubled area. For other values of k and γ, doubling the area would increase the volume by close to 2.5—thus somewhere below the ideal case of a fully three-dimensional volume.

Other area-volume scaling laws can be derived by combining other indicators, such as changes in length and topographic relief with area.

While at best an approximation for a single glacier, the accuracy improves when applied to a much larger number of glaciers. Estimates of global glacier volume using this approach differ due to differences in the particular scaling laws utilized.

The expense, logistics, and large-scale coverage required for ground-based glaciological studies and geodetic surveys, particularly in treacherous terrain, severely limit the number of field sites. Scaling laws that rely on empirical constants may not be universally valid. Furthermore, such methods neglect ablation losses due to calving. Therefore, remote sensing observations that encompass a much broader area and can detect ice motion now routinely supplement ground measurements. These instruments that take an eagle's-eye view—by aircraft and spacecraft—increasingly monitor the rapidly changing cryosphere. The next section briefly outlines various ways in which these eyes in the sky track how fast glaciers are wasting away.

EYES IN THE SKY—GLACIERS FROM AFAR

A large array of satellites silently circles the Earth, continuously collecting reams of data on the planet's swirling clouds, shifting ocean currents, sediment plumes spilling forth from river deltas, thick and dark clouds of volcanic ash spewing out of erupting volcanoes, crops nearing harvest time, new roadways, and cities springing up like mushrooms after a summer's rain. Satellites also train their eyes on the cryosphere, observing changes in the world's glaciers and ice sheets. Thanks to their ability to see large swaths of terrain at a glance, satellite observations vastly extend the knowledge gained from classic field studies, aerial photographic and ground geodetic surveys, or simple mathematical scaling laws. Among key glacier metrics that can be studied from afar are changes in length, area, and elevation, from which ice volume and mass change can be deduced. Other important glacier characteristics include surface reflectance, and hence albedo; surface temperature; and the ability to discriminate between different types of snow, ice, and debris cover.[35]

TRACKING GLACIER CHANGES

Instruments that eye the Earth from above employ interactions between electromagnetic waves, or gravity, and matter (see appendix B for additional information). Satellite imaging systems operate much like digital cameras, recording the intensity of reflected radiation in the visible to near-infrared portions of the electromagnetic spectrum, generally between 0.4 and 1.5 micrometers (1 μm = one millionth of a meter).[36] Materials absorb or reflect radiation in markedly different ways based on differences in their molecular makeup. For example, basalt, a dark volcanic rock, reflects little light throughout the spectrum, whereas chlorophyll absorbs the blue and red portions of the visible spectrum, reflecting strongly in the green (hence the verdant color of vegetation), and even more strongly in the near-infrared, beyond 0.7 μm. Fresh, fine-grained snow is extremely bright in the visible, but strongly absorbs wavelengths of around 1.4 and 1.9 μm in the near-infrared because of hydroxyl bonds in H_2O. The spectrum of coarse-grained, partially recrystallized snow resembles that of fine snow, but reflects less light. Glacier ice is darker yet (fig. 4.6).

Mapping different glacier snow/ice conditions or terrain types begins with the ability to recognize subtle differences in texture, brightness, or spectral characteristics, either by eye or by using automatic classification techniques. A trained glaciologist who manually identifies and maps glacial outlines or other features on aerial photographs can achieve higher accuracies than many existing computer algorithms; however, the process is time-consuming and thus more costly.[37]

The Landsat satellites, which have surveyed the Earth's surface since 1972, offer one of the longest available time series of satellite images (appendix B). The early Landsat satellites carried multispectral scanners (MSS) that spanned the visible to near-infrared in just four spectral bands. Later satellites added more spectral bands that provided more extensive coverage into the thermal infrared. The smallest swath of ground visible from space has also decreased significantly. Ground resolution has increased from around 80 meters (262 feet) on the earliest Landsats to 15 meters (49 feet) on Landsats 7 ETM+ and 8. The newest satellite imaging systems, such as SPOT 6 and 7 and WorldViews 1–4, now routinely spot objects 1 meter (3.3 feet) or smaller (appendix B).

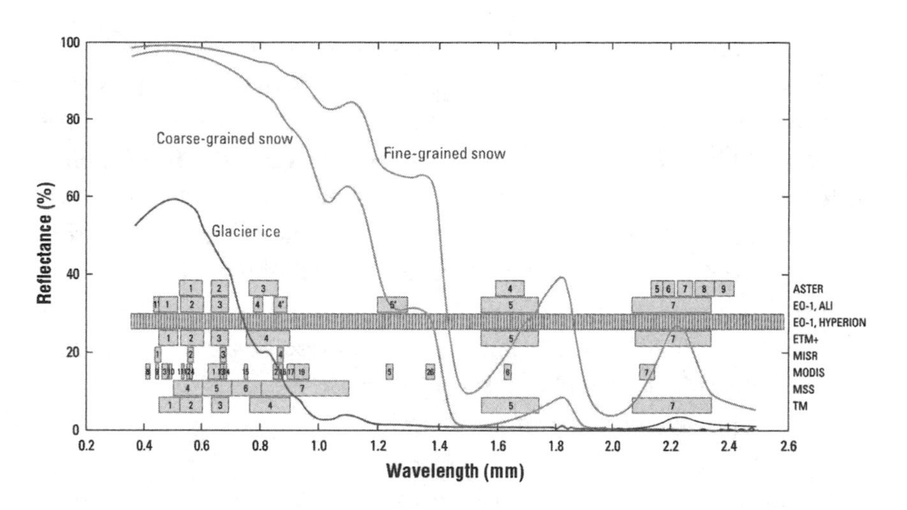

FIGURE 4.6

Visible, near-infrared, and short-wave infrared spectra of fine-grained fresh snow, coarse-grained snow (older and deeper), and glacier ice. The spectrum shows the variation in reflectance (brightness, or albedo) with wavelength. The numbers are the spectral channels used in the remote sensors listed on the right. (Williams and Ferrigno [2012], fig. 77, http://pubs.usgs.gov/pp/p1386a/images/gallery-2/full-res/pp1386a2-fig77.jpg)

Satellite sensors that peer beyond the visible and near-infrared regions have also greatly expanded our understanding of glaciers and other cryospheric constituents. The Landsat Thematic Mapper and Enhanced Thematic Mapper ETM+ and the Terra ASTER (Advanced Spaceborne Thermal Emission and Reflection Radiometer) carry multiple spectral bands, including a thermal infrared band, from 10.4 to 12.5 μm. The widened spectral range discriminates more clearly among fresh snow, various states of ice (fresh, dirty, or debris covered), and water.[38] The night scanning ability of thermal infrared bands (which detect emitted radiation, i.e., heat, rather than reflected light) compensates for their coarser ground resolution. Yet, as with visible and near-infrared sensors, they lack the ability to see beneath clouds.

On the other hand, microwave imaging systems, e.g., synthetic aperture radar (SAR), overcome these limitations by collecting data day and night regardless of cloud cover.[39] The SAR instrument operates as a type of imaging radar. In SAR, microwave pulses are beamed toward the Earth's surface at very short time intervals, and a two-dimensional representation of the ground is built up from the returned surface signals. SAR detects subtle differences in surface roughness and has also been applied to the study of sea ice (chap. 2). A modified form of SAR—interferometric synthetic aperture radar (InSAR)—records the phase of the signal, from which images acquired at different times or locations can be combined to form an interference pattern due to the phase differences. The resulting *interferogram* illustrates the phase changes resulting from ice motion between overflights, from which the glacier's speed can be deduced. InSAR has proved invaluable in tracking the velocity of glaciers, ice caps, and ice sheets.

Changes in glacier length, area, and motion can also be detected by comparing a sequence of aerial photographs, satellite images, and ancillary data of the same area spanning a known time period. Differences between digital elevation models (DEMs) over several years can be used to determine changes in glacier volume. DEMs are constructed from aerial and space stereophotogrammetry (e.g., ASTER) and topographic maps, as well as spaceborne radar and laser altimetry that measure changes in land elevation. But conversion of volume to mass yields only approximate results, due to uncertainties in ice/firn densities. Therefore, estimates of glacier mass balance increasingly rely upon other independent techniques, such as mass budget modeling and gravity measurements.

Ice thickness is a crucial index of glacier well-being. Satellite altimeters detect topographic changes resulting from changes in ice thickness. Satellite radar altimeters send out microwave pulses and record the returned signals from the Earth's surface. The time taken by the pulse to reach the Earth's surface and return at the speed of light gives the exact distance between the spacecraft and surface, after correcting for atmospheric and instrumental effects (appendix B). Repeated observations of the same area over time establish a trend in glacier or ice sheet elevation. Earlier radar altimeters were limited to smoother slopes, such as the interiors of the Greenland and Antarctic ice sheets. Technological advances, such as SIRAL (Synthetic Aperture Interferometric Radar Altimeter) on the European Space Agency's CryoSat-2, allows operation over more rugged terrain, which improves topographic mapping precision.

Laser altimeters, such as NASA's Ice, Cloud, and Land Elevation Satellite (ICESat-1), substitute radar with an intense, narrow laser beam (see appendix B). The IceBridge mission, as its name implies, bridged the gap between ICESat-1, which lasted from 2003 to 2010, and ICESat-2, launched in 2018. The IceBridge aircraft gathers data from various instruments including laser altimeters, radar sounders, and gravimeters, along with mapping cameras to survey sea ice extent, ice and snow thickness, and topography of the ice surface and base.

Not all remote sensors operate in the electromagnetic spectrum. The Gravity Recovery and Climate Experiment (GRACE), jointly operated by NASA and the German Aerospace Center, has measured detailed changes in the Earth's gravitational field since 2002.[40] GRACE consists of twin satellites that circle the Earth 15 times a day, detecting tiny gravitational variations along their orbits (appendix B). As the orbiting satellite feels a slightly stronger gravitational tug beneath, it accelerates relative to its orbital companion. Once past the anomaly, it slows down, whereas its twin accelerates upon reaching that spot. By combining the changing speeds with precise positions given by GPS, scientists construct detailed gravity maps of the planet.

Gravity is determined by mass. Natural gravitational variations occur because of differences in rock densities or topography. Some gravity anomalies detected by GRACE hint at buried mineral deposits containing large amounts of heavy metallic elements. However, widespread mass redistributions result from large-scale water transfers between land and sea, either natural or human caused. Natural causes encompass meteorological extremes of rainfall and drought. For example, the particularly strong 2010–2012 La Niña event ravaged Australia with record-breaking rainfall and floods. Global sea level briefly fell nearly 5 millimeters (0.2 inch) because of the excess water dumped on land.[41] Sequestration of water in large reservoirs or, conversely, excess groundwater mining also alter river runoff, ultimately affecting sea level.

Data gathered from GRACE and other orbiting satellites reinforce the accumulating evidence for ice losses on glaciers and, increasingly, ice sheets. The sea level contributions from receding glaciers and ice caps have been determined using GRACE data, after correcting spurious effects from changes in land water distribution and glacial rebound.

TABLE 4.1 Recent Sea Level Rise From Observations of Shrinking Glaciers (in mm/yr)

GLACIERS AND SMALL ICE CAPS			
SEA LEVEL RISE	**COMMENTS**	**METHOD(S)**	**REFERENCE**
0.58±0.15 (1950–2005)	Including Greenland glaciers	S	Leclercq et al. (2011)
0.84±0.64 (2003–2009)	Updated from Leclercq et al. (2011)		Marzeion et al. (2017)
0.41±0.08 (2003–2010)	Excluding peripheral ice sheet glaciers	G	Jacob et al. (2012)
0.63±0.23 (2003–2010)	Including peripheral ice sheet glaciers	G	Jacob et al. (2012)
0.59±0.07 (2003–2009)	Excluding peripheral ice sheet glaciers	G, L, Gl	Gardner et al. (2013)
0.71±0.08 (2003–2009)	Including peripheral ice sheet glaciers	G, L, Gl	Gardner et al. (2013)
1.37 (2001–2010)	Average of glaciological, geodetic data	Gl, Geo	Zemp et al. (2015)
0.53±0.09 (2002–2014)	Updated from Gardner et al. (2013)	multiple	Reager et al. (2016)
0.51±0.07 (2002/2005, 2013/2015)	Average of recent studies	multiple	Marzeion et al. (2017)

G = gravitational (GRACE); Geo = geodetic surveys; ground, air, or space mapping; Gl = glaciological methods, pit & stake approach; L = laser altimeter (ICESat); S = empirical area-volume scaling

One GRACE-based study concluded that shrinking glaciers and ice caps drove up sea level by 0.63 millimeters (0.02 inch) per year between January 2003 and December 2010 (including peripheral glaciers on ice sheets; table 4.1).[42] A separate study combined GRACE data with ICESat laser altimetry and classic ground-based glaciological methods.[43] During that period, glaciers yielded 0.72 millimeters (0.03 inch) per year (including peripheral glaciers; table 4.1). The glaciological method, in contrast to the other two techniques, may inflate ice losses, partly because ground measurements concentrated in regions with above-average melting rates that undoubtedly offered easier access.

Monitoring Glacier Speed

How fast a glacier moves—whether it is accelerating, slowing down, or creeping sluggishly forward—reveals another critical aspect of its behavior. The glacier's speed can be estimated by tracking known features in a sequence of aerial photographs or satellite images over a period of time. Suitable features include prominent ice pinnacles, crevasses, large boulders, or distinctive patterns of rocky debris that can be recognized on the entire set of images. All the images in the sequence need to be accurately overlaid and coregistered to a fixed reference point that appears on all images. Times of image acquisition should also be carefully recorded. Thus, the speed can be determined by subtracting the difference in distance between the same specific glacial features at two different times divided by the known time interval.

Interferometric synthetic aperture radar (InSAR) is increasingly being used to track glacier motion. This technique is capable of resolving glacier displacements as small as a few centimeters. InSAR assumes that the phase of the radar beam received by the instrument from a recognizable ground feature remains constant from one overflight to the next, assuming that the instrument remains in the same position. Subtracting the phase portion of the two radar images generates an *interferogram*—an image of the phase changes between the two overflights. Before applying this technique to measurements of glacier velocity, errors stemming from differences in topography and satellite positions between overflights must be removed. Topographic effects can be eliminated by acquiring DEMS from an independent source. Alternatively, interferograms from three satellite overflights can be merged to isolate the ice velocity component from changes in satellite instrument position.[44] Some of the fastest-flowing glaciers in Antarctica occur in the Amundsen Sea Embayment in West Antarctica. InSAR techniques, in conjunction with Landsat feature tracking prior to 1992, established their speeds.[45] Observing the flow of ice through a number of key "flux gates" near their grounding lines along with changes in ice thickness revealed a 77 percent increase in ice discharge since the 1970s, of which half occurred after 2003.

The above examples highlight the increasing utility of remote sensing in monitoring the rapidly changing cryosphere, including critical glacier indicators such as diminishing ice mass, thickness, and accelerations. Of equal importance to glacier studies are global databases that record historic trends in key vital statistics, as reviewed in the next section.

Global Glacier Databases

An accurate record of glacier evolution requires a detailed database of historic and current status. A comprehensive glacier inventory therefore proves an invaluable tool for estimating volume changes from DEMs and altimetry, determining glacier area/volume scaling relationships, and modeling future glacier mass changes. Richard Williams and Jane Ferrigno of the United States Geological Survey began assembling a global atlas of glaciers in 1978—mostly Landsat imagery, supplemented with other, more recent satellite data.[46] The atlas images, largely covering the period 1972 to1981, can be compared with more recent maps and satellite data from Terra ASTER, Landsat ETM+, and RADARSAT.

The National Snow and Ice Data Center (NSIDC) archives and distributes data on all aspects of the cryosphere, including glaciers. NSIDC, based at the University of Colorado in Boulder, is supported by multiple U.S. federal agencies, including the National Oceanic and Atmospheric Administration (NOAA) and and National Aeronautics and Space Administration (NASA). The World Glacier Monitoring Service (WGMS), established in 1986 and housed in Zurich, Switzerland, hosts a global repository of glacier data sets. WGMS collaborates closely with NSIDC, GLIMS, and other scientific organizations (see the following paragraphs) to conduct long-term monitoring of glaciers and regularly update the World Glacier Inventory (WGI). Every two years WGMS issues worldwide status reports that list changes in glacier thickness and cumulative mass balance.[47] In spite of considerable year-to-year fluctuations in ice thickness, their data dramatically underscore the cumulative downhill trend and rapid decline of glaciers in recent years, documented more thoroughly in the following section.

The Global Land Ice Measurements from Space (GLIMS) project uses advanced satellite remote sensing techniques to (1) acquire repeated imaging of the world's glaciers from ASTER at up to a 15-meter (49-foot) resolution spanning the visible through thermal infrared spectrum, (2) create a Web-based glacier database, and (3) analyze the images to determine the changes in glacier area.[48] While GLIMS draws most of its data from ASTER, supplementary Landsat and other satellite images, historic maps, and field-based "ground truth" verify the space observations. GLIMS actively cooperates with WGMS, and its findings expand on related efforts, such as the United States Geological Survey's *Satellite Image Atlas of Glaciers of the World*[49] and the WGMS World Glacier Inventory.

The Randolph Glacier Inventory (RGI) contains the currently most complete collection of area data for nearly 198,000 glaciers in 19 regions around the world.[50] Many studies now utilize the RGI database to generate area-altitude (hypsometry) data for each glacier, to improve regional and global estimates of glacier ice volume and mass balance and to predict the future evolution of glaciers.

Satellite, aircraft, and ground observations together with worldwide glacier inventories help us draw a picture of glaciers in distress. The next section vividly illustrates examples of glaciers in serious trouble and what this bodes for sea level rise.

WASTING AWAY—RECEDING GLACIERS

> *[Mer de Glace] . . . les glaces descendent, pareilles à un grand fleuve dont les ondes auroient été suspendues tout d'un coup par une force inconnue. . . . L'arcade de glace qui se forme à l'embouchure du glacier . . . est une des principales merveilles de la vallée de Chamonix. . . . Rien n'est plus frappant que le contraste des morceaux de glace écroulés, d'une blancheur pareille à celle de la neige, avec la couleur transparente du plus beau bleu foncé et d'aigue-marine de cette grotte enchantée.*
>
> *[Mer de Glace] . . . the descending ice, like a great river whose waves were suddenly suspended by an unknown force. . . . The arcade of ice forming at the mouth of the glacier . . . is one of the main marvels of the Chamonix valley. . . . Nothing is more striking than the contrast of crumbled pieces of ice, as white as snow, with the most beautiful transparent deep blue and aquamarine of this enchanted grotto.*

—Samuel Birmann, in *Souvenirs de la vallée de Chamonix* (1826)
(in Nussbaumer et al. [2007]; trans. V. Gornitz)

Glacial Pullbacks

Alpine glaciers, whose beautiful scenery now draws countless skiers, mountain climbers, and tourists, appeared more menacing to mountain farmers in centuries past. Folklore depicted glaciers as "nearly dragons, hanging on cliffs with open jaws and threatening to fall into the valley any minute; nearly snakes, winding themselves between the mountains, through narrow valleys."[51] Henry George Willink (1851–1938), an English lawyer, member of the Alpine Club, and artist, clearly inspired by this theme, drew the Mer de Glace (which he called Wilderwurm Glacier) in 1892 as a voracious dragon greedily devouring the hamlet of Le Bois.[52] The Mer de Glace, a 12-kilometer (7.5-mile)-long glacier on the northwestern side of Mont Blanc, highest mountain in the Alps (elevation 4,800 meters [15,780 feet]), currently spreads over 32 square kilometers (12.4 square miles), from 1,500 to 4,000 meters (2,950–13,100 feet) above sea level.[53]

Following a period of medieval warmth, winters in Europe had grown increasingly harsh by the 1550s during the Little Ice Age. Alpine glaciers advanced into the valleys, engulfing a number of small mountain hamlets in their path. In 1644, the Mer de Glace (then known as the Glacier des Bois) threatened the town of Chamonix. Alarmed residents summoned the bishop from Geneva, whose benediction appeared to ease the threat. But by the 1660s the glacier had begun to readvance, requiring additional blessings by the bishop in 1664 and 1669—this time to no avail.[54] Historical paintings and maps recorded numerous expansions and regressions of the Mer de Glace between 1600 and 1850, before its nearly continuous recession beginning in the 1860s.

Although each glacier, like the Mer de Glace, has a unique history shaped by local factors including elevation, slope aspect, temperature, winter precipitation, wind, and surging behavior, consistent trends emerge from piecing together regional and global histories. Since the 1860s, Alpine glaciers have retreated continuously, with a few minor temporary exceptions. Between 1850 and the 1970s the area of Alpine glaciers declined by 35 percent, and by 2000 only half of the original area survived.[55] A single extremely hot and dry summer in 2003 removed another 5–10 percent of the remaining ice. Several such recurrences could lead to even more extensive Alpine deglaciation than has already occurred.

The European Alps are not alone. Glacier National Park in Montana may soon lack the glaciers that inspired its name. Though it was covered by over 150 glaciers a century ago, only 25 survive today, and even these may be gone within a generation. As the ice disappears, the streams fed by melting snows dry up in summer, anglers catch fewer trout, alpine wildflowers diminish, and ranchers find less water for their livestock. Many widely separated glaciers across the globe have been receding: in southern Alaska (as we have already seen), the Canadian Rockies, the high Arctic, the Himalayas, the Andes, and New Zealand, and on Mt. Kilimanjaro. The glacier pullback has accelerated since around 2000, and now accounts for nearly a third of the current global sea level trend.

Rising temperatures threaten tropical glaciers, already on the brink of extinction. Lonnie Thompson, a glaciologist at Ohio State University, and his colleagues have measured the drastic recession of the snows of Mt. Kilimanjaro. To their dismay, around 85 percent of the ice cover on Kilimanjaro present in 1912 had vanished by 2007. Thompson predicts that at current rates, within one or two decades no ice fields will remain—for the first time in 11,700 years.[56] High in the Andes, the Qori Kalis Glacier on the Quelccaya Ice Cap in Peru has been retreating 10 times faster since 1995 than in the initial 15 years of Thompson's study, beginning in 1963.[57] Meanwhile, on the southern tip of South America, many Patagonian glaciers are retreating at record rates, in spite of heavy snowfall. Fourteen glaciers disappeared entirely between 1870 and 2011.[58]

From a Global Perspective

The thing most sensitive to climate change is a glacier. In the 1970s, people thought glaciers were permanent. They didn't think that glaciers would recede. They thought this glacier would endure. But then the climate began changing, and temperatures climbed.

—Qin Xiang, Chinese scientist[59]

High amidst the soaring peaks of the Qilian Mountains on the north side of the Tibetan Plateau, Qin Xiang and his colleagues climb the ice to document the evolving state of the Mengke Glacier. A mere six miles long and occupying nearly eight square miles, the glacier has shrunk twice as fast within the past decade as in the preceding 12 years. Here, as well as across much of the "Third Pole"—the Himalayas and Tibetan Plateau—dwindling glaciers could threaten Asia's water supply, as discussed further in chapter 9. Most of the mighty Asian rivers originate in these elevated mountain ranges, fed by glacial meltwater—the Ganges-Brahmaputra, Indus, Mekong, Yellow, and Yangtze Rivers, among others. Overall, Asian mountain glaciers are losing ice, although losses vary considerably across the region. Continued attrition of these glaciers could adversely impact food security for over 1.4 billion people downstream who depend on water resources from these major source rivers (chap. 9).[60]

Himalayan glaciers shed enough ice to raise sea level by 0.041 millimeters (0.002 inch) per year between 2000 and 2016.[61] The greatest losses occurred in the east and south, encompassing high peaks such as Mt. Everest, Annapurna, and Kangchenjunga, whereas the northwest Himalayas (Karakoram, Kunlun, and eastern Pamir) have remained stable or gained mass[62]—an exception examined more closely in the next section. The largest inputs of glacial runoff emanate from shrinking Himalayan glaciers feeding the Indus and Brahmaputra basins. Runoff over a large glacier source area accounts for the sizeable contribution from the Indus River. Mass loss from receding glaciers that flow into the Brahmaputra River also adds to the increasing runoff.[63]

Glaciers constitute an important yet increasingly endangered hydrological reservoir and a growing source of observed global sea level rise. Information pieced together from many widespread regions confirms the picture of worldwide glacier attrition. Although many glaciers have retreated since the 1860s, glacier mass losses have escalated significantly on a global scale, particularly within the past several decades. Lonnie Thompson, drawing upon years of personal experience, exclaims that "climatologically, we are in unfamiliar territory, and the world's ice cover is responding dramatically."[64] The glaciers' response is dramatic.

Mountain glaciers respond quickly to even minor climate fluctuations such as the warming of the last few decades. Some may argue that this simply reflects natural variability in the system. Nevertheless, many glaciers in various locations worldwide have been retreating since the

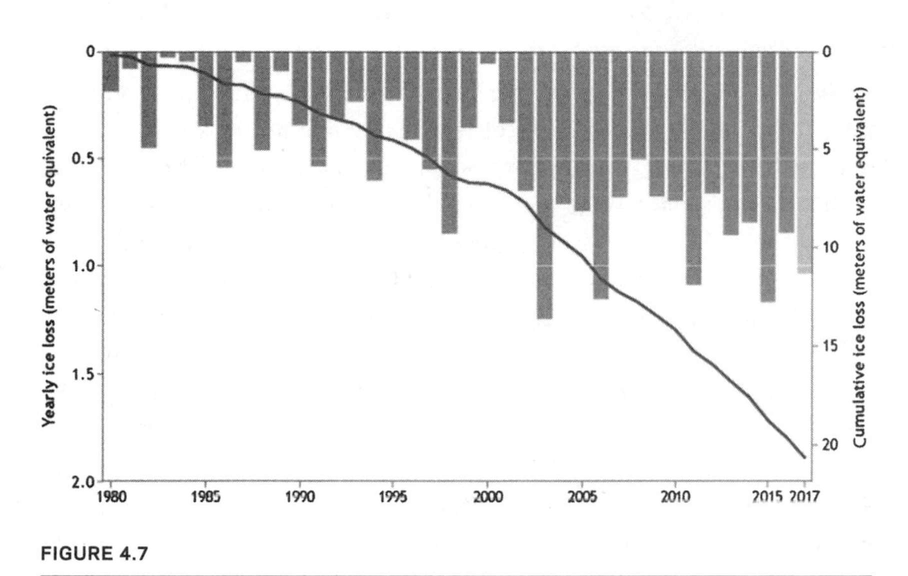

FIGURE 4.7

Cumulative loss of ice mass from 37 reference glaciers, 1980–2017. The bars show annual mass balance; the line shows cumulative annual balance (in meters of water equivalence). (NOAA, "2017 State of the Climate: Mountain Glaciers," August 1, 2018, https://www.climate.gov/news-features/featured-images/2017-state-climate-mountain-glaciers)

mid- to late nineteenth century. The recent mounting losses in glacier ice mass have been described as "historically unprecedented."[65] This raises a red warning flag, particularly if the trend is sustained (fig. 4.7). Glaciers hold sufficient ice to elevate the world's oceans by 0.35 to 0.52 meters (on average, 1.4 feet) if all melted and the water spread out uniformly. A meltdown is clearly under way, as summarized in table 4.1. Global ice losses since 2000 range from 0.5 to 0.8 millimeters (0.02–0.03 inch) per year in equivalent sea level rise. Diverse instrumentation, data sources, averaging techniques, and uncertainties in extrapolating from limited spatial coverage account for this broad spread. Nevertheless, while individual glaciers may still advance, the majority of sampled glaciers have been in retreat over the past century, and particularly within the last few decades. Glacier ice losses approach around a millimeter per year (see table 4.1).

What of the future? Will any glaciers survive global warming? A recent study, for example, suggests that a third of the Asian high mountain glaciers (by mass) could disappear if global temperature rises only 1.5°C (0.83°F) by 2100, relative to preindustrial conditions.[66] This estimate probably represents a lower bound, in that it is based on an optimistically low carbon emissions scenario (RCP2.6, appendix A). Another study concludes that the world's glaciers could add between 10 and 20 centimeters (3.9–7.9 inches) to the sea by 2100, depending on an assumed greenhouse gas emission scenario.[67] A more dire prediction foresees shrinkage of as much as 75 to 90 percent of western Canadian, U.S., Scandinavian, north Asian, Alpine, and tropical glacier ice volume by 2100.[68] Still more consequential would be losses of the much vaster ice stores held in Arctic, Alaskan, Asian, and ice sheet peripheral glaciers. Greenland could eventually play an even greater role, because polar amplification will likely warm the Arctic more than elsewhere.[69] But before turning our attention to the major players in terms of sea level rise—the ice sheets—we examine a contrarian region.

Bucking the Trend

Whereas glaciers from most regions of the world are silently wasting away, a few regions buck the global trend. The lofty Karakoram Range, nestled among Pakistan, India, and China high in the western Himalayas, presents such a case. The most glaciated area beyond the polar regions, the Karakoram boasts the highest concentration of 8,000-meter (26,250-foot) peaks on the planet.[70] Its snows feed the Indus and its tributaries, which deliver much-needed water to northern Pakistan and northwest India. In marked contrast to neighboring sections of the Himalayas and Tibetan Plateau (and elsewhere), Karakoram glaciers have remained stable or gained mass, a condition labeled the "Karakoram Anomaly." The anomaly encompasses neighboring glaciers in western Kunlun and eastern Pamir, as well.[71]

The resolution of the anomaly rests on contrasting atmospheric circulation regimes between the Karakoram and the central to eastern Himalayas and Tibet.[72] The entire region has generally warmed since the mid-1970s, although the Karakoram remains cooler than its neighbors.

However, westerlies that supply most of the Karakoram's winter precipitation have strengthened. Gains in winter snowfall, particularly at higher elevations, compensate for lower summer snowfall, neutralizing total annual snowfall loss. The rest of the Himalayas and Tibet, on the other hand, receive most of their precipitation from the Indian monsoon during summer, when most of any snowfall will melt. These trends are likely to persist in the future. Climate models suggest that the Himalayan glaciers would sustain losses ranging from a third to two-thirds of their present mass by the end of the century, depending on the extent of future warming. But, unlike the rest of the Asian glaciers, the Karakoram and eastern Kunlun regions would lose the least.[73]

Exceptions like the Karakoram and immediate surroundings notwithstanding, the fact remains that glaciers and small ice caps around the world are rapidly dwindling, and in some estimates, only slightly over half may survive by 2100.[74] Their shrinkage poses growing concerns for future water resources and food security for hundreds of millions of people in many parts of the world. Furthermore, glaciers furnish many benefits for hydropower, recreation, and tourism. Their relatively small total ice volume belies their importance to global welfare.

We now turn to the massive big gorillas—the enormous Greenland and Antarctic ice sheets. These two ice sheets combined hold the equivalent of 66 meters' (216 feet) worth of sea level rise (see table 1.1). Given the huge ice locker these two ice sheets represent and the serious global consequences of a large potential meltdown, as discussed further in chapter 9, the next two chapters focus on Greenland and Antarctica, respectively, to assess their current status, while a later chapter investigates their future prospects.

5

THE GREENLAND ICE SHEET

Three feet of ice does not result from one day of bad weather.

—Chinese proverb

WELCOME TO KALAALLIT NUNAAT (GREENLAND)

The Icelandic sagas relate how Norse settlers led by the Viking seafarer and adventurer Erik the Red colonized the southwestern tip of Greenland (known as the "Eastern Settlement") around 986 CE, lured there by Erik's glowing descriptions of a lush "green land." Soon thereafter, they established a second settlement on the west coast, farther north (the "Western Settlement"). Many historians have cited the Viking settlements as evidence that Greenland once enjoyed a more benign climate. The subsequent abandonment of the Norse colonies by around 1450 CE was therefore blamed on a deteriorating climate. Several markers of past climates indicate warmer conditions from around 800 to 1300 CE, around the time of the Medieval Warm Period.[1] Marine core data hint at less sea ice and more open water—conditions favorable for seafaring at that time. After the 1300s, colder temperatures set in and sea ice expanded.

The subsequent failure of the Norse settlements has been variously blamed on diverse factors including an unwillingness of the Vikings to adopt the hunting and fishing lifestyle of the native Inuit, overgrazing and soil erosion, disease, or hostilities. Nevertheless, archaeological remains show that in response to a worsening climate, the Norse settlers relied less on cattle and consumed more seafood, particularly seal meat.[2] By the end of the settlement period, 80 percent of their diet was marine based, including seal.

The demise of the Greenland settlements has always been puzzling. No signs of sudden disease, starvation, or warfare have been found, in spite of one medieval text that blames the *skraelings*—a derogatory term for native people—for destroying the Western Settlement in the 1360s. However, three factors, including climate change, may have contributed to the extinction of the colonies.[3]

Although largely self-sufficient, the Norse depended on trade with Europe for imports of grain, iron, wine, and other essentials in exchange for exports of luxury goods, such as walrus ivory and fur. Numerous ivory fragments at most archaeological sites suggests an important and lucrative walrus ivory trade that provided raw material for much of medieval Europe's ivory artifacts. But after Portugal and other nations acquired sub-Saharan African elephant ivory, demand for Greenlandic walrus ivory lessened.

Furthermore, trade with Norway and Iceland gradually diminished in the fourteenth century as regional cooling, increasing sea ice, and storminess made ocean voyages riskier. Finally, European population decline following the Black Death disrupted overseas trade. Access to imported metal tools and other non–locally produced goods eventually ceased. The Western Settlement was abandoned by around 1360, followed by the Eastern Settlement around 1450. Did any survivors migrate back to Iceland or Norway? No one knows for sure. While the demise of the Viking settlements in Greenland can be partially blamed on climatic cooling, the real reasons why they disappeared remain an unsolved mystery.

The Danes returned to Greenland in the early eighteenth century and asserted sovereignty over the island. Although still part of Denmark, Greenland today exercises a high degree of autonomy. Its population of around 56,000 is 89 percent Greenlandic Inuit and 11 percent European,[4]

Unlike the medieval Norse, today's Greenlanders face a warming climate. At Summit, a research station located near the center of the ice sheet close to its highest point, temperatures soared by six times the global average between 1982 and 2011, forcing the equilibrium line to migrate skyward and widening the ablation zone.[5] In 2016, Summit and

Kangerlussuaq, West Greenland, experienced their warmest spring on record. Summer temperatures at Nuuk, the capital city, were second only to those in record-setting 2012.[6] While winters and springs have warmed even faster than summers, higher summer temperatures pack a greater punch. The longer melt season expands the number and area of summer melt pools on the surface of the giant Greenland Ice Sheet. An exceptional 97 percent of the surface melted for a few days in July 2012 for the first time in nearly 130 years.[7] The ice sheet has thinned and darkened since the early 1990s, especially around its edges (see "The Darkening of Greenland" later in this chapter). A larger number of icebergs calve off tidewater glaciers and trigger more glacial earthquakes.[8]

Greenland's recent mounting contribution to sea level rise from melting ice portends ill for the future. Added freshwater inputs to the ocean could someday also initiate widespread changes in ocean circulation. In this chapter, we take a closer look at ongoing changes in the Greenland Ice Sheet, returning to a consideration of its future and its broader implications in chapter 8.

The Greenland Ice Sheet: Getting Acquainted

> *Nansen paced up and down the floe, gazing at the shark's-tooth coastal peaks of Greenland in the distance, with the edge of the ice sheet plainly visible in between. "It entices one far, far into the unknown interior," he wrote wistfully. "Oh well, our time will come yet."*
>
> —Roland Huntford[9]

In spite of Greenland's enticing name, a vast ice sheet buries roughly 80 percent of its land, occupying an area of 1.8 million square kilometers (695,000 square miles). The Greenland Ice Sheet (*Sermersuaq* [Greenlandic]; *Indlandsis* [Danish]), the world's second largest, extends nearly 2,400 kilometers (1,500 miles) across from north to south, and 1,100 kilometers (680 miles) across at its widest near 77°N. Ice spreads outward from two domes—3,300 meters (10,800 feet) high in the north, and 3,000 meters (10,000 feet) high in the south, with an average thickness between 2,000 and 3,000 meters (~6,560–10,000 feet). The remaining ice-free, largely mountainous land hugs the coast. Many of Greenland's glaciers end on land or in lakes, but the largest and fastest-moving glaciers enter the sea through narrow fjords, where they often calve numerous icebergs. Some of the best-known of these marine-terminating glaciers include Jakobshavn Isbrae in the west (previously encountered in chapters 2 and 4), as well as other swift-moving glaciers, such as Helheim and Kangerlussuaq Glaciers in southeast Greenland.

Buried under the thick ice lies an interior basin, part of which rests below sea level. When stripped of ice, a giant inland lake would emerge, with potential outlets to the sea at several points. Recent airborne ice-penetrating radar probes have further discovered a hitherto hidden "megacanyon" rivaling the Grand Canyon and deeply buried subglacial valleys underlying the Greenland Ice Sheet. These widespread ice-covered valleys penetrate much deeper below sea level and farther inland than previously suspected. The newly discovered subglacial topography hints at a potential Achilles' heel for the Greenland Ice Sheet[10] (see "Peering Beneath the Ice," later in this chapter).

BOX 5.1: MONITORING THE GREENLAND ICE SHEET FROM ABOVE

The sheer size, extreme climate conditions, and remoteness of most of the Greenland Ice Sheet warrant techniques that encompass much larger swaths of terrain than covered in traditional glaciological field studies. Some widely used methods include (1) climate-based calculations of mass balance, (2) repeated satellite or airborne laser and radar altimetry to detect elevation changes, and (3) measured gravitational variations to uncover mass changes (see chap. 4, "Eyes in the Sky"; appendix B). Ice sheets, like glaciers, respond closely with minor fluctuations in climate. Thus, regional-scale climate models can relate changes in the ice sheet's surface mass balance to oscillations in temperature, snowfall, and other climate characteristics. Many of Greenland's glaciers end in the sea. The volume of ice discharged across the grounding line of tidewater glaciers is derived from estimates of ice thickness, density, and surface velocity.

Changes in elevation are crucial markers of the ice sheet's state of health. Repeated measurement of the same area by satellite radar altimeters (e.g., CryoSat-2) establishes trends in glacier or ice sheet elevation. Airborne laser or satellite altimeters (e.g., NASA's IceBridge, ICESat) operate in a similar manner, but substitute an intense, narrow laser beam in place of radar (appendix B). Ice mass change is calculated from the elevation and volume change, together with estimates of snow/firn density.

Altimeters, however, measure all elevation changes, including uplift due to glacial isostatic adjustment (GIA) from the last ice age, as well as from

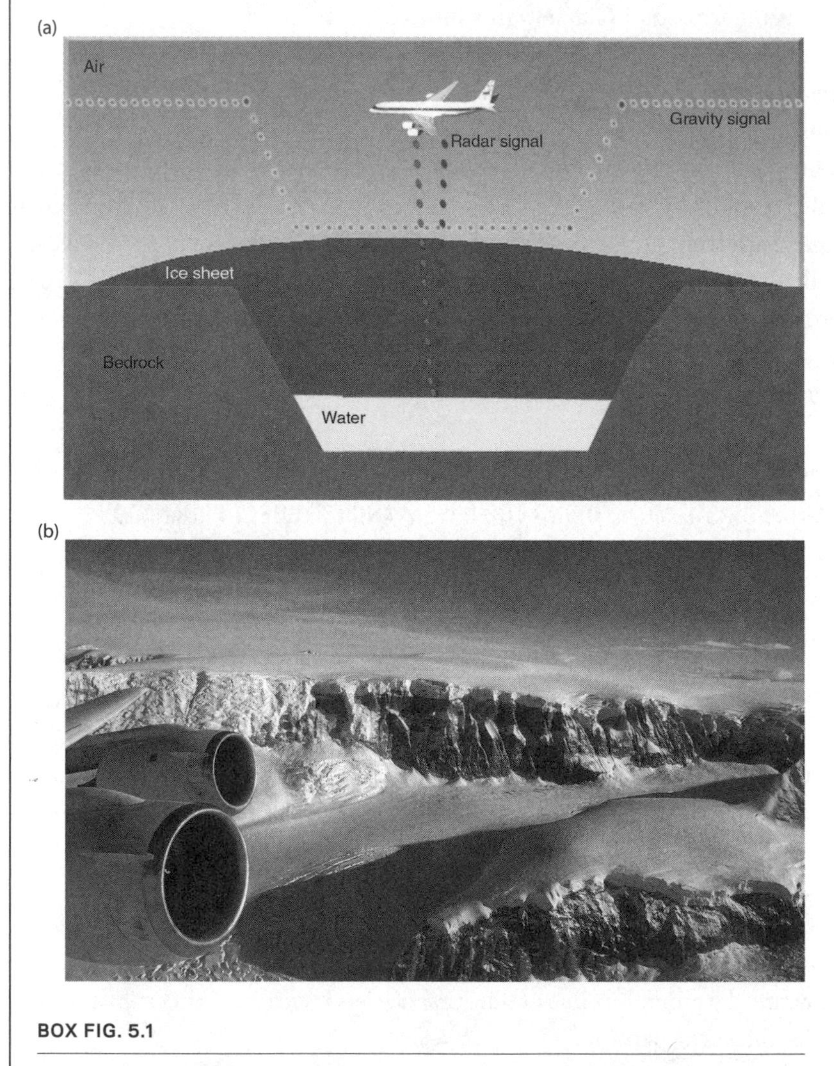

BOX FIG. 5.1

(*a*) Schematic diagram of NASA Operation IceBridge radar altimeter; (*b*) IceBridge aircraft flying over ice sheet. (NASA/Michael Studinger)

recent ice losses. The GIA contribution can be calculated from geophysical models that include rebound, gravitational, and rotational changes over time associated with redistribution of mass among ice sheets, land, and oceans.

Variations in the Earth's gravitational attraction also signal the changing mass of the Greenland Ice Sheet. Since 2002, the Gravity Recovery and Climate Experiment (GRACE) satellite mission has tracked detailed changes in the Earth's gravity (appendix B). However, as with altimeters, ice mass fluctuations need to be separated from changes in Earth's upper mantle and crust and in land water storage, using GIA models and GPS. GRACE data have added to the accumulating evidence for glacier and ice sheet attrition.

How fast a glacier or ice sheet moves—whether it is accelerating, slowing down, or creeping sluggishly forward—reveals another critical aspect of its behavior. The glacier's speed can be determined by tracking readily identifiable objects in a sequence of aerial photographs or satellite images over a period of time. Suitable features include prominent ice pinnacles, crevasses, large boulders, or distinctive patterns of rocky debris that reappear on the entire set of images. When the exact time the images were acquired is known, speed can be calculated by subtracting the change in distance of the same glacial feature at two different times and dividing by the elapsed time.

In addition to visible imagery, interferometric synthetic aperture radar (InSAR) can track glacier motion by examining phase changes in radar images over successive overflights (appendix B). InSAR, together with radar and laser altimetry, has proven a valuable tool in observing the speedup of outlet glaciers, described in this chapter.

LOSING ICE

The amazing aspect of Greenland glaciers is that (despite the specific variation in type, location specific configuration, etc.) their response has been as uniform and synchronous to global warming as has been observed. If this warming of the world persists long enough, the ice "banks" of Greenland will begin to fail.

—Mauri Pelto[11]

Growing "Bank" Deficits

The Greenland Ice Sheet sits in an area exposed to sharply shifting atmospheric circulation patterns and North Atlantic Ocean currents. Its location renders it especially sensitive to climate change. Since the 1990s, the Greenland Ice Sheet has begun to echo the story told in chapter 4 of the worldwide recession of mountain glaciers. Just as with personal finance, a glacier with a negative ice mass balance earns a lower "income" from snow accumulation than "expenses," or losses from melting ice and calving. In recent decades, Greenland's ice mass balance has shifted toward growing deficits. The Greenland Ice Sheet's sheer volume represents its main "asset." As higher temperatures extend northward into generally colder areas, will this initiate a drawdown on the "bank's" reserves?

Greenland Ice Sheet mass loss rates have doubled over the past quarter century. Average yearly losses between the 1980s and the 2000s range between 0.2 and 0.5 millimeters (0.01–0.02 inch) of sea level rise[12] (table 5.1). Losses appear to be accelerating: recent trends tend to exceed those spanning a longer time interval, although the trend of the last few years shows signs of slowing slightly (fig. 5.1).

Since the 2000s, Greenland has contributed between 0.4 and 1 millimeters (0.02–0.04 inch) per year to SLR. Roughly half these losses came from surface melting and runoff, and the balance came from ice calving and speedup in ice discharge at tidewater glaciers due to thinning of ice tongues or shelves (table 5.1). The attrition has now spread to parts of the ice sheet not long ago considered immune to retreat.

Not all regions shed ice uniformly, nor do these changes remain constant over time. Greenland's interior is actually gaining mass thanks to increased snowfall. However, losses initially concentrated in southern Greenland appear to have spread northeastward and northwestward since 2003.[13] The southeast and northwest sections of Greenland sustained the largest losses, with the southwest not far behind.[14] A team of glaciologists supplemented data they collated from airborne and satellite laser altimeters over northwestern Greenland with aerial photographs going back to the mid-1980s.[15] Their data revealed a "profound thinning" of the entire northwest ice sheet margin since the mid-1980s. Furthermore, ice losses have propagated northward. Closer inspection reveals that most regional

TABLE 5.1 Greenland Ice Melt Contributions to Sea Level Rise (millimeters per year)

GREENLAND ICE SHEET	METHOD (S)	REFERENCE
1980s–present		
0.33±0.08 (1993–2010)	Multiple	Intergovernmental Panel on Climate Change (2013a)
0.39±0.14 (1992–2011)	Multiple	Shepherd et al. (2012)
0.20±0.11 (1983–2003)	Geodesy; laser altimetry	Kjeldsen et al. (2015)
0.47±0.23 (1991–2015)	GRACE; mass budget model	van den Broeke et al. (2016)
Post-2000		
0.63±0.17 (2005–2010)	Multiple	Intergovernmental Panel on Climate Change (2013a)
0.58±0.10 (2000–2011)	Multiple	Shepherd et al. (2012)
0.68±0.05 (2003–2009)	Laser altimetry	Csatho et al. (2014)
0.68±0.08 (2000–2012)	Landsat 7 Enhanced Thematic Mapper+ and Terra ASTER	Enderlin et al. (2014)
0.42±0.09 (2000 2005)		Enderlin et al. (2014)
0.73±0.05 (2005–2009)		Enderlin et al. (2014)
1.04±0.14 (2009–2012)		Enderlin et al. (2014)
0.77±0.16 (2003–2013)	GRACE gravimetry	Velicogna et al. (2014)
0.40±0.04 (2003–2009)	Laser altimetry	Helm et al. (2014)
1.03±0.07 (2011–2014)	Radar altimetry	Helm et al. (2014)
0.51±0.05 (2003–2010)	Geodesy; laser altimetry	Kjeldsen et al. (2015)
0.65±0.24 (2000–2011)	GRACE; mass budget model	van den Broeke et al. (2016)
0.72±0.07 (2002–2015)	GRACE gravimetry	Forsberg et al. (2017)
0.74±0.07 (2003–2017)	GRACE gravimetry	Tedesco et al. (2017)

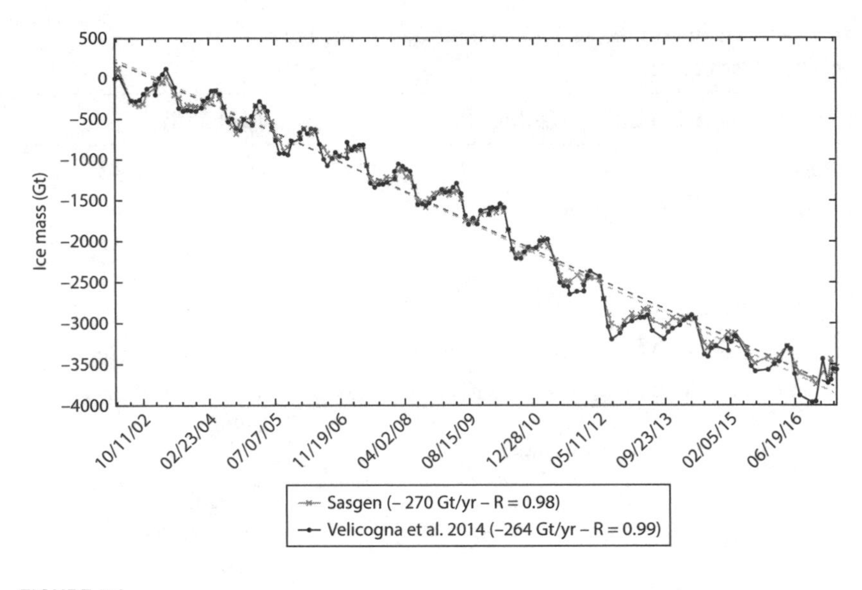

FIGURE 5.1

Ice mass loss on Greenland, from 2002 to 2017, from the GRACE gravimeter. (Adapted from M. Tedesco et al. [2017], fig. 3, https://arctic.noaa.gov/Report-Card/Report-Card-2017/ArtMID/7798/ArticleID/697/Greenland-Ice-Sheet/)

glaciers changed more rapidly in the period between 2005 and 2010 than previously. Marine-terminating glaciers accelerated and thinned after the ice tongues that had held them in check broke apart. Satellite and aerial altimeters also recorded a drop in ice-surface elevations. The reduced surface–mass balance accounts for a third of the regional ice losses.

A longer time perspective clearly illustrates how markedly glacier behavior varies over short periods and from one region to another. A study combining "glaciological research with a splash of Indiana Jones" extended the time coverage from the 1930s to 2010 by using a set of historic aerial and terrestrial photographs and satellite images over southeast Greenland.[16] The oldest photos, taken during an expedition led by the famed Danish explorer Knud Rasmussen, had been rediscovered gathering dust in a citadel outside Copenhagen. Other photos and satellite imagery came from declassified U.S. military sources. The photo analysis revealed widespread glacier recession along the edges of the ice sheet during two comparably warm periods—the first lasting from 1933 to 1943, the second from 2000 to 2010, separated by a cooler interim period. Rapid retreat during the recent warm spell affects more glaciers than before. But most of these are marine-terminating glaciers that are especially sensitive to sea surface temperatures, now almost 0.5°C (0.23°F) higher than during the 1930s. On the other hand, land-terminating glaciers had receded faster in the 1930s when they ended at lower elevations. Once they shrank and retreated to higher ground, their vulnerability to subsequent pronounced warming at lower elevations decreased. Given the sharply variable behavior of glaciers from one decade to another, or among different types of glaciers, these observations underline the need for caution in extrapolating current climate trends. Nonetheless, this analysis also demonstrates the high degree of sensitivity of glaciers to even short-term climate oscillations. Therefore, if the recent ocean warming trend of the past two decades persists, tidewater glaciers are likely to continue their swift landward withdrawal until they no longer abut the sea. As they reach higher ground, however, the retreat would decelerate.

Lowering Heights; Growing Wetter

Two additional signs reinforce the picture of thinning along the outer margins of the ice sheet and migration of surface melting to progressively higher elevations in summer. Satellite and airborne laser and radar altimeters from 1993 to 2014 detected a lowering of the entire western, southeastern, and northwestern margins of the Greenland Ice Sheet.[17] Surface elevations dropped faster between 2011 and 2014 than during the 2000s. The mass of ice lost during these few years corresponds to a sea level rise of nearly 1 millimeter per year (see table 5.1). The retreat of a few southeastern glaciers, including Helheim, decelerated for several years, but the short-lived slowdown soon ended and a faster pace resumed.[18] The thinning of Jakobshavn Isbrae now penetrates far into the interior. Helheim and Kangerlussuaq Glaciers, among others, have also thinned. Even the cold, formerly stable northeast has not escaped. The North East Greenland Ice Stream (NEGIS) has thinned as far as 100 kilometers (60 miles) upstream.[19] The interior of the ice sheet has maintained, or even gained, elevation due to added snowfall, in contrast to the peripheral lowering.

Ice is dwindling throughout much of the North Atlantic region, including Greenland, Iceland, and Svalbard. In these regions, orbiting Global Positioning System (GPS) satellites have detected an accelerating land uplift since the 1990s.[20] This uplift is caused by isostatic rebound due to recent ice melting, in addition to a lingering response to ice losses after the last ice age. Land quickly bounces back even after short-term ice melting, as we have seen around Glacier Bay, Alaska (chap. 4).

The surface of the ice sheet has also grown wetter in summer. (On July 11–12, 2012, nearly the entire ice sheet surface melted![21]) Melting ice collects in surface depressions, forming meltwater pools, or *supraglacial* lakes (i.e., on top of the ice) (fig. 5.2). A myriad of such lakes line the margins of the ice sheet. Satellites easily spot changes in meltwater lakes because of the sharp albedo contrast between water and bare ice or firn. Four decades of satellite tracking confirm that new lakes are forming at higher elevations than ever.[22] In general, lake elevations changed little between the 1970s and 2000, but subsequently many new lakes formed hundreds of yards uphill and tens of miles inland. Some of the most marked changes occur on Jakobshavn Isbrae, where lakes now appear near 1,900 meters

FIGURE 5.2

Meltwater pools on the surface of the Greenland Ice Sheet. (NASA Earth Observatory, "Ponds on the Ocean," July 12, 2011, photograph by Kathryn Hansen)

(6,230 feet) of elevation, roughly 30 kilometers (18.6 miles) farther up the ice sheet than before 2000. In the south, lakes even form above 2,000 meters (6,560 feet).

As lakes that overlie crevasses multiply, water pressure rises within the cracks, increasing hydraulic fracturing, or "hydrofracking." More water drains to the bottom and lubricates the basal ice, which then slides faster. Climate models estimate that by 2060 regional warming will lead to an approximately 50 percent expansion of the area over which surface meltwater lakes will be distributed. Nearly half of these new lakes will be large enough to drain to the bottom.[23] Opening up "fresh" terrain at higher elevations via newly formed hydrofractures and moulins could therefore theoretically direct additional meltwater to the bottom of the ice. This, in turn, could hasten the speed of ice over a greater area. Will enough "fresh" terrain open up at higher elevations?

Maybe not so soon.[24] Ice fracking requires both open cracks and an abundant supply of meltwater that forces its way to the bottom. Even though meltwater lakes now occur at elevations close to 2,000 meters (6,560 feet), these lakes are broad and shallow. However, bottom-reaching crevasses and moulins are even scarcer above ~1,600 meters (5,250 feet). These develop mainly over buried hills that stretch and break the ice. Thick high-elevation ice effectively buries such underlying topography, which inhibits development of fresh cracks or moulins. Instead, excess lake water overflows into surface streams that encounter moulins or crevasses much farther downstream. As new lakes develop at higher altitudes, increased volumes of excess meltwater would take this longer route to the base. This would mainly facilitate ice sliding at lower elevations. Very high-altitude ice would thin enough only after prolonged periods of sustained warming, such that hydrofracking could be initiated and route water directly to the ice sheet bed, thereby expanding the area of basal lubrication.

Furthermore, greater meltwater production by itself does not automatically speed up a glacier. Instead, the excess water may carve out broader channels or tunnels within the ice or at its base, enhancing the ability of water to drain out of the network quickly without necessarily speeding up the overlying ice (fig. 5.3).

FIGURE 5.3

Meltwater cascading down a moulin on the Greenland Ice Sheet (M. Tedesco, Columbia University/NASA GISS)

In the last few decades, Greenland has been losing nearly half its ice mass via discharge past the grounding line and calving. The ice sheet is growing more sensitive to both ocean and atmospheric warming. In the short term, at least, losses at tidewater glaciers may dominate. The next section takes a closer look at how these glaciers are evolving.

Fast-Flowing Outlet Glaciers

Fast-flowing, marine-terminating glaciers account for roughly half the ice lost from Greenland. Most of the recent losses have occurred in the northwest and southeast, from glaciers such as Jakobshavn Isbrae, Helheim, and Kangerlussuaq.[25] Jakobshavn Isbrae, one of the largest and fastest-streaming tidewater glaciers, flows into a deep fjord. It drains roughly 6.5 percent of the Greenland Ice Sheet. Up until the 1990s, Jakobshavn Isbrae may have been anchored to a 700-meter (2,300-foot)-deep underwater ridge before beginning to retreat.[26] Following disintegration of its floating ice tongue in the early 2000s and major calving events in 2008 and 2015 (chaps. 2 and 4), the glacier receded into a deeper part of the fjord. A rapid retreat between 2002 and 2015 set record speeds during the summers of 2012 and 2013.[27] Its grounding line now lies around 1,100 meters (3,610 feet) below sea level, with the glacier having receded a total of 12.5 kilometers (7.8 miles) between 1996 and 2016.

Beyond the current 1,100-meter-deep grounding line of Jakobshavn Isbrae, the trough drops farther to 1,600 meters (5,250 feet) below sea level, then gradually rises to 1,200 meters (3,940 feet) below sea level for at least another 50 kilometers (31 miles) inland before shallowing. The landward deepening bed of Jakobshavn Isbrae (and also of Helheim Glacier) represents a marine ice sheet instability (MISI), briefly introduced in chapter 2 (and discussed further in chapter 6).[28] Once initiated, the retreat will probably continue until the glacier passes beyond the deepest parts of the fjord later in this century.[29] For other glaciers, that point depends on their individual subsurface topographies. A glacier begins to decelerate once it loses contact with the ocean. Beyond that point, further recession upslope toward the heart of the ice sheet depends on surface melting, runoff, and frontal retreat.

The beds of many other speeding outlet glaciers also extend much farther inland below sea level than previously suspected (although not necessarily on reverse slopes). For example, the Zachariae Isstrøm glacier in the northeast, segments of which do rest on a reversed slope, has begun to recede. Because the NEGIS drainage network, of which this glacier is a part, extends deep into the interior, this could bode ill for the future stability of a large section of the ice sheet.[30]

Satellites tracking ice movement beneath the Greenland Ice Sheet coupled with airborne radar soundings have pieced together a detailed map of subsurface topography.[31] The mapping uncovered widespread deep valleys below the ice sheet, many of which lie below sea level. Many valleys also originate far inland and end at the sea. Out of 123 marine-terminating glaciers, "60 drain 88 percent of the ice sheet in area and are grounded below 300m depth at their termini, meaning they are deep enough to interact with subsurface warm Atlantic waters and undergo massive rates of subaqueous melting."[32] Under the right conditions, this could lead to a rapid meltdown that would affect a substantial portion of Greenland. These new findings spell out greater vulnerability of large parts of the ice sheet to rapid ice loss and retreat (see "Peering Beneath the Ice").

However, subsurface topography is just one factor in the recent retreat of Greenland's tidewater glaciers. The oceans also play a major role. The process works roughly as follows. In recent years, currents branching off the North Atlantic Current have delivered warmer, saltier ocean water to the ice-covered fjords of Greenland (fig. 5.4).[33] Most tidewater Greenland

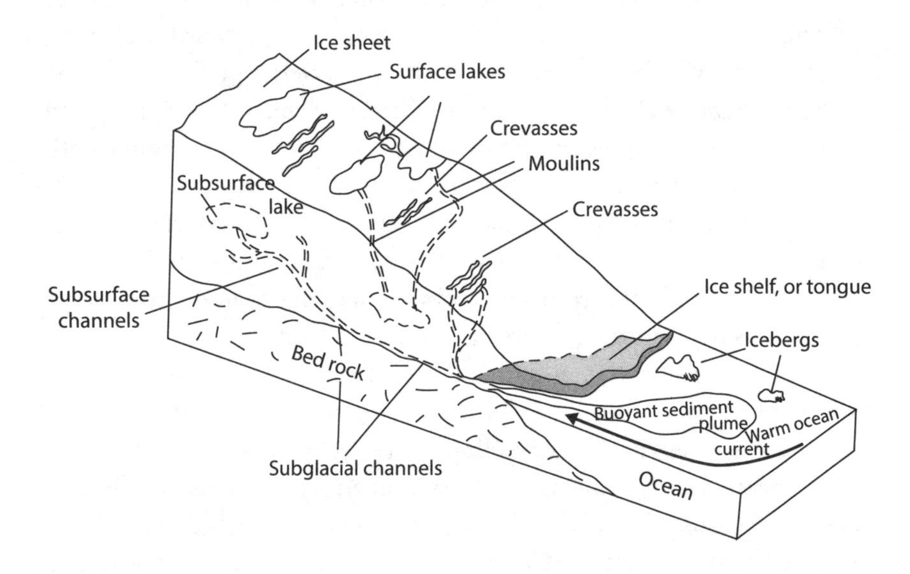

FIGURE 5.4

Processes contributing to retreat of Greenland tidewater glaciers.

glaciers enter the sea through these narrow fjords. In summer, warmer ocean water mixes turbulently with cold, fresh meltwater that drains from the ice surface and emerges at the glacier's mouth. As these waters mix, they generate a buoyant plume of water that rises to the surface and also undercuts the base of the ice cliff at the terminus. Warmer seawater can then penetrate beneath the ice and melt it from below. Caught in a positive feedback, increased stresses amplify calving and rifting that drive the grounding line farther landward. The glacier responds by accelerating, stretching, and thinning. The warm water may also hasten melting of icebergs and sea ice that otherwise clog the fjord.

A similar sequence of events befell Jakobshavn Isbrae. A protracted period of rising ocean temperatures after 1997 had weakened the base of its ice tongue.[34] Crevasses and hydrofractures multiplied as the ice tongue interacted directly with warming seawater. New rifts appeared on the surface, and calving rates soared with the glacier in contact with warmer seawater.[35] Friction lessened and resistance to flow weakened, particularly along the sides and base of the glacier. Ice moved faster with the reduced drag. The severely attenuated ice tongue finally crumbled in the early 2000s. Plunging forward, the glacier disgorged more icebergs in several dramatic calving events (chaps. 2 and 4).

Recent improved bathymetric mapping of several western Greenland fjords bolsters this scenario.[36] The bathymetry indicates deeper fjord troughs into which warmer, saltier Atlantic water can flow, mingle with colder subglacial water, and undercut ice at the grounding line, thereby increasing the calving rate.

In summary, regional air and ocean warming have thinned both the tops and bottoms of outlet glaciers and ice tongues, opened more crevasses and rifts near the terminus, escalated calving rates, and multiplied surface meltwater pools. Grounding lines retreat inland as tidewater glaciers accelerate.

Because of the extreme sensitivity of tidewater glaciers to ocean influences, these glaciers quickly adjust to changing external conditions. Today's rapid acceleration of many outlet glaciers may abruptly give way to natural climate fluctuations or feedback mechanisms that operate in the opposite direction. Still, glaciers would continue to shed ice for decades, even though much more slowly.

Some climate models predict that just a few degrees of global warming, if sustained over a longer period, would suffice to eventually destabilize Greenland irreversibly.[37] If this ever happened, sea level could theoretically climb as much as 7 meters (23 feet). Prolonged warming over centuries to millennia could generate a substantial future meltdown of the Greenland Ice Sheet, although total deglaciation would be extremely unlikely. How much of Greenland did actually deglaciate during the Last Interglacial, around 125,000 years ago, when sea levels were many feet higher than today? Chapters 7 and 8 will investigate whether or not Greenland ever did or ever could undergo a complete meltdown.

Icequakes

Rumbling noises and vigorous shaking accompany glacial earthquakes, or "icequakes," associated with rapidly speeding tidewater glaciers, such as Jakobshavn Isbrae, Helheim, and Kangerlussuaq. This came as a surprise to glaciologists, who long considered Greenland to be essentially aseismic, i.e., an area free of earthquakes. However, glacial earthquakes differ significantly from ordinary earthquakes that are set off when two masses of rock break and slide past each other.[38] The stronger an earthquake, the more energy it releases, and the longer it shakes. Typically, a moderate magnitude 5 earthquake would last a mere 2 seconds. Icequakes, on the other hand, normally last over 30 seconds, yet release energy equivalent to a magnitude 5 earthquake event.

Icequakes are another manifestation of Greenland's waning outlet glaciers. The annual number of these events has increased sixfold since the 1990s. Icequakes more or less coincide in timing with large calving events and abrupt glacier acceleration. Most active during late summer, icequakes cluster tightly around a number of Greenland's major tidewater glaciers. The closeness in timing between large calving events and icequakes strongly implies that calving triggers the seismicity. But not all major calving events generate icequakes. Calving style matters. An ice mass that rifts near the glacier terminus may "silently" break off a wide, tabular iceberg that does not capsize as it gently floats away from the glacier. This type of calving does not produce an icequake no matter how large the resulting iceberg(s). On the other hand, a massive ice mass that overturns and topples into the sea with a resounding splash generates an icequake, unleashing a small tsunami. The top of the capsizing iceberg pushes back against the glacier as it slumps forward, while water rushes in from below, exerting a buoyant force. The glacier front, initially pulled down, quickly bounces back like a spring in response to the reversing forces. Rushing away from the calving front, the detached iceberg(s) significantly reduce the resistance to flow provided by the formerly intact ice mass. The glacier, in turn, responds to the reduced back-pressure by picking up speed and moving forward.[39]

Exceptional or the New Norm?

Is the accelerated ice loss of the past two decades in Greenland a short-lived fluke or the actual onset of protracted warming—a ramped-up consequence of global warming? The 2017 Arctic Report Card, which updates the status of the Arctic cryosphere annually, finds no new record-setting lows since the summers of 2010 and 2012.[40] In spite of GRACE gravity data that show a slowdown of yearly mass losses since 2012, these mass losses continue the negative trend that began around 1990 (see fig. 5.1). Furthermore, average yearly Arctic temperatures since 2012 have climbed at double the rate of those in lower latitudes.

Observations point to marked year-to-year climate fluctuations that may temporarily mask the longer downhill trend. How much of the recent ice loss is simply the result of a short-lived warm spell due to natural climate variations rather than to global warming? Glacier retreat since the mid-1990s has coincided with the introduction of warmer water from the south flowing around the continental slopes of Greenland at moderate depths (500–1,000 meters; 1,640–3,280 feet).[41] Some of this water intrudes into the fjord troughs. But ocean currents and air circulation patterns are fluid; they can and do readily shift. Some have proposed that the North Atlantic Oscillation (NAO) may have influenced recent ice behavior. The NAO is defined by a variable atmospheric pressure difference between Iceland and Lisbon, Portugal (or, alternatively, the Azores). In a positive NAO phase, winds bring colder, stormier weather to Greenland. The NAO remained in a positive phase during the 1980s to mid-1990s, but has reversed since the mid-2000s to a near-normal or negative phase, leading to higher temperatures. Simultaneously, most of the upper North Atlantic Ocean has also warmed. The role of the NAO has been downplayed lately, in that it cannot explain a simultaneous warming in water masses of both subpolar and subtropical origin, which usually show opposite behavior. More likely, the recent ice losses stem from the influx of warmer southern ocean water into subpolar regions, as well as from increased surface runoff.[42]

The period since 2000 of above-average temperatures and diminishing ice remains too short, and year-to-year climate fluctuations too high, to confidently assert that recent trends signal the onset of a major Arctic meltdown. Nevertheless, the observations demonstrate a high degree of sensitivity to minor climate variations. However, if current climate trends do persevere, ice mass losses can be expected to escalate in the future. Chapter 8 will focus in greater depth on how much sea level is likely to rise as a result.

What could slow down, or even temporarily reverse, Greenland's ice losses? Ironically, icebergs may play an important role. Greenland sheds ice in two ways: as freshwater runoff, or as icebergs calving at the edges of tidewater glaciers. Freshwater from all sources reduces seawater saltiness, which in turn may slow down the Atlantic Meridional Overturning Circulation (AMOC)—an important part of the global ocean conveyor system (see chap. 1). Sluggish circulation of the Gulf Stream and North Atlantic Current—a branch of the AMOC—would convey less warm water and heat northward. While most of the Earth heats up, the region would contrarily cool off.[43] More significantly, a larger influx of icebergs, likely in a warmer world, would cool as well as freshen seawater as the bergs melted. The colder seawater would also chill the air above, leading to expansion of a sea ice cover in winter. These negative feedbacks would further lower air temperatures above Greenland and increase local snowfall. Greenland could thereby regain some of its lost ice, at least in the short term.

The ice sheet itself may harbor some additional negative feedbacks that could delay Greenland's contribution to sea level rise. For example, where exactly does the water that melts at the top of the ice sheet go? How much water that percolates into the ice actually makes it to the sea? In spite of much remaining uncertainty, some recent studies hint at an enormous, previously unsuspected water reservoir lurking beneath the ice.[44] At higher elevations, where a mix of the previous winter's snowfall and firn (partially compacted snow) mantle the uppermost layers, some of the meltwater runs off elsewhere, while the rest trickles down into the snow/firn layers. There, water confined within open-pore spaces refreezes. Joel Harper of the University of Montana, leading a team of geoscientists, collected field data in western Greenland during the summers of 2007 through 2009. They estimate that the firn holds a vast concealed water reservoir that could store between 322 and 1,289 gigatonnes (one billion metric tons).

Ground and airborne radar imagery that penetrates into the ice confirms the existence of such a massive meltwater aquifer.[45] Radar picked up a widespread, bright, radar-reflecting layer in south and southeastern Greenland, typical of a water table surface and matching the actual level of water observed in ice core drill holes. The top of the sub-ice water layer could be traced across 843 kilometers (524 miles) of southern Greenland. Tucked snugly within the firn, this water had survived the previous winters. High snowfall and melt rates characterize the region of the hidden aquifer. Blanketed by a thick layer of snow, which insulates the water from freezing in the winter, much of the water survives until the next melt season, when fresh meltwater recharges the aquifer. This hitherto concealed water storage tank could represent a significant holding area that could impede the journey of meltwater to the sea. But this storage tank may be somewhat leaky.

Unlike in western Greenland, where surface lakes and streams are common features during the summer, ice melt in the southeast penetrates into the firn layers. The region's heavy winter snowfall creates a thick insulating blanket that keeps the water liquid throughout the winter. However, the water doesn't just sit there. Crevasses lie downstream of the aquifer. The firn aquifer water likely feeds the nearby crevasses, filling them with meltwater. The weight of the water can then saw its way down the crack, allowing meltwater to reach the bottom within weeks to months, where it connects to subglacial channels.[46]

Clearly, meltwater on the Greenland Ice Sheet travels a circuitous route before reaching its marine destination. The next section traces some of these convoluted pathways as meltwater moves across and down into the ice, slowly making its way seaward, where it adds to the rising waters.

JOURNEY TO THE SEA

Greenland's Disappearing Lakes

A 2-kilometer (1.2-mile)-wide, 70-meter (230-foot)-deep crater suddenly appeared on the ice at the southwestern margin of the Greenland Ice Sheet in July 2011. This event marks the first sign that surface meltwater had drained rapidly into a subglacial lake under the ice. Glaciologist Ian Howat of Ohio State University suggests that "meltwater has started overflowing the ice sheet's natural plumbing system and is causing 'blowouts' that simply drain lakes away."[47] Michael Willis of Cornell and his team, in a separate study, closely monitored the changing volume of another subglacial lake in northeast Greenland over a 2-year period between May 2012 and March 2014. Carefully scrutinizing a series of satellite stereo images and altimeter data, the team members clearly saw how a mitten-shaped basin swelled as meltwater from surface streams cascaded down crevasses, filling a subglacial lake. As the lake filled to capacity and ultimately burst through, overlying ice sagged and formed a surface depression 75 meters (246 feet) deep.[48] The following melt season, fresh meltwater slowly recharged the subglacial lake. The ice surface bulged upward again as water saturated crevasses and lake. Once completely full, the lakes can empty abruptly. Willis exclaimed: "We're seeing surface meltwater make its way to the base of the ice where it can get trapped and stored at the boundary between the bedrock beneath the ice sheet and the ice itself. As the lake beneath the ice fills with surface meltwater, the heat released by this trapped meltwater can soften surrounding ice, which may eventually cause an increase in ice flow."[49]

During such events, water may cascade downward at rates rivaling Niagara Falls. These observations demonstrate that substantial bodies of water can remain trapped for considerable periods within the ice. Since the confined meltwater is generally several degrees warmer than the enclosing ice, it releases heat that melts surrounding ice and enlarges the confining cavity. Depending on the abundance of such trapped meltwater pools on or within the Greenland Ice Sheet, they could strongly influence the routing of water through subsurface channels and passageways and, hence, the motion of the ice sheet.

Ice Sheet Plumbing and Movement of Ice

We have just seen how significant volumes of meltwater can refreeze within the firn or remain stuck in holding pools within the ice for considerable periods. What happens to water that escapes these temporary or permanent traps? What is the nature of the ice sheet's internal water routing pathways, and how does the flow of water inside the ice affect the motion of the ice sheet?

Examination of surface basin fluctuations yields important insights into the movement of water inside the Greenland Ice Sheet. An array of 16 GPS stations at a site south of the Jakobshavn Isbrae carefully monitored surface displacements surrounding a lake between 2011 and 2013.[50] During each melt season, various premonitory clues signaled the imminent onset of a rapid drainage event. Some precursory signs included faster slippage of ice and/or localized ground uplift, while meltwater that poured down nearby preexisting crevasses and moulins swelled basal cavities. Water filling these preexisting crevasses creates tensional stresses within the ice strong enough to overcome the weight of the overlying ice. The stresses opened new hydrofractures under the lake, which propagated down to the bed, allowing the lake to drain out quickly.

Lakes are more likely to empty rapidly near abundant preexisting crevasses along the ice edge or at intermediate elevations. Although a warmer climate increases the population of lakes at high elevations, lakes will overflow onto a much smoother ice surface crisscrossed by far fewer crevasses. An overflow of meltwater would run off farther along surface streams before disappearing down crevasses or moulins into preexisting crevasses downstream.[51] Any enhanced basal lubrication would preferentially benefit the mid- to low-elevation parts of the ice sheet.

Other escape routes for water exist. Shortly after the extraordinary surface melt event in July 2012, Lawrence C. Smith of UCLA and his team of glaciologists immediately installed a host of instruments to measure water flow on the ice, at moulins, and at a river connecting subglacial water that emerges from the ice edge to the sea.[52] Their instrumentation included specially designed, remotely operated boats to collect data from slow-moving streams, lakes, and rapid currents. High-resolution satellite imagery that supplemented ground measurements revealed extensive

FIGURE 5.5

Greenland Ice Sheet, meltwater streams on the ice surface. (NASA/Maria-José Viñas)

surface river drainage networks dissecting much of the lower sections of the ice sheet. "It's the world's biggest water park, with magnificent and beautiful—but deadly—rushing blue rivers cutting canyons into the ice," Smith said.[53]

The surface streams are well organized into parallel, dendritic, and radial patterns that only loosely follow local topography, often leaving depressions unfilled (fig. 5.5). All streams eventually end in moulins. Water that cascades down these vertical shafts in the study area supplies nearly all of the meltwater reaching the ocean. This enormous runoff dwarfs the volume of water contained in surface lakes or depressions, indicating a well-organized internal plumbing system. Nevertheless, the observed volume of water exiting the ice sheet falls short of regional climate model predictions based on temperature and precipitation data. Water stored in firn/ice or in subglacial lakes likely explains the discrepancy.

A buildup of high water pressures at the bed of a glacier propels it forward. This condition eases gliding along its bed. The ice should therefore accelerate toward lower elevations where additional melting can take place. But recent observations show that the flow of water through ice depends not only on how much water courses through open conduits, but also on variations in water supply and extent of conduit connectivity, which varies seasonally.[54]

At the end of winter, small cavities form on the ice bed in the lee of obstacles and connect to other cavities via narrow, tortuous passageways. As with mountain glaciers, the transport of water through this linked cavity network remains sluggish and inefficient until spring meltwater reaching the base gradually builds up water pressure. By early summer, the increasing water pressure eventually overwhelms the drainage and partially lifts the ice off its bed. Friction between the ice and its bed diminishes, speeding up the ice flow—initially. However, the warmer water also melts surrounding ice and enlarges cavities and channels. As the season progresses, these evolve into larger channels that connect into a dendritic (branched) network resembling river drainages. This well-connected, efficient, channelized drainage network can now remove large volumes of water. As a consequence, water pressures at the base drop and the ice soon slows down. Therefore, a steady influx of water into well-interconnected channels does not automatically lead to ice acceleration.

Surface water, however, usually arrives at the base in pulses—after a heavy rainfall or abrupt drainage of surface lakes (as discussed earlier), or even from daily summer melting. These high pressure spikes force water from channels into adjacent narrow openings, causing short-lived ice accelerations. For example, massive surface runoff after heavy late-summer rains in 2011 in western Greenland provoked a widespread, although temporary, acceleration in ice flow.[55] At that time, the internal ice plumbing had been transitioning to a less efficient winter style. Rainfall has generally increased over the past 30 years, particularly late in the season and at higher elevations. Late-season rainfall will probably become more prevalent in the next few decades and could contribute to further ice loss.

Ice flow early in the melt season remains fairly constant from year to year. However, during exceptionally hot summers, in spite of attaining a greater peak speed, the ice does not maintain this rapid pace for long.

As basal water pressures build up, the initially inefficient drainage of interconnected cavities quickly transforms into an efficient channelized drainage that conveys water away and slows down overlying ice flow.[56]

A series of borehole measurements installed from the edge of the ice sheet to the interior opens up further windows into the movement of water within the ice sheet. At the ice sheet margin, the borehole instruments recorded large daily water pressure swings, in contrast to steady high pressures with little variation inland. Near the ice edge, the subglacial channel network may have grown more efficient than in the interior, where a more sluggish system persists. A relatively flat inland topography and rapid sealing of narrow passageways by ice creep may inhibit formation of a well-channelized network.[57]

Glaciologists also detect striking differences in ice drainage behavior a few kilometers apart. One study found that daily water pressure variations in moulins closely correlate with ice motion, in contrast to pressure variations in boreholes.[58] The moulins, unlike boreholes, connect directly to an efficient channelized network at the bed that allows the ice to respond rapidly to changes in water pressure. Toward late summer, however, the ice slowed down, while moulin water levels and channel pressure remained steady. Meanwhile, borehole water pressures dropped, as an initially poorly linked basal drainage below gradually evolved over the summer into well-connected channels that quickly removed subsurface water.

A Look Below

Airborne radar that "sees" all the way to the bottom of the ice, coupled with instruments that detect subtle gravitational variations, has discovered a totally unforeseen icescape. This buried icescape also influences how the ice moves. The ice-penetrating radar delineates extremely contorted ice masses along the base of the ice sheet in many places, instead of the flat near-surface layers. The ice units bend into severely distorted layers, tight folds, and upwarped, fingerlike intrusions that invade and deform overlying flat ice layers deposited over successive years of snowfall (fig. 5.6). These latter structures closely resemble the folding and deformation seen in salt domes and *diapirs*—soft, plastic salt or rock material that squeezes

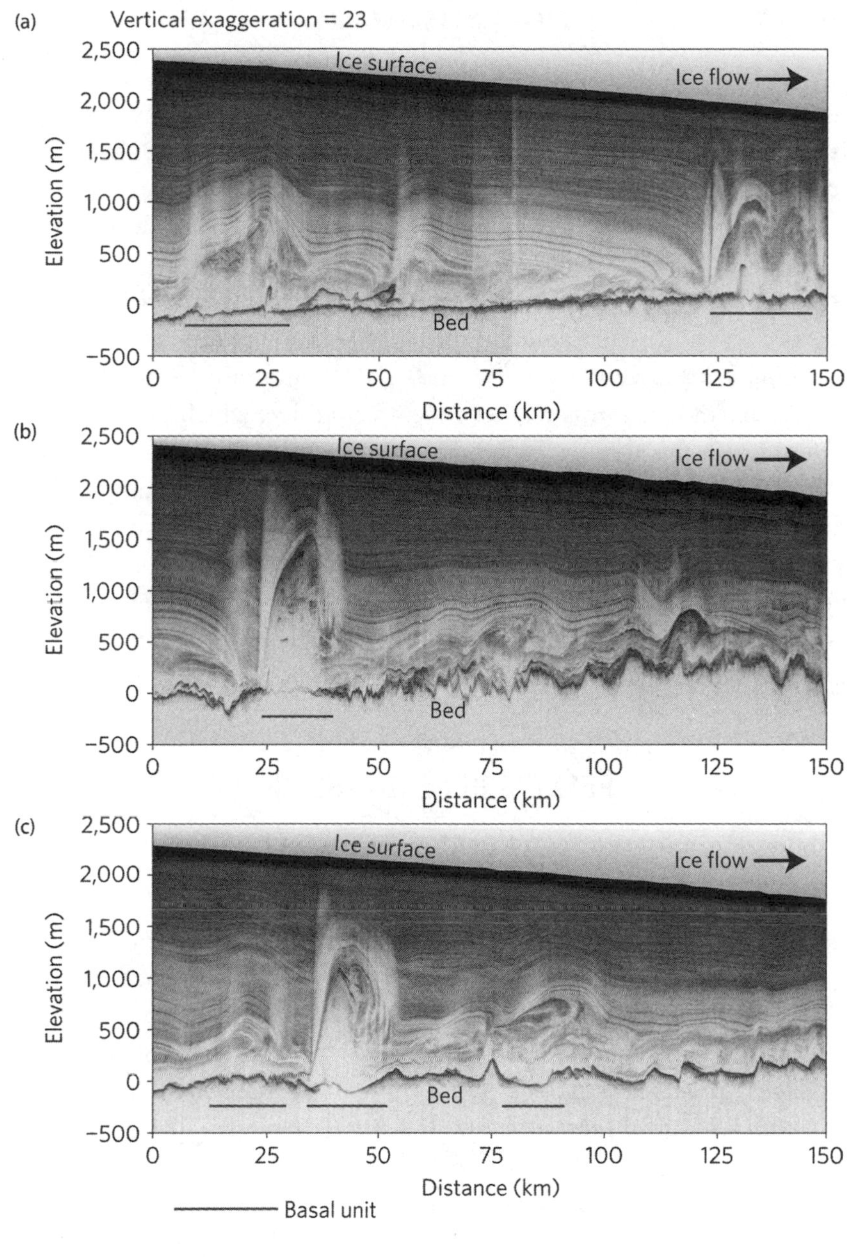

FIGURE 5.6

Severely deformed and contorted ice layers at the bottom of the Greenland Ice Sheet. (Bell et al. [2014], fig. 3. © 2014 Macmillan Publishers Ltd. Reprinted by permission from Springer Nature)

up into denser, harder rocks. The chaotic ice masses may also have refrozen from meltwater and become cemented to the base of the ice sheet. The meltwater was formed by geothermal or frictional heating.[59]

The discovery of "freeze-on," or refrozen meltwater, in deep-seated Greenland ice (as in Antarctica—more in the next chapter) bears on the important issue of future ice sheet stability. As meltwater freezes, it releases heat to the surrounding ice and softens it. The refrozen ice holds more impurities, such as water droplets, sediment grains, and other dissolved constituents. This weakens the accreted ice, which subsequently deforms into the sharply folded and overturned layers detected by the radar. The weakened basal ice allows ice to flow faster, as has happened at the Petermann Glacier in northwestern Greenland.[60] Petermann Glacier, which has calved several massive icebergs since 2010 (see fig. 1.5), connects to a subsurface canyon that extends across a large part of northern Greenland (see the following section). The subglacial canyon may someday provide a convenient conduit to the sea for basal water.

PEERING BENEATH THE ICE

The Grand Canyon of Greenland

A discovery of this nature shows that the Earth has not yet given up all its secrets. A 750 km canyon preserved under the ice for millions of years is a breathtaking find in itself, but this research is also important in furthering our understanding of Greenland's past. This area's ice sheet contributes to sea level rise and this work can help us put current changes in context.

—David Vaughn, British Antarctic Survey

A great gorge cuts through the middle of northern Greenland, buried under the ice[61] (fig. 5.7). The canyon meanders some 750 kilometers (460 miles) north from Summit, central Greenland's highest point, to the fjord at the snout of Petermann Glacier on the north coast. The megacanyon is nearly 800 meters (2,600 feet) deep and 10 kilometers (6 miles)

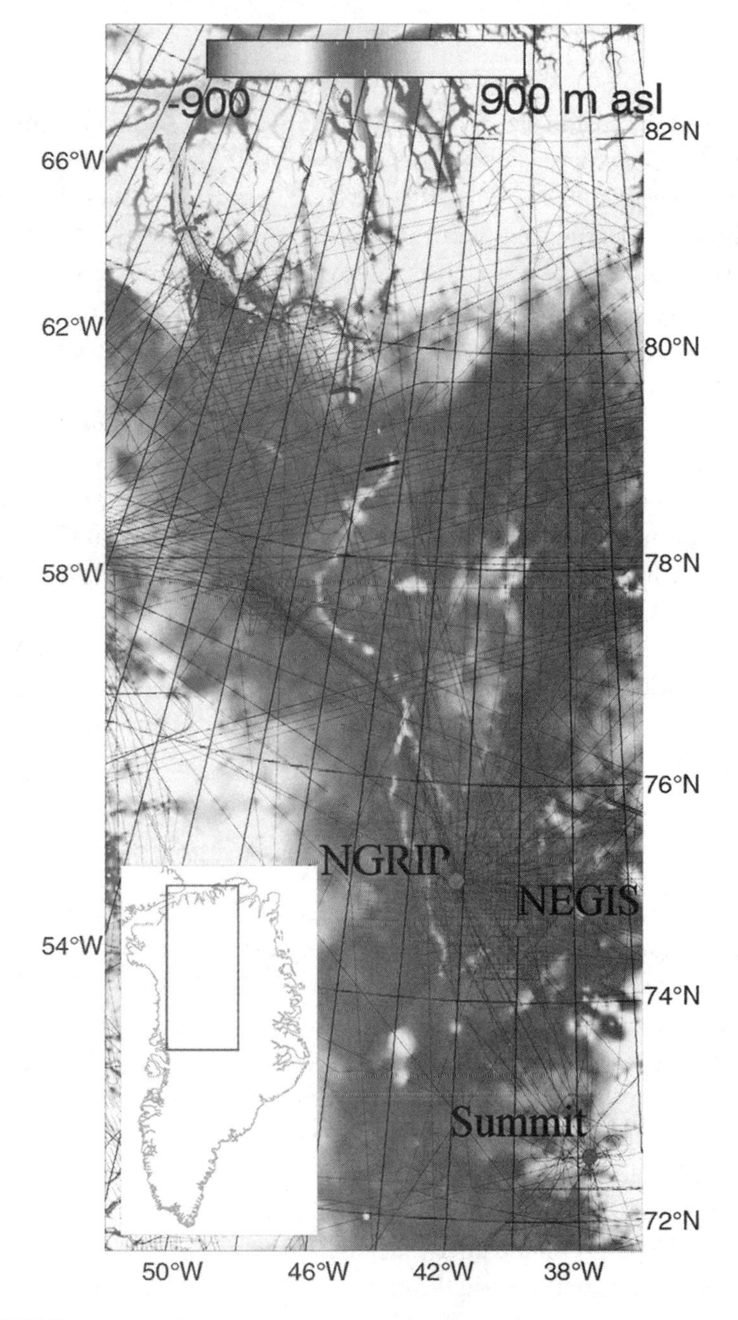

FIGURE 5.7

The hidden "Grand Canyon" at the bottom of the Greenland Ice Sheet, comparable in dimensions to its famous Arizona counterpart. (Bamber et al. [2013], fig. 1)

wide—comparable in size to Arizona's famous Grand Canyon, although not as deep. Discovered by NASA's Operation Ice Bridge ice-penetrating radar between 2009 and 2012, this remarkable feature proves that much remains to be learned about our planet.

Rivers, rather than ice, appear to have gouged out much of the giant ravine, judging from its morphology. While the canyon's exact age has not been determined, it presumably predates the buildup of ice on Greenland, making it at least 3.5 million years old. Scientists studying this feature believe that it has served as an effective meltwater roadway to the sea even since before the ice sheet formed. Efficient drainage along the buried canyon may explain the lack of large subglacial lakes in this part of Greenland. An arm of this canyon extends from Summit in central Greenland to the northern margin of the ice sheet, reaching the fjord that drains Petermann Glacier. Ongoing ocean warming within the fjord and outflow of basal water may further thin and weaken the ice tongue protecting the glacier, leading to more calving and grounding line retreat. Continued undercutting of basal ice could eventually open an entry point for warmer ocean water to penetrate far into the interior.

A Weak Underbelly

In spite of the recently discovered subglacial "Grand Canyon" and the basal ice freeze-on phenomenon, much of Greenland's subsurface topography remains hidden from view. Mathieu Morlighem of the University of California, Irvine, and his colleagues set out to learn more about this concealed world.[62] The research team combined ice-penetrating radar observations, bathymetry, and satellite detection of ice motion to calculate ice thicknesses and velocity (see appendix B). They found many glacially carved, U-shaped valleys that lie below sea level and extend much farther inland than expected. These features often coincide with fast-flowing outlet glaciers (see "Fast-Flowing Outlet Glaciers," earlier in this chapter).

Ice is funneled to the ocean through a narrow set of escape hatches along the margins of the Greenland Ice Sheet. Although ice grounded below sea level occupies a mere 8 percent of the total length of these potential escape hatches, they control 88 percent of all ice discharged from Greenland.[63] Some of these troughs penetrate tens to hundreds of kilometers inland, as in the case of Petermann and Jakobshavn Isbrae. Using newer, higher-resolution data, Morlighem and his team found extensive areas of ice grounded 200 to 300 meters (656–984 feet) below sea level in deep, narrow fjords with potential access to warmer Atlantic water.[64] In fact, over two-thirds of the marine-terminating glaciers in western and northern Greenland connect to the ocean via such ocean water–accessible pathways. The new subglacial maps expose the weak underbelly of the ice sheet that is potentially susceptible to warmer ocean water.

The subglacial Jakobshavn Isbrae drainage network, also laid bare by ice-penetrating radar, sprawls across around 440,000 square kilometers (169,900 square miles), or 20 percent of Greenland's land area.[65] Like Greenland's "Grand Canyon," these channels, carved by flowing rivers, predate the ice sheet buildup. The ancient inherited buried landscape constrains the flow of ice. The surface velocity jumps up sharply toward the outlet where several major subglacial channels converge. This hitherto-concealed topographic element that speeds up the ice may be one factor facilitating the retreat of Jakobshavn Isbrae, described in the previous paragraph.

Ice-penetrating radar and ice core drilling have recently discovered a widespread region of northeast-central Greenland where the ice sheet is melting at its base. Melting has been attributed to anomalously high levels of heat that emanate from the Earth's interior. Some researchers believe that the unusual heat flow stems from Greenland's passage over the Iceland "geologic hotspot" some 80 to 35 million years ago.[66] Furthermore, Greenland's longest ice stream, NEGIS, originates in this area, near the center of the ice sheet. It forms part of a dense network of subglacial pathways. As mentioned earlier, some of the overlying ice has already shown signs of thinning and increased speed.

For those outlet glaciers whose beds lie below sea level, such as Jakobshavn Isbrae or parts of the NEGIS drainage, warmer ocean water that enters the fjord may initiate a swift inland glacial retreat, until the glacier bed no longer rests below sea level. However, the beds of some other currently fast-moving outlet glaciers, like Helheim and Kangerlussuaq, rise above sea level going inland. These glaciers will be less susceptible to continued rapid ice sheet drawdown. Nevertheless, the widespread occurrence of deep, far-inland-penetrating subglacial channels renders the ice sheet more sensitive to rising sea level than previously suspected.

THE DARKENING OF GREENLAND

On a field trip to Greenland in 2014, Jason Box, a glaciologist with the Geological Survey of Denmark and Greenland, was utterly shocked to see so much black ice. He noted that the ice is not merely slightly dark; instead, "it's record-setting dark." He further remarks that "in 2014 the ice sheet is precisely 5.6 percent darker, producing an additional absorption of energy equivalent to roughly twice the U.S. annual electricity consumption."[67] His photos of the ice sheet show a heavily crevassed surface coated by a grimy, dark-gray layer that looks just like the dirty snow on the ground after a few days in New York City. Why should we worry about "dark snow" (fig. 5.8)? Why is Greenland growing progressively darker, what's the dark stuff made of, and why should we care? The answers to these questions bear important

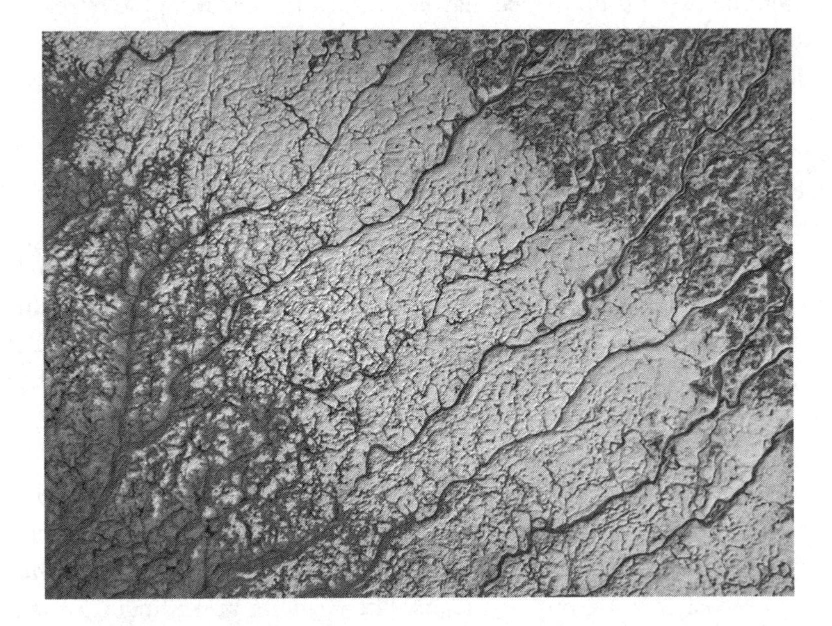

FIGURE 5.8

Dirty ice on the southwest Greenland Ice Sheet (*upper right*), near Kangerlussuaq, July 24, 2015. Summer ice melt is accumulating in meltwater lakes (*lower left*). The band of dark ice (*upper right*) is covered by surface impurities. (M. Tedesco/Lamont-Doherty Earth Observatory, Columbia University. Used by permission)

implications for the future of the Greenland Ice Sheet and how much it may add to the rising seas.

The Greenland Ice Sheet has been darkening each summer since the late 1990s, with a record low set in the summer of June–August 2012. In particular, the fringes of the ice sheet are growing darker. Several factors lead to the albedo decline (lowering reflectivity or brightness). Among these are rising air temperatures, greater surface melting that spurs enlargement of snow grains, exposure of more bare ice patches, increased growth of algae, and dark impurities trapped in ice. The balmier summers and extended melt season enlarge the area and number of darker surface meltwater pools, also reducing albedo.[68] Surface lakes, now mostly confined to the ablation zone at lower altitudes, may extend to higher altitudes in the future.[69] Once-exceptional events, such as the near-total, though short lived, surface melting of the ice sheet in July 2012 may someday become commonplace.

A darker surface absorbs more of the sun's heat, which enables more and more snow or ice to melt and expose additional low-albedo bare ground. This initiates the now-familiar positive snow/ice-albedo feedback loop (see chap. 1). Fresh snow is very bright; its albedo is around 80 percent, while that of bare ice is only 30 to 40 percent. As snow melts and refreezes, grains grow larger—and also darker. Melting snow also uncovers more bare soil, which further lowers albedo. Dark impurities encased in ice concentrate on the surface as it melts. Algal growth on ice also promotes darkening, while deposition of airborne particles adds to the mix. This enhanced surficial pileup of dust, algal, and soot particles creates the black ice that so shocked Jason Box. Climate models that calculate future albedos examine the effects of expanded bare ice area and increasing ice grain size, but not the increase in impurities covering what ought to be white ice. What are these dark particles, and where do they come from?

The lowest albedos are confined to a dark, narrow band on the western and southwestern margins of the Greenland Ice Sheet. This dark strip contains more dust than surrounding, brighter ice. Its location correlates closely with the length of the melt season, days of exposed ice, and size of ice grains.[70] The dark dust in this region (often called "cryoconite") could come from either (1) an increase in recent dust deposition or (2) accumulation of particles from older ice layers that have melted recently.

Microscopic and chemical examination of impurities in ice from several sites along a transect crossing the dark band shows similar grain shapes and mineral assemblages.[71] The sharp, angular, scratched grains suggest glacial transport. The lack of rounding further implies a relatively local source, as does the overall mineral composition and chemistry that are consistent with the ancient igneous or metamorphic rocks that underlie much of the Greenland Ice Sheet.

Powerful Icelandic volcanic eruptions, such as that of Eyjafjallajökull, which disrupted European air travel for several weeks in April of 2010, or of Grimsvötn in May 2011, could have also supplied a host of dust grains. However, the Greenland summer darkening anomaly began years before the 2010 volcanic activity. Furthermore, the mineral and chemical makeup of the sampled grains differs from that derived from Icelandic volcanoes or eastern and central Asian deserts, ruling out these sources.

Therefore, the now-exposed layers of ice probably originated at higher elevations, once closer to the center of the ice sheet. These older layers, subsequently buried under multiple layers of younger ice, slowly slid toward lower elevations at the western margin where they now form outcrops. As the old ice thaws, embedded dust concentrates at the surface. The dust grains may even date back to the last ice age, when conditions were drier and windier. On the other hand, increased industrial pollution, forest fires, wind-borne volcanic ash, and even microbes that thrive in melting snow could also amplify the darkening.

As the Greenland Ice Sheet darkens, the albedo feedback will play a growing role in ice loss by amplifying a cascading chain of events. A darker surface helps thaw more summer ice, which enlarges meltwater pools and increases surface runoff and subsurface drainage. This paves the road for smoother basal ice lubrication, swifter-moving ice sheets, and more iceberg calving.

This chapter has highlighted evidence of recent ice attrition in Greenland caused by the growing warmth of the surrounding air and ocean. Although year-to-year or even decadal temperatures can and do fluctuate significantly, in the long run Greenland is likely to warm up further, causing it to lose greater amounts of ice. These losses will be amplified by the snow/ice-albedo feedback, and to a still-uncertain extent by instabilities created by submarine topography of tidewater glaciers. However, before we ponder the fate of Greenland, we first turn to the big gorilla in the room—the planet's largest ice sheet: Antarctica.

ANTARCTICA

The Giant Ice Locker

A FROZEN WORLD

The sky gradually lightens to the day's one hour of twilight, shifting in invisible stages from a star-cluttered black pool to a dome of glowing indigo lying close overhead; and in that pure transparent indigo floats the thinnest new moon imaginable, a mere sliver of a crescent, which nevertheless illuminates very clearly the moonlight glittering on the snow, gleaming on the ice, and all of it tinted the same vivid indigo as the sky; everything still and motionless; the clarity of the light unlike anything you've ever seen, like nothing on Earth.

—Kim Stanley Robinson, *Antarctica* (1998)

Antarctica is indeed a place unlike anywhere else on Earth—the closest thing to an alien planet. It is a barren world of snow, ice, bare rock, sky, incredible shapes, and few colors—a stark land of overwhelming beauty—a rhapsody in sparkling, brilliant white; somber grays; browns; coal black; and the most intense shades of cerulean, turquoise, teal blue, sapphire, cobalt, and ultramarine blue. The continent abounds in sharp contrasts—ice pinnacles sculpted into fantastic shapes by wind and waves floating in a dark, slate-blue, foam-flecked sea; massive aquamarine-tinged, flat-topped ice shelves girding a mountainous interior . . . and everywhere, a land buried under so much ice that only the highest mountain peaks protrude above vast white plains extending as far as the eye can see—an endless carpet of ice crisscrossed by gaping crevasses and fast-moving ice streams.

A few hardy species claim the icy continent as their permanent or temporary home—penguins and migratory birds, with seals, dolphins, whales, and fish stocking its coastal seas. Green is a rarity in Antarctica. It lacks trees or shrubs; only a few robust plants can survive its frigid temperatures. Certain well-adapted grasses, flowering plants, mosses, liverworts, and lichens find a precarious foothold, particularly on the warmer Antarctic Peninsula, nearby islands, and some coastal regions. But what the continent lacks in biodiversity, it makes up for in numbers.

Antarctica is the terrible fabled *Terra Australis Incognita*, a land of mists and legend, hinted at by ancient Greeks and imaginative Renaissance geographers, whose existence remained speculative until 1820 when land was first sighted by a Russian expedition led by Fabian Gottlieb von Bellingshausen, and shortly thereafter by British captain Edward Bransfield and several American sealers. A brisk sealing and whaling industry soon arose, lasting into the 1960s when declining stocks of many whale species limited profitability. Belatedly, the International Whaling Commission placed a moratorium on further commercial whaling in 1982, although several nations including Japan, Russia, and Norway remain opposed. Some nations issue so-called "scientific permits" allowing limited whaling to their citizens.

Scientific exploration began in earnest in the late nineteenth century, culminating in the race to the South Pole, which the Norwegian explorer Roald Amundsen reached on December 14, 1911, followed a month later by Robert F. Scott and four others (see chap. 1). Within the decade, several other major scientific expeditions embarked, intent on peeling the veils of ignorance from this hitherto-unknown continent. Ernest Shackleton led the last major expedition of the "Heroic Age of Antarctic Exploration" in 1914–1917. While adverse conditions prevented his expedition from fulfilling its goal, it still stands as a lasting testament to human endurance and survival.

Although Antarctica, unlike all other continents, boasts neither independent nations nor bustling cities nor major seaports nor highways, it stands as a unique scientific laboratory protected by international treaty from environmental degradation and exploitation. The Antarctic Treaty regulates relations among the participating parties, now numbering 53.[1] Among other things, the treaty ensures "in the interests of all mankind that Antarctica shall continue forever to be used exclusively for peaceful purposes and shall not become the scene or object of international discord." The treaty therefore prohibits military activity, nuclear explosions, and the disposal of nuclear waste, instead promoting scientific research, free exchange of data, and protection of the natural environment. It further stipulates that activities should be organized so as to limit any adverse environmental impacts, including those resulting from activities involving mineral resources other than for scientific purposes.[2] Around 30 nations maintain permanent or summer research stations on the continent, including the McMurdo and Byrd Stations (United States), Vostok Station (Russia), and Dumont d'Urville Station (France).

Antarctica is a land of extremes. Straddling the South Pole, it is the Earth's coldest continent, far colder than Greenland and nearly entirely covered by ice. The marked climatic contrast between the two poles lies in their reversed distributions of land and sea. Whereas land surrounds the Arctic Ocean, the ocean encircles Antarctica. The Arctic Ocean moderates the polar climate to some extent, while the higher average Antarctic elevation and extensive snow and ice cover intensify the extreme cold. Furthermore, the vastly expanded sea ice surrounding Antarctica in winter blocks off any ocean warmth.[3] From its September maximum of ~16 million square kilometers (6.2 million square miles), sea ice shrinks to a mere 2 million square kilometers (0.77 million square miles) in February. This seasonal fluctuation greatly exceeds that of the Arctic. However, unlike the Arctic, which has been losing sea ice since the 1980s, Antarctic sea ice extent has held steady or increased slightly (see chap. 2).

Antarctica thus holds the record for the lowest temperature on Earth, −89.2°C (−128.6°F) at the Vostok Station in July 1983. Winter temperatures on the Antarctic Plateau typically plummet to −70°C (−94°F), while summer temperatures remain below freezing. Milder temperatures prevail along the coast, ranging between around −16°C (3.2°F) in winter and −2°C (28°F) in summer. The interior is also the driest part—a "white desert," in spite of its vast reservoirs of frozen water. Precipitation over the polar plateau averages a mere 5 centimeters (2 inches) per year in water equivalence, while coastal areas receive up to 10 times that. The extreme cold and arid climate make Antarctica a veritable Mars on Earth—the closest analog to the surface of the red planet. Antarctica endures the world's strongest, fiercest winds, reaching 327 kilometers (199 miles) per hour at Dumont d'Urville Station in July 1972.[4] Cold, dense air hugging the surface rushes rapidly downslope off the polar plateau toward the continental edge, setting up intense, gale-force winds.

Because nearly the entire continent lies south of the Antarctic Circle, most is plunged into total darkness for several months, with a near-total blackout half the year at the South Pole (which does, however, receive a month of dim twilight). The prolonged darkness only reinforces the frigid conditions. However, conversely, half a year later, the sun never sinks below the horizon.

While Antarctica long remained the last continent to feel the weight of a human footprint and the last to be fully explored, it holds vital keys to our future. Its fate and ours are thus intimately intertwined, as this chapter relates. A key question revolves around the future stability of the West Antarctic Ice Sheet, with its potential to raise global sea level by several meters. A mounting body of evidence builds a case for an eventual breakdown if the Earth continues to warm up, as many climatologists predict.

AN ENDLESS CARPET OF ICE

And now there came both mist and snow,
And it grew wondrous cold:
And ice, mast-high, came floating by,
As green as emerald.
And through the drifts the snowy clifts
Did send a dismal sheen:
Nor shapes of men nor beasts we ken—
The ice was all between.
The ice was here, the ice was there,
The ice was all around:
It cracked and growled, and roared and howled,
Like noises in a swound!

—Samuel Taylor Coleridge, "Rime of the Ancient Mariner" (1798)

Antarctica spans some 14 million square kilometers (5.4 million square miles), including ice shelves—roughly 1.3 times larger than Europe. Almost 90 percent of the continent lies buried beneath the Earth's largest and thickest ice sheet, averaging around 2,000 meters (6,600 feet) thick. This ice mass would elevate sea level by 58 meters (191 feet) if it all melted (see table 1.1). Its thick ice mantle makes it the world's highest continent, with an average elevation of about 2,200 meters (7,200 feet). The highest mountain, Mt. Vinson, reaches 4,892 meters (16,050 feet) above sea level, somewhat higher than Mont Blanc in the French Alps (fig. 6.1). However, stripped bare of ice, the average height would drop to a little above 460 meters (~1,500 feet). Antarctica would also shrink in area, as parts of West Antarctica, in particular, would turn into an island archipelago.[5]

Although divided into two parts by the Transantarctic Mountains and often treated separately, both the East Antarctic Ice Sheet (EAIS) and West Antarctic Ice Sheet (WAIS) are covered by a continuous mass of ice.

FIGURE 6.1

Mt. Vinson (Vinson Massif), Antarctica, is the highest mountain in Antarctica. (NASA/Michael Studinger)

Only the highest mountain peaks (known as nunataks), stick out above the ice. East Antarctica is considerably larger and higher in elevation than its western counterpart and endures some of the coldest and driest conditions on Earth. On the other hand, the Antarctic Peninsula, which extends northward into the Southern Ocean, enjoys a comparatively milder climate than elsewhere on the continent.

Ice spreads outward toward the sea from the highest elevations on Dome A near the center of East Antarctica, and Domes C and F, also in East Antarctica. Approaching the coast, ice discharges into ice streams and outlet glaciers. Ice streams, like ribbons of ice, flow many times faster than surrounding ice (fig. 6.2). Crawling at a leisurely pace of a few centimeters to a few meters per year near the summit of the ice sheet, the ice gains speed, from tenths of a kilometer to several kilometers per year, upon entering the swifter streams. Streams form over topographic lows and between areas of slower-moving ice. Because of this differential motion, highly crevassed shear zones develop along the edges of the streams. Ice streams also occur over soft, deformable sediments such as glacial till, weaker ice, or a well-lubricated bed. Many ice streams originate far inland. Like tributaries of temperate-climate rivers, smaller ice streams coalesce into larger ones. On the East Antarctic Ice Sheet, for example, three major ice streams (Lambert, Mellor, and Fisher Glaciers) merge into what has been described as "[one of] the world's larger glacier[s]."[6] They also account for the majority of ice exiting Antarctica. On the West Antarctic Ice Sheet, ice also departs via outlet glaciers such as the Pine Island, Thwaites, and Kohler Glaciers, as well as numerous ice streams that feed into fringing ice shelves such as the Ronne-Filchner, Ross, and Getz Ice Shelves. Narrow valleys or topographic highs constrain outlet glaciers before they enter the sea. Most, however, end in floating extensions—as ice tongues or larger ice shelves.

Aprons of Ice

Ice shelves represent floating marine extensions of the ice sheet (see fig. 2.2). They encircle roughly three-quarters of the shoreline of Antarctica, covering an area of 1.56 million square kilometers (0.6 million square miles).[7] The two largest include the Ross Ice Shelf, at almost 473,000 square

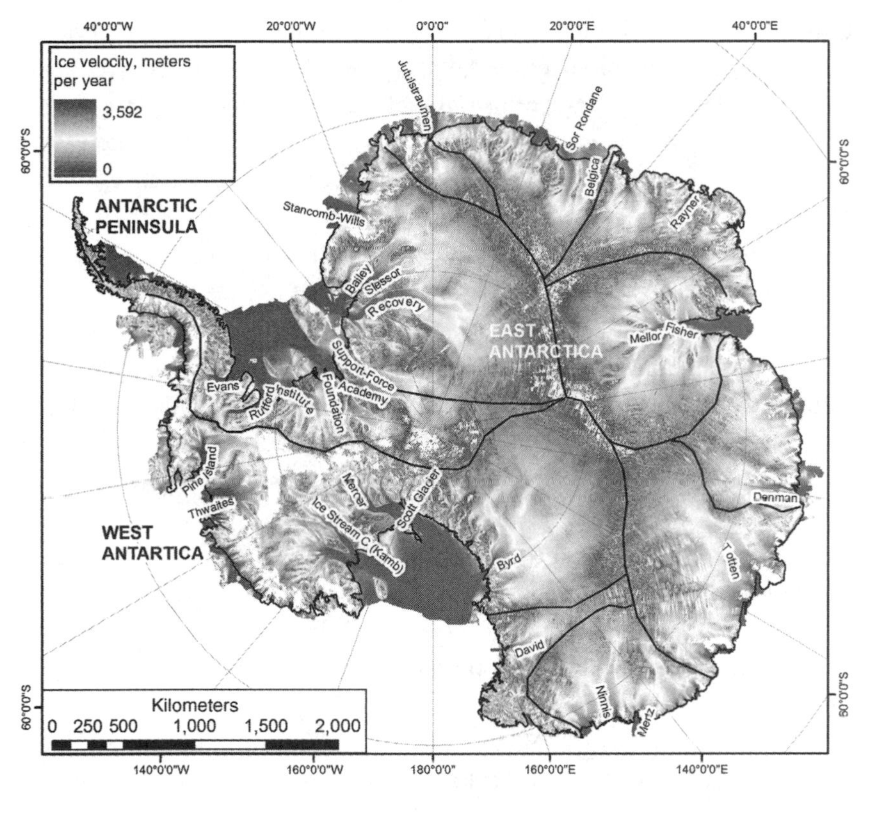

FIGURE 6.2

Ice streams on Antarctica are relatively fast-moving rivers of ice that feed into the major ice shelves. Dark lines are ice divides that separate ice drainages, analogous to river watersheds. (Rignot et al. [2011], https://earthobservatory.nasa.gov/images/51781/first-map-of-antarticas-moving-ice)

kilometers (182,630 square miles), and the Ronne-Filchner Ice Shelf, at 422,000 square kilometers (162,930 square miles).[8] Others include Larsen C, Lambert-Amery, Riiser-Larsen, and Shackleton. While these still hold fast, some ice shelves have not fared as well.

At least 10 ice shelves on the Antarctic Peninsula have disintegrated or shrunk in a steady southward progression since the 1950s (see table 2.1). What about the even more extensive ice shelves on the mainland, beyond which lie massive glaciers? Many of these have also begun to thin, including a handful in colder, more stable East Antarctica. Because ice shelves help to stabilize the Antarctic ice sheet and thereby control its mass balance, early warning signs of potential instability raise serious concerns, as discussed in a later section. Large icebergs often split off tongues of tidewater glaciers or ice shelves. In Antarctica, the land of extremes, these icebergs can be enormous. Iceberg B-15, the world's current record holder at 10,800 square kilometers (4,170 square miles)—dwarfing the island of Jamaica—calved off the Ross Ice Shelf in March of 2000. It subsequently broke up into several smaller pieces, the largest of which was named B-15A. Another megaiceberg, A-38, cracked off the Ronne-Filchner Ice Shelf in October 1998. It measured 7,500 square kilometers (2,900 square miles)—larger than the state of Delaware.

Despite such headline-grabbing events, calving is a normal process, the result of stresses on the exposed ice shelf by currents, tides, waves, collisions with other icebergs or moving pack ice in winter, or hydrofracturing. Summer meltwater pools spread over the surface of ice shelves and press down on water entering crevasses, until the cracks are wedged apart and propagate, bit by bit, all the way to the base. Icebergs then finally split off with a loud, resounding splash.

In late September of 2017, another large iceberg of about 185 square kilometers (72 square miles) broke off Pine Island Glacier.[9] The newest iceberg was larger than another berg that had calved earlier that year, though smaller than one in 2015 and the giant Iceberg B-31 (nearly 700 square kilometers; 250 square miles) that calved in November of 2013. At the time, NASA glaciologist Kelly Brunt noted: "The detachment rift, or crack, that created this iceberg *was well upstream of the 30-year average calving front of Pine Island Glacier, so this is a region that warrants monitoring*"[10] (italics added). Starting in January 2013, an array of instruments had been eying its progress, including IceBridge aircraft and the Moderate Resolution Imaging Spectroradiometer (MODIS) on board the NASA Terra and Aqua satellites. B-31's calving attracted widespread attention because Pine Island Glacier—the source of the massive iceberg—is often labeled "the weak underbelly of the West Antarctic Ice Sheet." Rifting also preceded the 2015 calving event. But this rifting was unusual in that cracks had developed at the center of the ice shelf near the grounding line and propagated toward the margins, causing calving farther up-glacier than previously observed. This new rifting style may be associated with warming Antarctic shelf bottom water.[11] Continued formation of such crevasses farther landward could eventually undermine the shelf and lead to its demise.

Once set adrift in the polar seas, some Antarctic bergs run aground in embayments or drift with local currents, while others get swept up in the in the Antarctic Circumpolar Current and drift northward where they disintegrate and ultimately melt. A similar fate awaits B-31. In April 2014, the giant berg was spotted slowly drifting out to sea. Nevertheless, Pine Island Glacier and its neighbors in the Amundsen Sea sector of the West Antarctic Ice Sheet continue to be carefully watched for signs of instability (see "The West Antarctica (WAIS)—on Shaky Ground," later in this chapter).

Hidden Mountains and Volcanoes

Beneath the seemingly endless continental carpet of ice, girded by floating ice shelves and drifting icebergs, lies a hidden world—the vanished world that once existed before the deep freeze. Hints of its existence emerge in the nunataks—the mountaintops or ridges that project above the ice. But only in recent years have we been able to peer under this frozen mantle to study a previously unknown landscape of mountains, valleys, islands, lakes, deep embayments, and volcanoes.

First detected in the 1950s, the true nature of the Gamburtsev Mountains became much clearer in 2011 as ice-penetrating radar, gravimeters, and magnetometers revealed hitherto-concealed features. The Gamburtsevs rival the Alps in size but are totally buried beneath the Antarctic ice. Their story began a billion years ago when a mountain range formed as a result of continental collisions. Over time the original mountains wore down and only the roots were preserved. Between 250 and 100 million years ago, when dinosaurs roamed the Earth, rifting broke apart the supercontinent of Gondwanaland. One of the pieces that became Antarctica drifted toward the South Pole. These tectonic forces "rejuvenated" Antarctica, such that part of the land pushed upward and new mountains formed. By 34 million years ago, ice had begun to build up on the East Antarctic Ice Sheet. The Gamburtsevs may be close to the location where the ice sheet first began to grow. The ice sheet, when fully formed, protected the mountains from further erosion, preserving their youthful appearance.[12] The ice is now so thick that in places it quietly flows uphill over the buried mountain peaks.

Bedmap2 yields much new information on the hidden physiography of Antarctica. This project compiled all available information on ice-surface elevation, ice thickness, and subglacial topography.[13] Airborne ice-penetrating radar, satellite laser and radar altimetry, satellite imagery, gravity measurements, and even people on dog-drawn sleds supplied the raw data.

Most of the Bedmap2 data come from radio echo sounding (RES). This technique exploits a portion of the electromagnetic spectrum in which emitted radio waves travel freely through both ice and air. When radio waves encounter boundaries between materials with different physical properties, the speed of wave propagation changes, setting up reflections. Radar antennas on aircraft or on the ground detect reflections from the base of or within the ice sheet. Since the velocity of the radar pulse is known, the thickness of the ice can be calculated by measuring the time elapsed between "echoes" emanating from air/ice and ice/bedrock surfaces. A cross section of the ice sheet and underlying bedrock can be built up from a successive series of such echoes recorded as the aircraft flies over the ice sheet.

This massive effort resulted in the most detailed map yet of the southernmost continent's bedrock (fig. 6.3). "It's like you've brought the whole thing now into sharp focus. . . . In many areas, you can now see the troughs, valleys and mountains as if you were looking at a part of the Earth we're much more used to seeing, exposed to the air," said Dr. Hamish Pritchard from the British Antarctic Survey, a coauthor of the Bedmap2 report published in 2013. Among other features, Bedmap2 discovered a trough in the northwestern part of East Antarctica that is one of the largest on the continent, penetrating 650 kilometers (400 miles) into the interior of the ice sheet (fig. 6.3). Another major rift, the Lambert Rift, extends seaward to the east of the Gamburtsev Mountains. On the new map, the Transantarctic Mountains continue into the Ellsworth Mountains of West Antarctica. The map also illustrates sharp topographic differences between East and West Antarctica. East Antarctica features much more rugged terrain than does West Antarctica. Much of the latter would be reduced to a series of islands without its ice cover. The new map furnishes

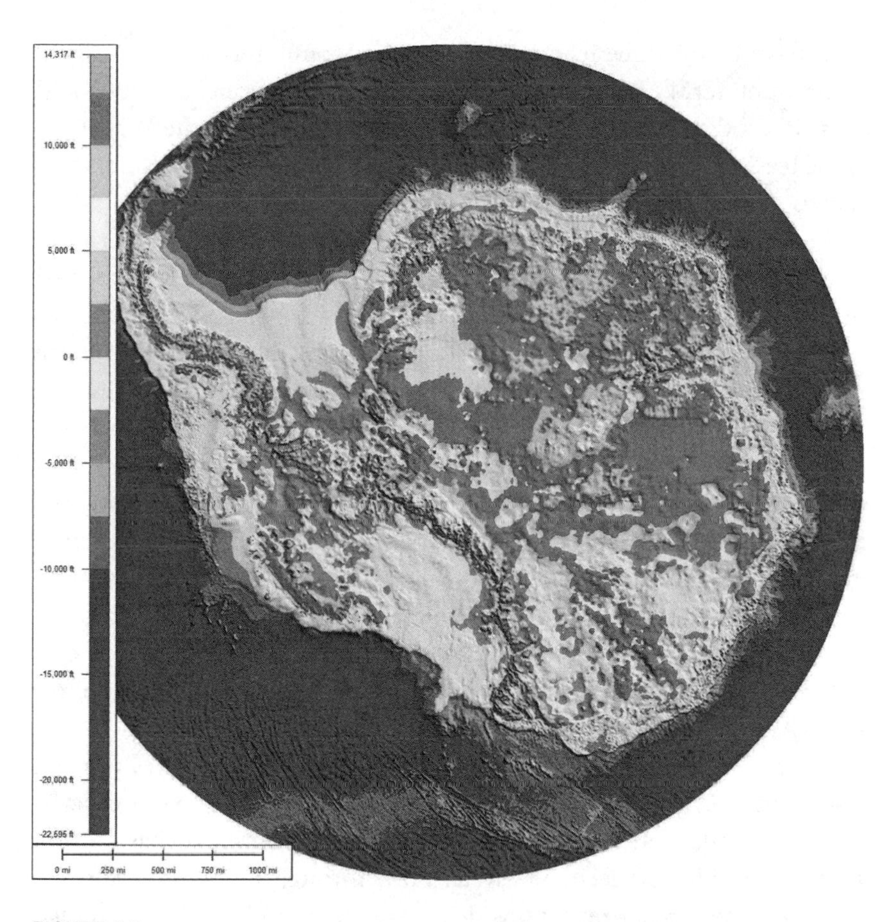

FIGURE 6.3

Antarctica stripped of its ice cover, showing bare bedrock. Note the large areas that would become submerged. ("Antarctic Bedrock," prepared from Bedmap2, March 13, 2008, https://commons.wikimedia.org/wiki/File:AntarcticBedrock.jpg; courtesy of Wikimedia Commons)

scientists with an important tool with which to study how ice flows across Antarctica and how it might react to future climate change.

Antarctica is also home to a number of volcanoes, some buried under ice, others exposed, and a few still active. Mount Erebus, the southernmost active volcano in the world, is the second highest in Antarctica after Mt. Sidley, and stands at 3,794 meters (12,448 feet) elevation.[14] It shares

Ross Island with three inactive volcanoes—Mount Terror, Mount Bird, and Mount Terra Nova. The volcano erupted most recently in 2005 and 2014. Recently, 138 volcanoes have been mapped beneath the West Antarctic Ice Sheet, identified as such from Bedmap2 data on the basis of their shapes. Of these, 91 were previously unknown.[15]

Seismicity on the WAIS further suggests magma sloshing around under the ice.[16] Earthquake swarms, detected in 2010 and 2011, displayed behavior more typical of seismic activity related to volcanism and movement of magma than to quakes associated with tectonic activity or ice movement. The quakes originated in an area of subglacial mountains with magnetic anomalies characteristic of volcanic origin. Furthermore, a prominent seismic reflecting layer interpreted as volcanic ash was draped over a volcano-like subsurface feature. The ash layer was estimated to have been deposited as recently as 8,000 years ago. While geologists doubt that any local subsurface volcano will break through to the surface any time soon, geothermal heat generated by magmatic activity could melt the base of the ice sheet. The decreased pressure on subterranean magma due to a thinning ice sheet could spur volcanism that would further destabilize the WAIS.

Other geothermal sources were discovered beneath Thwaites Glacier, also on the WAIS. Airborne radar surveys detected higher-than-expected subsurface radar-reflecting signals typical of large quantities of basal water. Combined with a model of how water flows through a subglacial network of canals, these radar reflections denote an area of higher-than-average geothermal heat flow from the Earth's interior.[17] The location lies near a buried rift structure and associated volcanism. Thwaites Glacier may thus be sitting on a "stovetop." Don Blankenship, one of the scientists on the project, exclaimed: "And then you plop the most critical dynamically unstable ice sheet [WAIS] on planet Earth in the middle of this thing." Thwaites Glacier (like its neighbor, Pine Island Glacier) has been retreating within the last two decades, mainly due to incursion of warmer ocean water beneath its ice shelf. The weakened ice shelf has relaxed its tight grip on the glacier, allowing it to move forward and speed up. Thwaites Glacier may therefore be exposed to a double whammy—underground as well as at its marine edge. This glacier definitely bears more watching (see "The West Antarctica (WAIS)—on Shaky Ground").

The Dry Valleys—Mars on Earth

Mars looks more and more like Antarctica every day. . . . Both Mars and Antarctica are cold deserts. They experience intense [UV] radiation, and liquid water is only briefly and ephemerally stable in special locations and at the warmest part of the year.

—Joe Levy, geologist

Across McMurdo Sound, opposite Ross Island at the eastern edge of the Ross Ice Shelf, lies a series of ice-free valleys—otherwise a rarity in Antarctica. Narrow gaps in the topography have slowed down the flow of ice from the East Antarctic Ice Sheet. Much less snow accumulates in the Dry Valleys than on nearby Ross Island. Furthermore, strong westerly winds in winter scour the valleys, removing surface snow. The fierce winds also polish large rocks and cobbles, often carving them into bizarre shapes called *ventifacts*. As in the Arctic, permafrost develops in the subzero temperatures (see, e.g., chap. 3). Polygonal ice wedges and patterned ground are widespread. Small glaciers nestle in cirques along the ridge crests of surrounding mountains, a few spilling over onto the valley floors. Loose sand, gravel, and occasional sand dunes cover the ground. Several lakes fill valley low spots, covered by ice most of the year.

The Dry Valleys rouse great scientific interest as a terrestrial analog for extraterrestrial life. The search for life on Mars motivates much of NASA's exploration of the Red Planet. Earth and Mars share many features. Although Mars today is extremely cold and dry, its surface bears many traces of past water: branched river valleys, deltas, lakebeds, and hydrous minerals. Even today, both poles are covered by H_2O ice caps, and more ice lies buried under the soil. The coldest temperatures in the Dry Valleys approach those of Mars. Thus, the study of life in the Dry Valleys guides scientists in their search for possible martian life. A few hardy microbes manage to survive in the icy water and in the barren soils despite the inhospitable environment. Mostly single-celled organisms, these include cyanobacteria, diatoms (a type of silica-shelled marine or freshwater algae), mosses, lichens, and many species of bacteria. Cyanobacteria (formerly called blue-green algae) are among the oldest forms of life, going back some 3.5 billion years.[18] Their photosynthetic activity may have led to the buildup of Earth's oxygen atmosphere starting around 2.5 billion years ago. Diatoms live in a wide range of environments, including the oceans, lakes, brackish lagoons, sea ice, and soils, and even on wet rocks. Their abundance, widespread distribution, and sensitivity to the environment make them useful markers of changes in temperature and water chemistry, including salinity.

Blood Falls, a plume of saltwater, flows over the Taylor Glacier, in the McMurdo Dry Valleys, onto an ice-covered lake.[19] Iron oxide, mingled with salts, give Blood Falls its rusty red color. Microbes that metabolize both sulfate and ferric iron live in the water. Such an unusual ecosystem also offers a relevant Mars analog, since both types of minerals coexist in rocks on the Red Planet.[20] Not only do the frigid, hyperarid climate and absence of higher life forms in the Dry Valleys represent the closest martian analog on Earth, but Antarctica also yields a veritable treasure trove of other bits of extraterrestrial planetary bodies.

A Meteorite Bonanza

> *Meteorite hunting is not for wimps. The best places to look are also the coldest and windiest. You need very old ice, and you need wind, lots of it, strong and unrelenting. Antarctica fits the bill.*
>
> —Mary Roach

Meteorites constitute virtually our only samples of extraterrestrial materials. Nearly all meteorites found on Earth are pieces of asteroids; a handful may originate on the moon or Mars. Because most date back to when the solar system formed, they offer important insights about the birth of the planets and their moons. Antarctica turns out to be one of the best places on Earth to hunt for these interplanetary objects.[21] The cold climate and aridity preserve freshly fallen meteorites from weathering, or from rusting in the case of iron meteorites. Quickly buried under snow and ice, meteorites travel downslope with the ice sheet toward the ocean. But the Transantarctic Mountains obstruct this flow and trap numerous pieces along their flanks. Powerful winds that rush down from the South Pole strip the ice bare by sublimation and abrasion. Researchers on snowmobiles or on foot quickly spot these dark rocks sitting on the blue ice like ripe plums ready for the picking (fig. 6.4). Over the years, more than 50,000 specimens have been recovered from Antarctica, far more than from the deserts of northern Africa, Asia, and Australia combined. This vastly expanded collection has proven a bonanza for cosmochemists, planetary geologists, and astronomers studying the origins of our solar system. But, returning to Earth, the thick ice sheet holds many other surprises, as the next section will show.

FIGURE 6.4

An abundance of meteorites lie on the Antarctic ice, waiting to be collected by scientists. (NASA)

Note: Not all these rocks are meteorites; however, the experts can tell the difference.

Buried Lakes

Mountains and volcanoes are not the only terrain features the ice conceals. Antarctica harbors hundreds of subglacial lakes. Of these, Lake Vostok, at 15,700 square kilometers (6,060 square miles), is the largest and best-known. Situated near the center of the East Antarctic Ice Sheet, close to the Russian Vostok Research Station, the lake may have been sealed by ice around 15 million years ago. However, motion of the ice sheet may be renewing the water every 13,300 years.[22]

The possibility that the lake harbors a unique ecosystem has spurred interest in drilling 3.7 kilometers (2 miles) beneath the thick ice seal to find out. Despite concerns about contamination by drilling fluids or surface microbes, the Russians pierced the ice in 2012 and again in January 2015. A Russian-led team reported in 2013 that they had identified 3,500 gene sequences in ice accreted to the underside of the ice sheet, just over the lake.[23] The vast majority derived from bacteria and the remainder from eukaryotes (organisms with cell nuclei). Most of these species resembled primitive organisms living in lake sediments, brackish water, seas, soils, and even deep-sea thermal vents. Presumably, most of these life forms inhabit the lake as well.

Almost 400 subglacial lakes have been discovered in Antarctica, thanks to radar (fig. 6.5).[24] The subsurface topography of the ice sheet can be reconstructed by combining ground-penetrating radar data with surface elevation measurements from satellite altimeters. Slight variations in density or crystal orientation, or presence of impurities such as volcanic ash, produce distinctive radar reflections that outline the ice layers. If no escape route for water exists, ice entering a lake will float over water. This produces a flat, featureless surface like that of an ice shelf. Thus, many (though not all) flat radar reflections within the ice correspond to buried lakes. Since ice meets no resistance when flowing over water, it accelerates and thins on the upstream side of the lake. The reverse occurs when ice encounters bedrock again farther downstream. Subtle changes in surface elevation may therefore reveal movement of water between lakes or within an interconnected hydrological system.

Seismic profiling and gravity measurements have also helped detect and map subglacial lakes. Seismic surveying somewhat resembles that

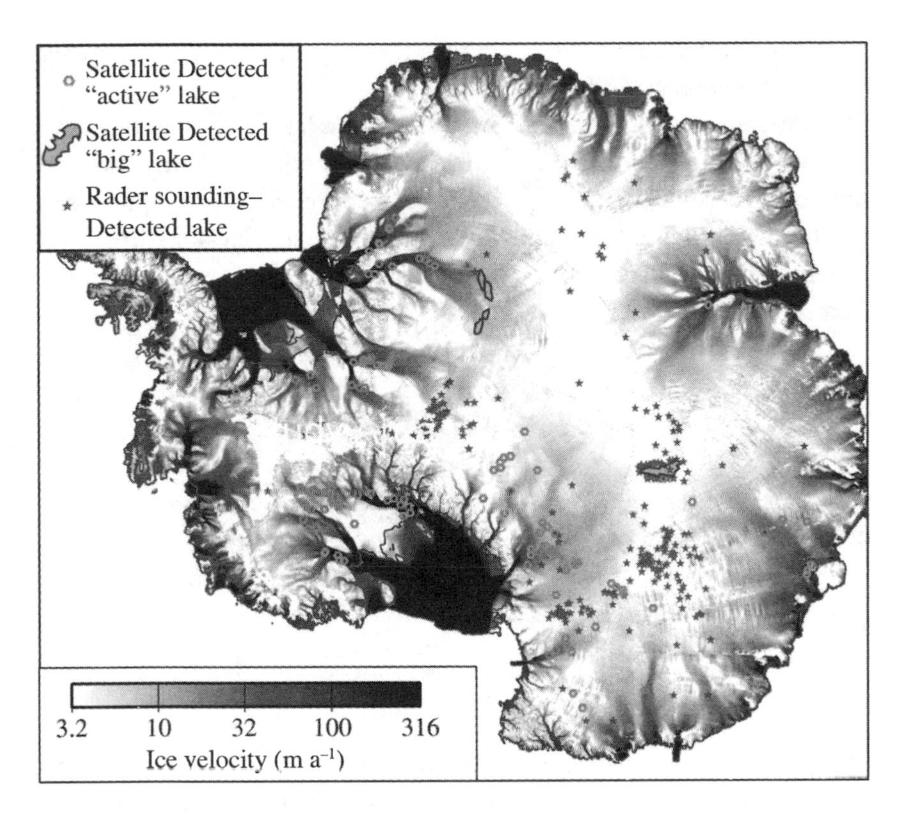

FIGURE 6.5

Distribution of lakes buried beneath the Antarctic ice sheet. (Frank Pattyn, Sasha P. Carter, and Malte Thoma, "Advances in Modelling Subglacial Lakes and Their Interaction with the Antarctic Ice Sheet," *Philosophical Transactions of the Royal Society A* 374 [2015], doi:10.1098/rsta.2014.0296)

of radar sounding, but substitutes sound waves for radio signals. Seismic mapping takes time and effort, acquiring only a single data point for each "shot." (A small explosive charge is detonated in order to trigger the seismic waves.) A number of large lakes, including Lake Vostok, have been studied in this manner. Gravity measurements have been used to a lesser extent. If the geometry of the overlying ice sheet is known and taken into account, the depth of the lake can be inferred from the remaining gravity signal. This approach identified two separate basins in Lake

Vostok, one larger and deeper than the other, separated by a 40-kilometer (25-mile)-wide rocky ridge.

No mere geographical curiosities, the buried lakes of Antarctica have become objects of intense study. They hold important information on unique ecosystems adapted to darkness, extreme cold, and high pressures. Furthermore, they reveal elements of a complex subglacial hydrological system that may affect how fast the ice sheet moves—as explained in the next section.

MOVING ANTARCTIC ICE

> *Abundant liquid water newly discovered underneath the world's great ice sheets could intensify the destabilizing effects of global warming on the sheets. Then, even without melting, the sheets may slide into the sea and raise sea level catastrophically.*
>
> —Robin E. Bell, "The Unquiet Ice"

The immense ice blanket over Antarctica slowly wends its way seaward along faster-moving corridors of ice, or ice streams, that convey some 90 percent of the ice to the ocean. Like branches on a tree, smaller ice streams join larger ones that eventually reach the sea. Inland ice streams generally flow over buried valleys and occasionally glide over subglacial mountains. During the International Polar Year, 2007–2009, an array of satellites equipped with interferometric synthetic aperture radar (InSAR) instruments surveyed the entire Antarctic continent and collected emitted microwave radiation reflected back off the Earth's surface. Eric Rignot of the University of California, Irvine, and his colleagues used changes in the resulting pattern of reflections to map the motion of ice streams across Antarctica. They determined that ice crawled a few centimeters a year at ice divides but sped up to several kilometers per year on fast-moving glaciers and ice shelves.[25] They also recognized the importance of bottom slipperiness in moving Antarctic ice. In the fast-flowing regions, ice flows over soft, squishy sediments or over water. Ice streams also gather speed as they approach major ice shelves, such as the Ross, Ronne-Filchner, and Avery Ice Shelves.

The speed of outlet glaciers and ice streams ultimately governs ice losses from Antarctica. Several factors, in turn, influence ice stream velocity. While smoothness or roughness of the substrate are important factors, water plays a major role in "greasing the skids."[26] Unlike in Greenland, where surface melt drains to the base via an internal plumbing system (see fig. 4.5), water beneath the Antarctic ice sheet is primarily generated by pressure melting, friction, and geothermal heat (see "Hidden Mountains and Volcanoes," earlier in this chapter). Surface temperatures over much of East Antarctica average −50°C (−58°F); temperatures under 3–4 kilometers (2–2.5 miles) of ice are near −2°C (28°F)—close to the melting point of ice at that pressure. Friction generated by ice moving over its bed also increases melting; so does motion over a geothermal hot spot.

The discovery of hundreds of lakes beneath Antarctic ice has revolutionized thinking about the ice sheet's internal plumbing.[27] Ice-penetrating radar reveals that many lakes can form interconnected entities in which water actively flows from one lake to another. Many such "active" lakes cluster around fast-flowing ice streams (compare fig. 6.5 in the previous section with fig. 6.2). However, these features may change quickly as lakes fill up or discharge water. For example, the ERS-2 satellite altimeter detected an abrupt drop in elevation over a known subglacial lake near Dome C in East Antarctica. Soon thereafter, elevations rose farther downslope. These successive changes hint at a network of interconnected subglacial tunnels through which water flows from one location to another.[28] How do the subglacial lakes affect the motion of overlying ice? Basal friction vanishes as ice flows over water. Furthermore, lake water that refreezes at the base of the ice releases heat, which warms up the ice. The partially thawed, softened ice then accelerates until it hits bedrock again before refreezing. Buried lakes and places of high geothermal heat flow are therefore "potentially future nucleation sites for the initiation of fast-flowing ice streams."[29]

Not all ice on the East Antarctic Ice Sheet originally fell as snow. Radar probes demonstrate that nearly a quarter of the ice at the base of the East Antarctic Ice Sheet consists of accreted ice, refrozen from meltwater.[30] As in Greenland (chap. 5), the freeze-on process can severely distort and upwarp overlying ice layers, which effectively destroys the bottommost ice stratigraphy. This complicates efforts to core the oldest ice layers in order to reconstruct ancient climates (chap. 7). Refreezing of meltwater releases heat to the enclosing ice, which weakens it, allowing it to deform and accelerate. The extensive distribution of accreted ice implies the existence of a long-lived internally draining hydrological system. Simple calculation of lengths and ice velocities suggests that freeze-on may have persisted for tens of thousands of years, spanning much of the last ice age to recent times. Thicker and more widespread than previously suspected, this different form of glacial ice may render the Antarctic ice sheet more vulnerable to change than most models assume.

The giant frozen gorilla may be stirring fitfully deep inland, but most of the action is taking place along the margins of the ice sheet, as the following sections will demonstrate. Mounting evidence points to a growing imbalance within the past two decades between gains of ice from snowfall and losses from ice outflow past the grounding line and calving. Dynamic thinning may be a sign of trouble ahead. It implies that ice moves faster and sheds more mass at its terminus than it did previously. The most dynamic portions of the ice sheet are West Antarctic outlet glaciers and their floating extensions. As is increasingly the case in Greenland, warmer ocean water encroaching under ice shelves has been implicated as a major culprit.

THE GIANT GORILLA AWAKENS

What's Up with Antarctic Ice?

> *It is the first time we can say that if you are looking at the entire ice sheet, it is losing mass. . . . We can now see Antarctica melting. We have a number for the ice sheet. It's a big step toward understanding how the sea level is going to change.*
>
> —Isabella Velicogna

Is Antarctica's endless winter beginning to relax its gelid grip? Has the growing warmth that has pushed back the cryosphere elsewhere on Earth reached the southernmost continent yet? The Antarctic ice locker that long preserved a record of ancient climates may be starting to defrost. As this section suggests, the icy giant may be slowly awakening from its long slumber.

Most experts believe that Antarctica is currently unloading more ice than it gains. Furthermore, losses appear to have increased since the 1990s, when satellite observations of Antarctica's ice began. Several other subsequent reports echo similar findings (table 6.1). While the East Antarctic Ice Sheet in general has remained stable, or expanded, Antarctica's ice sheet is in decline.

In 2015, H. Jay Zwally, a NASA glaciologist, and his colleagues declared that, to the contrary, Antarctica piled on more ice than it shed—yielding a surplus of 82 billion metric tons per year from 2003 to 2008.[31] Other recent studies have measured deficits of between 92 and 160 billion tons per year since 2003 (see table 6.1). Zwally does concede ice losses on some coastal sections of West Antarctica—notably the Amundsen sector, including the Pine Island and Thwaites Glaciers —and less so on the Antarctic Peninsula. Yet the East Antarctic surplus far outweighs these losses. Lest climate change skeptics rejoice, Zwally does not expect this snow and ice buildup to last for long. "If the losses of the Antarctic Peninsula and parts of West Antarctica continue to increase at the same rate they've been increasing for the last two decades, the losses will catch up with the long-term gain in East Antarctica in 20 or 30 years—I don't think there will be enough snowfall increase to offset these losses."[32]

Other scientists express doubts. Theodore Scambos, a researcher at the National Snow and Ice Data Center, says: "I think the [Zwally] study is just plain wrong, far too inconsistent with other lines of evidence, and not worth the public's attention." Ian Joughin of the University of Washington observed that the Zwally team's findings differed from previous studies that used the same data sources.

Conflicting findings may stem from using different methods or time intervals (see table 6.1). For instance, the Zwally team's data cover only up to 2008, although ice losses, especially in West Antarctica, have stepped up considerably since then. Because of Antarctica's vast size, remote location, and extremely harsh climate, satellite remote sensing provides most of our information. Radar or laser altimetry and GRACE gravimetry are the two most commonly used satellite techniques.[33] Altimeters measure changes in surface elevation. These data, plus knowledge of area and snow or firn density near the ice sheet surface, yield volume estimates. Since scientists do not always know snow/firn densities precisely, they must make

TABLE 6.1 Recent Ice Losses on Antarctica and Contributions to Sea Level Rise (mm/yr)

ANTARCTIC ICE SHEET	ANTARCTIC PENINSULA	WEST ANTARCTIC ICE SHEET	EAST ANTARCTIC ICE SHEET	REFERENCE
0.20±0.15 (1992–2011)	0.06±0.04	0.18±0.07	−0.04±0.12	Shepherd et al. (2012)
0.22±0.10 (2005–2010)	0.10±0.03	0.28±0.05	−0.16±0.09	Shepherd et al. (2012)
0.27 (0.37 to 0.16) (1993–2010)	—	—	—	Intergovernmental Panel on Climate Change (2013a)
0.41 (0.61 to 0.20) (2005–2010)	—	—	—	Intergovernmental Panel on Climate Change (2013a)
0.31±0.06 (2003–2012)	0.07±0.01	0.32±0.02	−0.07±0.04	Sasgen et al. (2013)
0.44±0.13 (2010–2013)	0.06±0.05	0.37±0.07	0.01±0.10	McMillan et al. (2014)
0.18±0.12 (2003–2013)	0.09±0.01	0.32±0.02	—	Velicogna et al. (2014)
0.25±0.03 (2003–2014)	0.07±0.01	0.33±0.02	−0.15	Harig and Simons (2015)
−0.31±0.17 (1992–2001)	0.02±0.03	0.04±0.06	−0.38±0.14	Zwally et al. (2015)
−0.23±0.07 (2003–2008)	0.08±0.01	0.07±0.04	−0.38±0.08	Zwally et al. (2015)
0.25±0.14 (2002–2015)	—	—	—	Forsberg et al. (2017)

Note: One GT ice (1 billion metric tons) per year = 362.5 millimeters per year sea level rise equivalent.

certain simplifying assumptions that introduce uncertainties in calculating mass changes. The GRACE gravimeter measures ice sheet mass fluctuations directly. But these must be corrected for redistribution of mass due to glacial isostatic rebound of land after the last deglaciation. However, different glacial isostatic adjustment (GIA) models often yield divergent outcomes. A third method, the mass budget method (also known as the *input-output method*), relies on climate data to calculate the difference between accumulation (i.e., snowfall) and ablation (see chap. 4). Because the extreme Antarctic cold limits surface melting, ablation consists mainly of ice discharge at the grounding line and iceberg calving. Regional climate models and in situ observations provide accumulation rates, while ablation rates are calculated from ice motion and grounding line thickness. Interferometric synthetic aperture radar (InSAR) provides ice velocity data, while airborne radar estimates ice thicknesses. Some studies combine data from multiple sources. The scientific discord therefore hinges on reconciling such technical differences.

Setting aside the scientific debate, all parties agree that the Antarctic Peninsula and West Antarctic Ice Sheet have been losing mass since the early 1990s, enough to raise sea level by up to 0.44 millimeter (0.02 inch) per year—which corresponds to roughly 15 percent of the current global trend (see table 6.1).

The West Antarctic Ice Sheet (WAIS)—on Shaky Ground

Is the West Antarctic Ice Sheet rapidly approaching a point of no return? Could loss of ice along its margins set off an irreversible collapse, eventually pushing up sea level by several meters? A map of Antarctica stripped bare of its present ice cover, which in places is more than 2,000 meters (6,600 feet) thick, shows that much of the WAIS is "grounded" (i.e., rests on bedrock) below sea level (see fig. 6.3). Portions of the ice sheet margin overlie subglacial lakes and channels. With the WAIS stripped of its ice cover, seawater would engulf large areas of it and surround an archipelago of islands.

At present, many West Antarctic ice streams and glaciers (and a few in East Antarctica) near the grounding line rest on slopes that tilt landward (i.e., have reverse, or retrograde, slopes). The marine ice sheet instability (MISI) hypothesis proposes that an ice sheet or glacier grounded on a

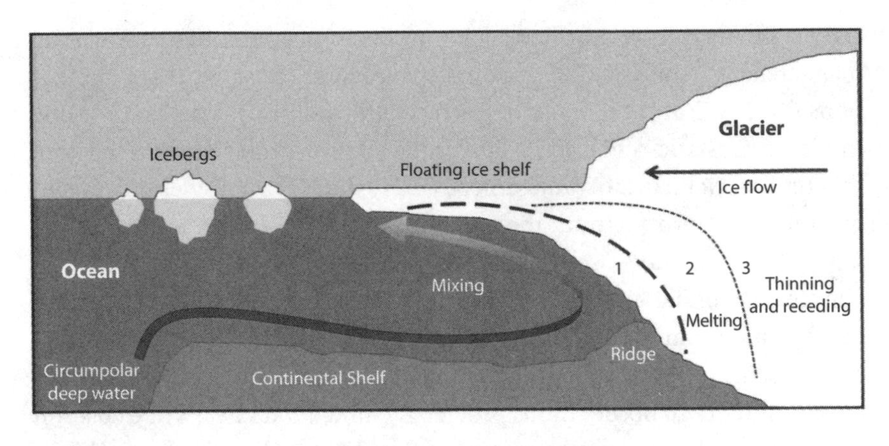

FIGURE 6.6

Marine ice sheet instability in Antarctica: (1) The ice sheet or glacier is grounded on a bedrock ridge; (2) warm Circumpolar Deep Water circulates beneath the ice shelf, melts it from below, and erodes the grounding line; (3) the grounding line retreats beyond the ridge downslope, the ice shelf continues to thin, and the glacier accelerates forward. (Modified from Bethan Davies, "Is the West Arctic Ice Sheet Collapsing?," AntarcticGlaciers.org, May 13, 2014, http://www.antarcticglaciers.org/2014/05/west-antarctic-ice-sheet-collapsing/)

reverse slope is inherently unstable[34] (fig. 6.6). Ice thickness at the grounding line is a key factor controlling the forward motion of the glacier. As its bed deepens inland from the grounding line, the glacier thickens. Once a retreating grounding line reaches the thicker section, ice begins to accelerate. Moreover, ice melts at a lower temperature under the greater pressure of a thicker glacier. Increased basal melting at the grounding line reinforces the effects of warm ocean-water incursion. In a cascading positive feedback loop, the accelerating glacier discharges more ice across the grounding line, which retreats farther downslope. The process continues until the bed slope levels out or steepens landward. As discussed in the following paragraphs, this particular topographic configuration is particularly sensitive to rising oceanic temperatures. Although several scientists proclaim that such an instability already exists, most experts doubt that a catastrophic collapse of the WAIS would occur within this century. Rather, they foresee a much slower yet inevitable breakdown playing out over several centuries or longer as temperatures continue to climb.

The West Antarctic Ice Sheet could, however, be issuing ominous warning signals of incipient instability. Glaciers entering the Amundsen Sea sector, such as Pine Island, Thwaites, Smith, Kohler, Pope, and Haynes, drain a third of the WAIS and are among the fastest-flowing ones in Antarctica. These glaciers discharged 77 percent more ice in 2013 than in 1974, with a third of the increase occurring between 2003 and 2009.[35] Another study concludes that Amundsen Sea–draining glaciers contributed an average of 0.23 millimeter (0.06 inch) per year to sea level rise from 1992 to 2013, with an increasing trend from 2003 to 2011.[36]

Is a "disaster . . . unfolding—in slow motion," as a popular science magazine boldly declares, with the headline "West Antarctic Ice Sheet Is Collapsing"?[37] Perhaps not yet, but some day . . .

Eric Rignot and fellow glaciologists mapped the continuous and rapid retreat of the grounding lines of the Pine Island, Thwaites, Smith, and Kohler Glaciers between 1992 and 2011, with InSAR radar data from the European Remote Sensing satellites (ERS-1 and 2). Most spectacularly, the grounding lines of the Smith and Kohler Glaciers separating floating from land-based ice had shifted landward by 34–37 kilometers (21–23 miles).[38] Pine Island and Thwaites Glaciers have also been thinning and retreating in recent years. These two glaciers and their neighbors hold the equivalent of 1.2 meters (4 feet) of sea level rise. All four glaciers rest on reverse beds with no immediate subglacial obstacles to rein in their landward migration. Rignot and colleagues concluded that "this sector of West Antarctica is undergoing a marine ice sheet instability that will significantly contribute to sea level rise in decades to centuries to come." Other glaciologists voice similar concerns. For example, Ian Joughin of the University of Washington, Seattle, and his team note that the grounding line for Thwaites Glacier is now held in check by a 600-meter (1,970 foot)-deep sill, with only some tens of kilometers separating it from a deep interior basin.[39] Although occurring slowly at present, "an early-stage collapse" may have already started, and would accelerate once the grounding line reaches the deeper sections. This could occur in as little as a few centuries if the shelf base continues to melt at a high rate. Furthermore, as noted above, the Thwaites Glacier overlies a geothermal hot spot, which could also melt ice deep under the ice sheet. Meanwhile, the grounding line of the Pine Island Glacier has already retreated some tens of kilometers and now enters a region with a reverse slope, opening the door to its possible irreversible retreat far inland.[40]

Among Antarctic glaciers surveyed between 2010 and 2016, 22 percent of the grounding lines on the West Antarctic Ice Sheet retreated faster than 25 meters (82 feet) per year (59 percent in the Amundsen Sea sector alone). This compares with a 9.5 percent retreat rate on the Antarctic Peninsula, and only 3 percent on the East Antarctic Ice Sheet.[41]

Grounding lines have also retreated along the WAIS's Bellingshausen Sea coast over the last four decades, except for ice streams feeding one ice shelf.[42] Will such widespread pullbacks inevitably trigger a complete WAIS meltdown, provoking a 3.3-meter (10-foot) jump in sea level?

Some advanced ice sheet and ice-ocean computer models support just such a dire outcome. If present rates of melting beneath West Antarctic ice shelves continue for as little as 60 years, Johannes Feldmann and Andres Levermann of Potsdam University warn, not only would the Amundsen Sea sector eventually destabilize, but so would the entire marine-based West Antarctic Ice Sheet.[43] Deep inland basins connect Thwaites Glacier, in particular, to marine-based ice beneath the West Antarctic Ice Sheet.[44] Once the grounding line on Thwaites Glacier retreats beyond an "island" of relative stability (a 600-meter [1,970 feet] sill that holds it back), the door opens to a deeper, wider calving front that would precipitate rapid surging and thinning of the glacier. Because Thwaites connects to deep interior subglacial basins, this could conceivably "empty the weak underbelly of the West Antarctic Ice Sheet."[45] The grounding lines of nearby Pine Island Glacier and its neighbors are also receding. If these trends persist for centuries and glaciers keep thinning, the perturbations, in a gelid analogue of stream piracy,[46] would eventually spill over, propagate into adjacent catchments that feed into the presently stable Ronne-Filchner and Ross Ice Shelves, and march toward their grounding lines. After centuries or millennia, most of the WAIS would ultimately disappear.[47] Thirteen thousand years from now, the surviving remnants of the ice sheet would attain a new equilibrium state, having raised sea level by some 3 meters (10 feet).

Access of warm Circumpolar Deep Water (CDW) to the glacier grounding line will also affect how quickly the WAIS can destabilize. Recent improved mapping of the adjacent seafloor reveals more clearly defined topographic features that control entry of the warmest, deepest CDW layers to the ice shelves. While several troughs are shallower, narrower, or more topographically variable than previously suspected, others are instead deeper. The new data will greatly benefit modeling of ocean and ice sheet interactions.[48]

Ice Shelves in Trouble

> *It's melting from above and below and crumbling at the edges. Antarctica is in trouble. Its frozen edges, or ice shelves, are disappearing into the ocean faster than we thought . . . and the process is accelerating. The most rapidly melting ones are likely to be gone within a century.*
>
> *As this happens, the ice sheets sitting over Antarctica's land—which hold the equivalent of 60 meters of sea level—will speed up their descent into the ocean, causing it to rise globally.*
>
> —Michael Slezak, *New Scientist* (April 4, 2015)

The previous section underscores the critical keys to the stability of the WAIS held by ice shelves. Ice shelves act as a "safety band" constraining the motion of ice streams as they approach the sea. Once ice shelves weaken or break apart, their ability to restrain or hold back the advancing ice diminishes. This opens the door for glaciers and ice streams to quicken their pace across the grounding line and disgorge more icebergs. Furthermore, as indicated above, the reverse slopes underlying many WAIS ice streams and glaciers near their grounding lines set the stage for a runaway marine instability, with serious repercussions for future sea level rise.

Although still intact, West Antarctic ice shelves, especially those lining the Amundsen and Bellingshausen Sea coasts, have thinned and melted within the last 10–15 years.[49] In particular, ice shelves in the Amundsen Sea Embayment, fed by glaciers including the aforementioned Pine Island, Thwaites, and Smith Glaciers, have been melting faster than elsewhere. The ocean has gained 282 to 310 billion metric tons of water per year over the past two decades by eating away at ice shelves.

Deteriorating Ice Shelves

Ice shelves expand by ice crossing the grounding line, snowfall, and basal freezing of sea water. Conversely, they shrink by calving icebergs, by melting from top or bottom, and by wind stripping surface snow. Melting from below now accounts for over half of Antarctica's ice attrition.[50] Within the past two decades, favorable winds have steered somewhat milder water into a number of submarine troughs that traverse the continental shelf. This water, a modified form of Circumpolar Deep Water, circulates around the continent at intermediate depths at a few degrees above the freezing point.[51] The water has warmed by 0.1° to 0.3°C (0.2–0.5°F) per decade under the Amundsen Sea and Bellingshausen Sea ice shelves.[52] There, westerly winds induce upwelling and steer warmer CDW beneath the ice shelves, causing greater basal melting there than elsewhere. In other regions, such as off the Ross and Ronne-Filchner Ice Shelves, strong easterlies instead force CDW down to greater depths, which limits warm-water incursions beneath ice shelves.

However, other forces also attack ice shelves. Flexing of the ice shelf at its ocean edge by waves or tides stresses the ice, which initiates multiple small surface cracks. Water draining summer melt pools enters crack tips under pressure and eats deep into the ice, gouging out the cracks and crevasses (hydrofracturing).[53] Seawater trapped in firn pores furnish additional inputs of water. Bottom-reaching cracks can then calve off iceberg slabs.

Ice cliffs may tumble into the sea once an ice shelf disintegrates. Cliffs fail as deep crevasses open up near the grounding line or thinned ice shelves reduce buttressing. Crevasses also develop in other zones of high shear stresses, for example, where the ice shelf rubs against valley or bay walls, or against slower-moving ice. Thinner ice shelves and faster-moving tributary glaciers in such marginal zones may tear cracks in the ice that expand into full-fledged rifts, whose presence may precondition the ice shelf to further rifting and breakup in a positive feedback loop. The growing frequency of such events in a warming climate will hasten ice sheet shrinkage.[54]

A different kind of rift, this one of geologic origin, may possibly also contribute to a pullback of the West Antarctic Ice Sheet one day. Such rifts

BOX 6.1: BURIED RIFTS AND RETREATING ICE SHELVES

Ground-penetrating radar that "sees" under the ice discovered a narrow basin underlying the Ferrigno Ice Stream in the Bellingshausen Sea sector of the WAIS.* This depression belongs to a major fault system that runs from the interior of the ice sheet to the sea. The rift further intersects a glacially eroded submarine trough at an angle offshore, which probably originated along a preexisting tectonic lineament. The direction of ice flow and, more importantly, that of ice thinning matches that of the proposed rift basin. A similar rift system also underlies the Pine Island Glacier. Warmed ocean water thus gains convenient access into submarine troughs on the continental shelf and thence into the onshore subglacial rifts, where additional ice erosion occurs. These pathways of geologic origin, buried under ice, could therefore exacerbate any potential marine ice instability.

* Bingham et al. (2012).

in solid rock, now buried under the ice, were created by tectonic plate motions. These are described further in box 6.1.

"Upside-down rivers," or channels etched into the base of the shelves' ice by warm ocean water currents, present another threat to the ice shelves' survival.[55] Satellite laser altimeters detect these ice-bottom channels by the shallow, winding depressions they leave on the surface of the ice shelf. (Because ice is thinner over a basal channel, it floats lower than adjacent thicker, channel-free ice. The surface depressions are therefore mirror images of the water-etched channels below.) Not too surprisingly, their abundance peaks on West Antarctic ice shelves, where CDW incursions have intensified basal melting (see "Deteriorating Ice Shelves," earlier in this chapter). The vast majority of these channels form by intrusion of warmer ocean water, including channels that start at grounding lines but otherwise bear no obvious relation to outflow of subglacial meltwater.

Satellite images illustrate a number of cases where ice-bottom channels have focused crevassing and even caused splits in the ice. The formation of multiple channels can therefore weaken the ice structure, adding yet another link to the cascading chain of events threatening ice shelves: warm ocean water incursions under ice shelves, increased basal melting, etching of subice channels, ice shelf thinning/breakup, escalating ice discharge and calving rates, crumbling ice cliffs, ice sheet retreat, and accelerated sea level rise.

This simple picture of ice shelf wastage is clouded by the myriad ways in which meltwater can reach the sea. Aerial photos and satellite imagery from 1954 to 2014 show meltwater exiting the ice sheet and coalescing into streams that drain across the ice shelves, often terminating in cascading waterfalls along the edges of the shelves.[56] By efficiently removing surface meltwater, these streams prevent pooling of water into meltwater lakes and help buffer the ice shelves against damage by hydrofracturing. On the one hand, steeper slopes, low-albedo rock, and bare ice exposure favor stream development and shelf stability, while on the other hand, a flatter topography and thick snow cover promote surface ponding that would lead to shelf instability. Which of these opposing processes would dominate in the future remains highly uncertain at this point.

In contrast to West Antarctica, where warming ocean water has undermined ice shelves, the higher summer air temperatures and overall milder Antarctic Peninsula climate have spurred the retreat of a number of ice shelves through proliferation of surface meltwater ponds, as well as hydrofracturing (chap. 2, and "Aprons of Ice" and "Deteriorating Ice Shelves," earlier in this chapter). While atmospheric influences are currently less important elsewhere in Antarctica, this may change in the future as global temperatures rise.

Releasing the Floodgates

Does the rapid surging of tributary glaciers on the Antarctic Peninsula after collapse of their ice shelves foreshadow analogous behavior on the much larger West Antarctic Ice Sheet (WAIS)? Mounting evidence links WAIS ice shelf thinning to retreat of grounding lines and speedup of ice streams or glaciers. After analyzing recent Antarctic ice sheet losses, H. D. Pritchard of the British Antarctic Survey and his colleagues put it emphatically and succinctly: "Ocean-driven ice-shelf thinning is in all cases coupled with dynamic thinning of grounded tributary glaciers . . . and the majority of Antarctic ice-sheet mass loss . . . It is reduced buttressing from the thinning ice shelves that is driving glacier acceleration and dynamic thinning. This implies that the most profound contemporary changes to the ice sheets and their contribution to sea level rise can be attributed to ocean thermal forcing that is sustained over decades and may already have triggered a period of unstable glacier retreat."[57] Other recent studies corroborate these conclusions. The fastest-changing ice shelves—Getz, Dotson, Thwaites, Pine Island, Abbot, and Venable in the Amundsen Sea and Bellingshausen Sea sectors of the WAIS—are the very shelves attached to glaciers with the fastest-retreating grounding lines that unleash the most ice: e.g., Kohler, Smith, Haynes, Thwaites, and Pine Island Glaciers.[58] As Pritchard and many others observe, warming ocean waters aided by favorable winds set the ball in motion. But the loss of ice shelves, abetted by the marine ice sheet instability, will keep the ball rolling.

Floating ice shelves are held in place by small islands, submerged obstacles, and friction of ice scraping against the sides of a fjord or embayment. These "pinning points" exert a pressure, or backstress, that resists the advance of glaciers or ice streams. But as thin or highly fractured ice shelves become "unpinned," they lose this buttressing support. The outcome is like releasing the floodgates of a barrage on a river.

Nevertheless, not all parts of an ice shelf exert equal buttressing capabilities. Calculating the forces that determine the stability of an ice shelf, a group of researchers led by Johannes Jacob Fürst of the Centre National de la Recherche Scientifique (CNRS) in Grenoble, France, found that only 13 percent of the ice shelf area around Antarctica constitutes a "safety band," or, in other words, comprises "passive ice shelves" that will have a minimal effect on ice velocity farther upstream if lost.[59] Once this safety band disappears, whether through ice shelf thinning or calving, the surviving ice shelves becomes even more vulnerable to processes that undermine buttressing. The most vulnerable ice shelves—those with the smallest safety bands—lie in the Amundsen and Bellingshausen Sea sectors, which also sustain the highest shelf-thinning rates. Many of these ice streams overlie sunken subglacial basins or valleys that plunge deep into the heart of the West Antarctic Ice Sheet. This potential marine ice sheet instability setup could therefore lay the groundwork for the deicing of much of the WAIS.

But will the loss of ice shelves inevitably lead to irreversible marine ice sheet instability? Are such alarmist visions of an imminent or long-term WAIS meltdown premature? The next section looks into some counteracting processes that could slow down or even neutralize such a dire outcome.

Maybe Not So Fast!

An ice stream resting on bedrock that slopes inland (i.e., a MISI setup) may not always be unstable. Its stability depends on multiple factors, such as how readily ice flows over the evolving surface characteristics of the floor, variations in bedrock slope or ice stream width, ice accumulation rates, and strength of ice shelf buttressing at the grounding line. The grounding line may survive on a reverse slope for a while under favorable conditions. For example, a narrow subglacial trough slows down retreat, because first, the ice stream experiences greater drag along the sides of the pinched-in pathway, and second, advancing ice from the hinterlands piles up, thickens, and steepens the surface slope, which may temporarily stall retreat, assuming rates of ice thinning or sea level rise do not change.[60]

Furthermore, ice shelf melting involves a complex web of reactions that affect the grounding line position and ice stream flow. In spite of advances in modeling of ice shelf and grounding line behavior, it is still difficult to predict outcomes of multiple interactions that stabilize or destabilize a reverse slope, or whether loss of an ice shelf would inevitably trigger a marine instability.[61]

Two important negative feedbacks, generally overlooked in extreme scenarios of a runaway ice meltdown, may also rein in the rise of global sea level. The first feedback involves adjustments of the Earth's lithosphere to a diminished ice load and weakened gravity.[62] The Earth's lithosphere rebounds after an ice burden disappears (called "glacial isostatic rebound"), thereby lowering local sea level. Glacial rebound acts partly in an instantaneous, elastic manner and partly in a much slower, sluggish, or viscous fashion, lasting centuries to millennia. Because West Antarctica overlies a less viscous (i.e., more fluid) region of the mantle, ice losses result in a quicker, higher, and more localized uplift associated with a deeper fall in sea level near the grounding line, which could impede further ice recession.

The gravitational feedback causes water to flow away from the shrinking ice sheet due to its weakening gravitational pull, which further cuts back local sea level rise. A fall rather than rise in sea level near the grounding line hinders the two main driving forces that lead to collapse of the ice sheet. The drop in sea level prevents warmer CDW from reaching the base of ice shelves, thus impeding bottom thawing and fending off ice shelf breakup. Second, the grounding line would migrate to a higher elevation as the land uplifts and local sea level drops, thus reversing the march toward a marine ice sheet instability. Moreover, the once-inland-slanting slope would flatten, and even slant seaward due to glacial rebound. Computer models that include both feedbacks acting in concert find that these offsetting feedbacks significantly cut anticipated Antarctic ice losses, as compared with losses without these offsets.[63] These offsets become most effective for modest atmospheric carbon dioxide increases and assumption of a less viscous or rigid upper mantle, like that of West Antarctica. Very high CO_2 levels will probably overwhelm the effectiveness of these two geophysical feedbacks. Nonetheless, their existence teaches us to treat claims of an imminent onset of a marine ice sheet instability with a note of caution.

THE EAST ANTARCTIC ICE SHEET (EAIS)

Does only the West Antarctic Ice Sheet lie on shaky ground? How stable is the East Antarctic Ice Sheet? Far more massive than the West Antarctic Ice Sheet, the East Antarctic Ice Sheet holds a mass of ice equivalent to a sea level rise of 52 meters, versus 5 meters (171 versus 16 feet) for the WAIS.[64] Unlike its western neighbor, gains in EAIS snow and ice accumulation still exceed losses from ice shelf melting or iceberg calving. Nevertheless, a few fringing ice shelves have thinned between 2003 and 2008, notably Shackleton, Amery, and Totten Ice Shelves, as well as the Wilkes Ice Shelf.[65]

What potential exists for a marine ice sheet instability in East Antarctica? Bedmap2, which mapped the subglacial surface of Antarctica in unprecedented detail (see "Hidden Mountains and Volcanoes," earlier in the chapter), discovered lengthy subglacial rifts, large segments of which are grounded far below sea level. More recent airborne surveys over the East Antarctic Ice Sheet with radar and gravitational and magnetic sensors have detected a subglacial, 1,100-kilometer-long canyon—the longest in the world and nearly as deep as the Grand Canyon.[66] These subglacial characteristics may influence the future strength of large portions of the EAIS.

Like many ice shelves in West Antarctica, the Totten and Moscow University Ice Shelves in the Wilkes region of the East Antarctic Ice Sheet abut a large, marine-based portion of the ice sheet. Totten Glacier in East Antarctica drains a large catchment—the Aurora Subglacial Basin—which holds enough ice to raise global sea level by 3.5 meters (11.5 feet) if melted—comparable to the amount for the WAIS.[67] A fifth of the Totten Glacier catchment lies well below sea level, covering a much vaster area than previously suspected. The Totten Glacier Ice Shelf has thinned more than other East Antarctic ice shelves, bathed by mild, modified Circumpolar Deep Water that enters cavities under the ice.[68] A recently discovered subglacial channel appears to slice through a ridge between the main Totten Glacier Ice Shelf cavity and an adjacent cavity.[69] This passageway could convey warmer water into cavities beneath the Totten Glacier Ice Shelf, the nearby Moscow University Ice Shelf, and Reynolds Trough, facilitating an eventual ice shelf pullback over a broad region.

The Totten Glacier grounding line further retreated 1–3 kilometers (0.6–1.9 miles) between 1996 and 2013.[70] Sections of the grounded edge of the ice shelf show topographic features that favor a rapid retreat, but rises in slope farther upstream would hinder subsequent regression. While no specific conduits to rapid upstream retreat were found, ruling out any immediate threat of marine instability, the Totten Glacier does share several features with West Antarctic ice shelves. The Totten Glacier Ice Shelf catchment drains an ice volume comparable to that of the WAIS. A significant portion of the basin is grounded below sea level, and yet-to-be-discovered passageways may exist that allow warm ocean water access into deep, downward-sloping cavities toward the interior of the ice sheet. The large sea level rise potential and potential MISI setup underscore the need for more-detailed subglacial bathymetric data for this area.

The Wilkes Basin, located west of the Transantarctic Mountains, holds the largest volume of marine ice on the East Antarctic Ice Sheet that overlies subglacial troughs. The Wilkes Basin, although still stable, is drained by the Cook and Ninnis Ice Streams, with a bathymetry similar to that of the Totten and Moscow University Ice Shelves. What would happen if warm water could enter the ice shelf's underside? Two German researchers devised a computer model simulation to find out.[71] In their simulation, warm water pulses of 1°–2.5°C (34°–36.5°F) invaded ice shelf cavities over periods lasting between 200 and 800 years. A small ice mass called the "ice plug" controls the stability of the marine-based ice. Once this plug melts beyond a threshold of a more than 80-millimeter sea level equivalent, the entire basin begins to destabilize.

The self-sustaining retreat eventually causes sea level to rise 3–4 meters (10–13 feet) over 10,000 years. But the instability sets in only if the grounding line can reach the deep inland steepening canyons beyond the ice plug. This topographic setup closely resembles that of Pine Island Glacier, which may already be experiencing unstable retreat (see "The West Antarctic Ice Sheet (WAIS)—On Shaky Ground," earlier in this chapter). While these finds raise tantalizing questions, the exact pathways of warmer water into subglacial basins await further clarification.

A LOOK BACK AND AHEAD

The foregoing shows that even Antarctica, the coldest continent on Earth, may no longer be immune to the forces that undermine the cryosphere elsewhere. Although we may not yet fully understand all relevant processes, an emerging picture suggests that parts of Antarctica may be susceptible to the great thaw also gripping the Arctic and the high mountains. But a much longer record, one that stretches far beyond the historic period of instrumental records, gives us a better long-range context with which to judge the true significance of the observed twentieth- and early twenty-first-century ice attrition. By peering into the distant past, we gain a deeper understanding of the processes governing ice sheet growth or decay. This knowledge strengthens our ability to interpret recent changes, anticipate plausible scenarios of ice sheet instabilities, and improve predictions of future sea level rise. The next chapter looks back at the long journey of ice buildup in Antarctica and, more recently, in Greenland. Major meltdowns also punctuated the road to an ice-house world. What can these past episodes of large-scale, rapid ice sheet meltdowns teach us that would shed light on the future of the two major ice sheets? How likely is it that such a rapid thaw will occur within the next century or two? What can our reading of the past record and computer models foretell about the future? After answering these important questions to the extent possible given our present understanding, in the final chapter we explore the human consequences of the oncoming deglaciation and transition to a more aquatic existence.

7

FROM GREENHOUSE TO ICEHOUSE

A VANISHED GREENHOUSE WORLD

Antarctica was once covered with tropical forests . . . The . . . story starts in much warmer climes . . . in the early Eocene. Back then, the climate was subtropical, the verdant landscape dominated by palms and trees such as the monkey puzzle. By the early Oligocene, around 31 to 33 million years ago, the palms and monkey puzzles had disappeared. They gave way to more temperate species, including Huon pines . . . living fossils that still thrive in New Zealand and Tasmania. For trees, the transition from the Oligocene to the Miocene 23 million years back was the beginning of the end . . . But the end for all greenery came around 12 million years ago, when even the tundra disappeared.

—Andy Coghlan, *New Scientist* (2016)

Alligators, primitive hippopotami, and primates once roamed a lush, subtropical, swampy, forested, ice-free Arctic during the early Eocene Epoch, 56 to 50 million years ago (appendix C). Near the poles, surface oceans were a mild 20°C (68°F), tropical surface waters a sultry 35°C (95°F).[1] Atmospheric CO_2 soared to two to four times the pre–Industrial Revolution level of 280 parts per million (ppm).[2] In the absence of ice-covered poles, sea level climbed up to 70 meters (230 feet) above that of the present.

Since that climax of Cenozoic warmth, the Earth's climate has slowly but inexorably descended into an "icehouse" world, beginning with a fairly abrupt cooling event around 34 million years ago. This signaled the first significant ice accumulation in Antarctica. Nevertheless, small, isolated glacier and ice fields may have graced the southernmost continent even earlier. The icehouse world has lasted until the present, but that may be beginning to change, as the next chapter will show. Some day we may even return full cycle back to the hothouse Eocene. But before contemplating such a return, we first explore the origins of the icehouse climate that produced the cryosphere.

DESCENT INTO THE ICEHOUSE WORLD

Birth of the Antarctic Ice Sheet

A 34-million-year event halfway through the Cenozoic Era marks a major climate transition—the initial entry into an icehouse world. Atmospheric CO_2 levels declined below critical thresholds of roughly 750 to 600 ppm.[3] Large ice masses built up in East Antarctica, precipitating a 50- to 60-meter (164- to 197-foot) drop in sea level, while the surrounding ocean cooled by several degrees.[4] But even at this early stage, the nascent ice sheet demonstrated signs of mobility. Cores drilled into sediments beneath the ice cover on the Ross Ice Shelf record multiple advances and retreats of the West Antarctic Ice Sheet (WAIS) between 23 and 14 million years ago. Atmospheric CO_2 levels fluctuated between around 280 ppm during cold periods and more than 500 ppm during warm ones, within this time interval.[5] The climate ameliorated briefly around 23 million years ago and again 17 to 14 million years ago.

Between 34 and 23 million years ago a lowland cold temperate forest and shrub vegetation similar to modern Tasmania and New Zealand still covered the Wilkes Land margin of East Antarctica. At higher elevations, southern beech and conifers flourished in a cold tundra shrubland.[6] The woody sub-Antarctic vegetation occupied a partially ice-free environment until 14 million years ago. Further cooling pushed the surviving vegetation to lower elevations until only tundra survived. After 14 million years ago, glaciers overrode the last remnant plants and turned Antarctica into a white desert. Once frigid polar conditions set in, Antarctica remained locked in the icehouse, even when climates ameliorated elsewhere.

Burial of the Gamburtsev Mountains

The Gamburtsev mountains are probably older than 34 million years and were the main centre for ice-sheet growth. Moreover, the landscape has most probably been preserved beneath the present ice sheet for around 14 million years.

—Sun Bo, *Nature* (2009)

The Gamburtsev Mountains, a range 2,400 meters (7,900 feet) in elevation at the center of the present East Antarctic Ice Sheet, lie entombed beneath 1,600–3,000-meter (5,250–9,800-foot)-thick ice.[7] Discovered in the 1950s, their true nature only became apparent decades later, after ice-penetrating radar and other modern probes exposed their hidden traits (see also chap. 6). The well-preserved mountains still retain topographic features inherited from a period before the ice came, while other features represent progressive stages of the ice buildup. Deep valleys form a dendritic network like that etched by rivers in mountainous terrain. Glaciers later took advantage of these preexisting troughs. They also sculpted cirques high up near the peaks and produced hanging valleys that overlooked the main valleys, forming a landscape not unlike that of the European Alps (see fig. 4.2). The ice gradually built up in three main stages. Initially, ice only occupied cirques—hollows along the mountain crests. But as ice continued to pile up, glaciers spilled into the valleys and gouged their flanks and floors into their characteristic U-shape. Over tens of millions of years, more and more ice accumulated until it completely buried the mountains.

These stages mirror the growing ice buildup in Antarctica. Andy Coghlan, writing in *New Scientist*, depicted an ice-free Antarctica during the peak Eocene greenhouse, in which palm trees, araucarian conifers,[8] and beech forest thrived in a subtropical coastal climate, while temperate forests occupied higher elevations. As Antarctica chilled, Alpine-style cirques and mountain glaciers gradually appeared at the highest elevations. By 34 million years ago, temperatures had cooled enough for glaciers to expand and coalescence into the first massive ice sheet on East Antarctica. Nevertheless, summer temperatures still remained above freezing. Beyond a critical threshold, 14 million years ago, when hyper-cold conditions set in, the ice sheet expanded to near-present dimensions, frozen to its bed. This blocked significant glacial erosion, which helped to preserve the preexisting topography.

Further Cooling

During the Pliocene Epoch, 5.3 to 3 million years ago, the WAIS expanded and contracted repeatedly.[9] These climate cycles, lasting roughly 40,000 years, marched in step with the changing tilt of the Earth's rotation axis—the obliquity cycle (see box 7.1). Sediments were deposited in the Ross Sea of Antarctica during warmer periods of open water with little or no sea ice. Some accumulated later beneath an ice shelf, as happens today. Sediments were absent when an advancing ice sheet rested directly on the seafloor. However, the region enjoyed a prolonged period of benign climate and ice-free open water between 3.6 and 3.4 million years ago, when the WAIS probably deglaciated completely.

BOX 7.1: THE ASTRONOMICAL THEORY OF THE ICE AGES

Milutin Milankovitch (1879–1958), a Serbian geophysicist and mathematician, proposed that an ice age starts when high-latitude summers remain cold enough to prevent the previous winter's snowfall from melting completely. The tilt of the Earth's axis (the *obliquity*)—the angle between the rotational axis and the perpendicular to the plane of the orbit (the *ecliptic*)—controls the amount of solar energy that reaches the top of the atmosphere. Currently 23.5°, this angle varies between 22°and 24.5° over a 41,000-year cycle. The greater the tilt when the Earth's axis points more directly at the sun, the more intense the rays that bathe high latitudes in summer and the more snow and ice melt. Once water freezes in winter, colder temperatures no longer matter. Spring and summer thaw is more important. Conversely, a smaller tilt angle produces cooler summers and milder winters, though not mild enough to melt snow or ice. Thus, onset of glaciation is favored by a low tilt angle combined with maximum distance from the sun at the Northern Hemisphere's summer solstice (i.e., cooler northern summers). The obliquity effect is strongest at higher latitudes, and therefore the amount

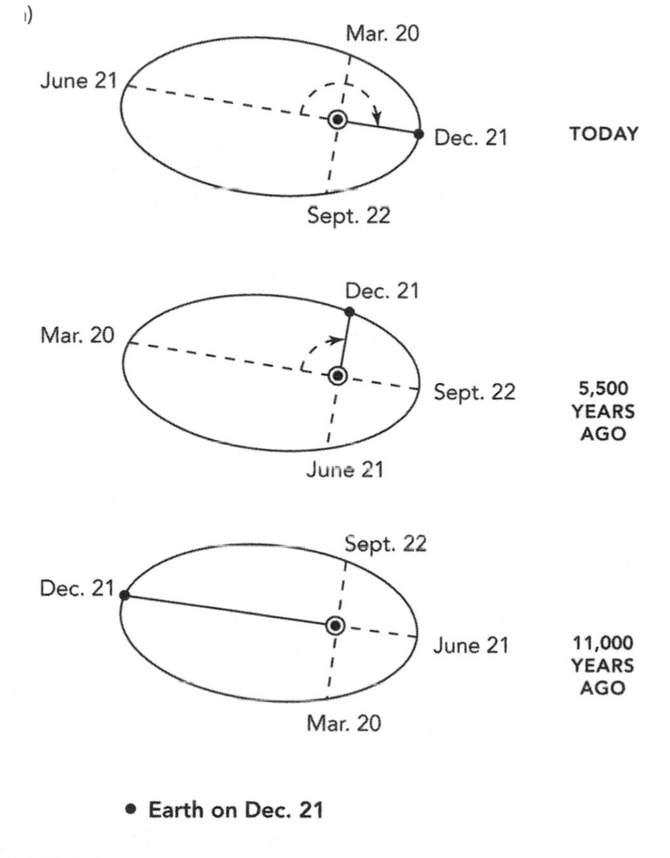

BOX FIGURE 7.1A

Precession of the equinoxes. *Top*: Perihelion (closest approach to the sun) occurs today near the Northern Hemisphere winter solstice. *Bottom*: Perihelion 11,000 years ago occurred at the summer solstice. (After John Imbrie; fig. 4.6 in Gornitz [2013])

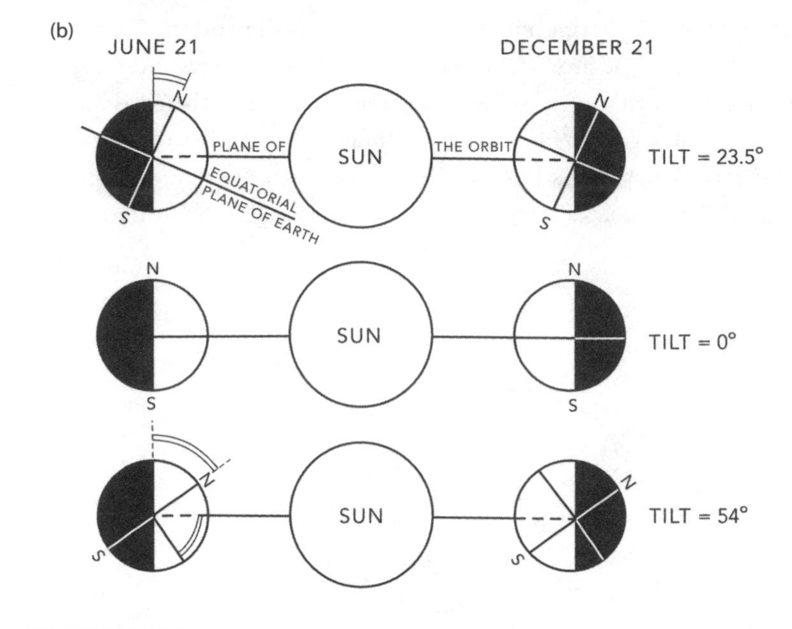

BOX FIGURE 7.1B

Obliquity (tilt of axis) and latitudinal distribution of sunlight. *Top*: Today (tilt angle 23.0°). *Middle*: Near-zero tilt: Even latitudinal distribution favors an ice age. *Bottom*: High tilt angle favors a deglaciation. (After John Imbrie; fig. 4.7 in Gornitz [2013])

of solar radiation at 65°N is considered critical to the start (or end) of a glacial cycle. *Precession of the equinoxes* combines two elements: (1) axial precession—a wobble like that of a spinning top, caused by the gravitational pull of the moon on the Earth's equatorial bulge, which repeats every 26,000 years; and, more importantly for past climates, (2) elliptical precession, which determines the season when Earth is closest or farthest from the sun (*perihelion* or *aphelion*, respectively). This varies over a 22,000-year cycle. *Eccentricity* (100,000-year variations in the circularity of the Earth's orbit around the sun) produces only minor changes in the total solar radiation input. However, a greater orbital elongation combined with favorable alignment of obliquity and precession may nudge the Earth enough to push it into or out of an ice age. The recipe for an ice age is outlined schematically in the following sections.

Recipe for Starting an Ice Age

Low Northern Hemisphere tilt angle (obliquity) + aphelion at June 21, summer solstice (precession) + highly elongated orbit (greatest eccentricity)

Recipe for Ending an Ice Age

High Northern Hemisphere tilt angle + perihelion at June 21 + highly elongated orbit

Initially greeted with great skepticism, the astronomical theory—the pacemaker of the ice ages—received a tremendous boost after the discovery of quasi-cyclical peaks at 100,000, 42,000, and 23,000 years in oxygen isotope ratios of tiny marine fossils. These periods correspond to the eccentricity, obliquity, and precession cycles, respectively. But the dominance of the 100,000-year eccentricity cycle, which should have been the weakest, remains puzzling. Clearly, other feedback processes must amplify the orbital signals; among these are the ice-albedo feedback and the greenhouse gases carbon dioxide (CO_2) and methane (CH_4), which vary in sync with the ice ages.

This period of peak Pliocene warmth offered fleeting memories of a fading greenhouse world. Atmospheric CO_2 levels were comparable to today's (~400 parts per million), while global temperatures were around 2–3°C (3.6–5.4°F) higher. Sea level during warm intervals could have climbed up to 20 to 25 meters (66–82 feet) above present levels.[10] However, the upper estimates remain uncertain because of as-yet-unresolved issues concerning regional land uplift, glacial isostatic adjustments, and gravitational attraction.[11]

The subsequent cooling and onset of continental-scale Northern Hemisphere glaciation around 3 million years ago set the stage for a full icehouse world. The latter part of the succeeding Quaternary Period—the last 2.6 million years of Earth history—featured the periodic waxing and waning of both polar ice sheets.

Looking Closer

Some temperature changes have occurred without carbon dioxide changes, and some carbon dioxide changes have occurred without temperature changes because other factors were more important. But carbon dioxide has been important, and there is no good way to explain the ice-age cycles without appealing to the importance of the carbon dioxide.

Richard Alley, *The Two-Mile Time Machine* (2000)

Over the 65-million-year-long Cenozoic Era, the slow breakups and collisions of tectonic plates have shaped the Earth's protracted descent into an icehouse world. Some key milestones include the breakup of Pangaea, the collision of the Indian plate with Eurasia, the rise of the Himalayas and Tibetan Plateau, the closure of the Tethys Ocean, and the opening and closing of critical ocean "gateways." These latter tectonic events reconfigured ocean circulation patterns and enabled ocean currents to transport heat and moisture to the poles. Falling atmospheric CO_2 levels also intensified the global cooldown (see "The Icehouse World," later in this chapter).

Starting 180 million years ago, the supercontinent Pangaea began to break apart, opening the North Atlantic Ocean between southeastern North America and West Africa (fig. 7.1). This was followed by the opening of the South Atlantic around 140 million years ago, and the separation of Africa, Australia, Antarctica, and India. India slowly drifted north toward Eurasia, with which it collided around 50 million years ago. Antarctica moved toward the South Pole. The Tethys Sea that once had separated Africa from Eurasia began to close as several Southern Hemisphere plates drifted gradually northward. The collision of the Indian plate with Eurasia led to the uplift of the Himalayas and the Tibetan Plateau.

This massive tectonic reorganization not only altered the topography, but induced major changes in the world's climate and ocean circulation. During the early Cenozoic greenhouse, the Earth basked under high temperatures and elevated atmospheric CO_2 levels, but later, the planet cooled as CO_2 levels began to drop. How? The rise of the Tibetan Plateau (close to 5 kilometers, or 16,000 feet) intensified the summer monsoons.[12]

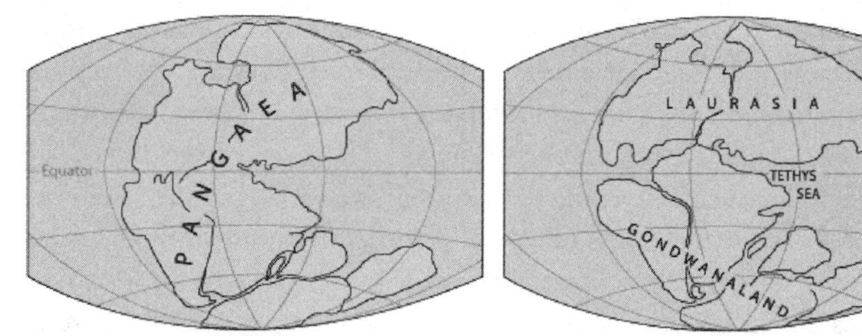

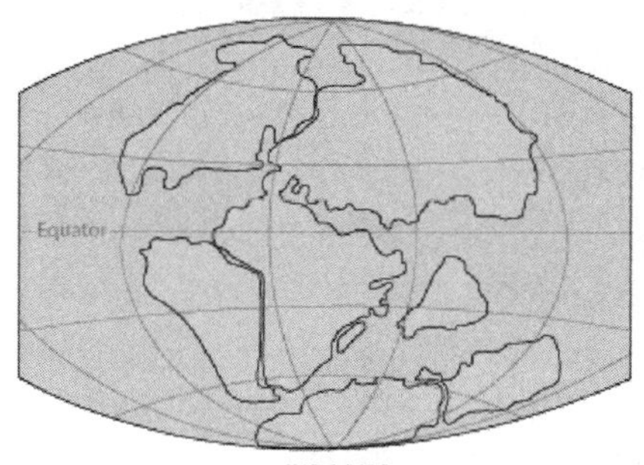

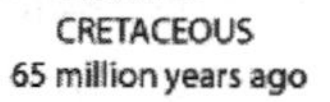

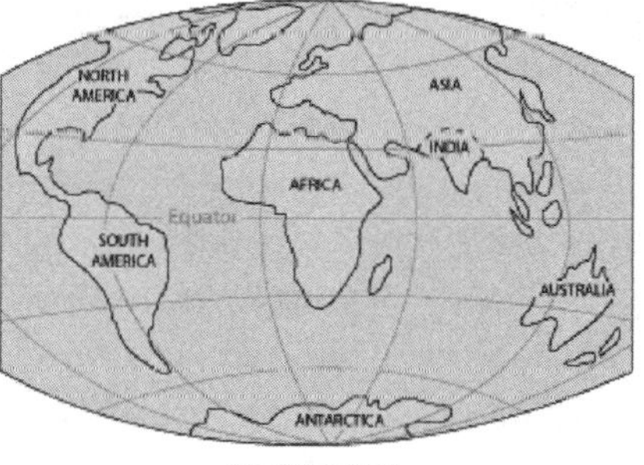

FIGURE 7.1

Breakup of supercontinent Pangaea and the changing positions of the tectonic plates. (USGS)

Nevertheless, a stronger Asian monsoon may not have sufficed to jump-start the growth of large ice sheets in both hemispheres. Marine geologists and paleoclimatologists Maureen Raymo and William Ruddiman therefore proposed that mountain building and faulting delivers fresh silicate rocks to the surface and exposes them to weathering by atmospheric CO_2.[13] Carbon dioxide reacts chemically with silicate minerals in rocks and transforms them into carbonates and silica. This chemical weathering sequesters the greenhouse gas CO_2, thereby cooling the Earth, causing expansion of ice, and lowering sea level. Carbon-rich organic matter in delta, estuary, and continental shelf sediments thus faces exposure to erosion and weathering. Oxidization of the organic matter regenerates CO_2, which offsets some of the cooling. Furthermore, the same tectonic forces that pushed up the Himalayas also create volcanoes, which release fresh CO_2 into the atmosphere. This replenishes some of the CO_2 consumed by weathering. Thus, volcanic activity saves the Earth from freezing. Climate plays a delicate balancing act between these opposing forces.

Geoscientists Dennis Kent and G. Muttoni believe alternatively that the collision of the Indian plate with Eurasia around 50 million years ago brought a vast expanse of 65-million–year-old basaltic lavas—the Deccan traps sitting atop the Indian plate—to the hot, humid tropical zone. There, intense tropical weathering removed atmospheric CO_2, as described in the previous paragraph.[14] The "silicate weathering machine" was in now high gear! This atmospheric CO_2 drawdown was reinforced by eruption of Ethiopian basaltic lavas 30 million years ago, after the Deccan traps had exited the tropical zone. Volcanically active Indonesia, Borneo, and New Guinea furnished additional basaltic rocks for chemical weathering. Either way, by weathering of the high Himalayas or of volcanic rocks, the atmospheric CO_2 levels dropped, the climate cooled, and the Earth continued its descent into the icehouse world.

The moving tectonic plates not only raised lofty mountain ranges, but opened and closed key ocean "gateways" that significantly altered ocean circulation with important climatic consequences. With the closure of Tethys and expansion of the Atlantic, ocean circulation gradually reoriented from a mainly east-west to a north-south direction. Although various dates have been proposed for the initial opening of the Drake Passage between Antarctica and South America in the south, a deep enough passageway had developed by 34–30 million years ago to allow currents to flow freely eastward.[15] The Tasmanian Passage between Antarctica and Australia also opened around 34 million years ago.[16]

These gateway changes may have sparked the development and strengthening of the Antarctic Circumpolar Current (ACC) (fig. 7.2; see also chap. 6). The formation of the powerful ACC may have effectively led to isolation of Antarctica from warmer waters and cooling of the Southern Ocean, which may have set the stage for the massive ice buildup in Antarctica. Much later, another gateway, the Central American Seaway that once separated North and South America, began to shoal around 4.6 million years ago, and finally closed within the last three million years. This further restricted east-west ocean circulation and intensified the Gulf Stream, which by delivering more warmth and moisture to high latitudes would have fed the growth of Arctic ice sheets.[17]

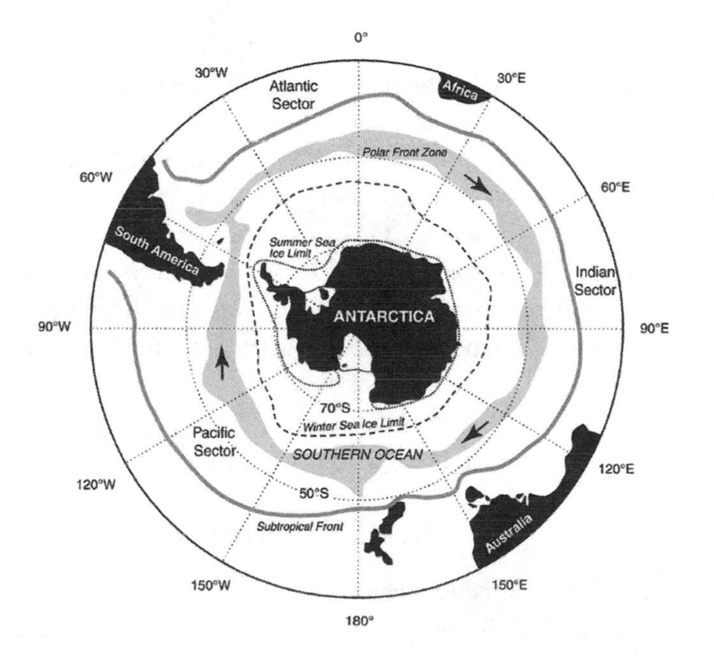

FIGURE 7.2

The Antarctic Circumpolar Current. (After B. Diekmann [2007])

However, the poleward drive of heat and moisture may not have sufficed. The ACC may have intensified well after an initial Antarctic ice buildup. Computer simulations of past climate and ice sheet behavior further suggest that the ACC (and other gateways) strengthened along with the lowering of CO_2 levels, which increased cooling and glaciation.[18] Diatoms, a type of free-floating marine algae (see box 7.2), would also have proliferated, nourished by the upwelling of nutrients in a vigorous ACC. Photosynthesis by diatoms would have extracted CO_2 from the atmosphere, as would have burial of organic debris in deep-sea sediments.

Marked changes in CO_2, ice sheet volume, and sea level closely accompany major Cenozoic climate transitions.[19] During the torrid Eocene greenhouse, CO_2 levels reached beyond 1,000 ppm; sea level was 60 to 70 meters (197–262 feet) higher than the present level. Computer simulations show significant ice mass accumulations only when CO_2 drops below 750 to 600 ppm.[20] The models also suggest that enough CO_2 declined by the Oligocene, 34 million years ago, to grow an East Antarctic Ice Sheet and lower sea level by 50–60 meters (164–197 feet). During a brief climate respite between 17 and 14 million years ago, CO_2 surpassed the present-day value of ~400 ppm.[21] But by the mid- to late Pliocene, around 3 million years ago, CO_2 levels attained near-modern values, although temperatures still remained 3–4°C (5–7°F) higher globally, even 10°C (18°F) higher near the poles.[22] Sea level may have climbed over 20 meters (66 feet) above that of the present.[23] However, CO_2 had to plunge still lower—to near-preindustrial levels (~280 ppm)—before massive ice sheets appeared in the Northern Hemisphere by around 2.8 million years ago.[24] The development of major ice sheets at both poles marks the entry into a full icehouse world.

THE ICEHOUSE WORLD

When Greenland Turned White

Greenland really was green! However, it was millions of years ago. Greenland looked like the green Alaskan tundra, before it was covered by the second largest body of ice on Earth.

—Dylan Rood, Imperial College London

Naming "Greenland" may have been clever real estate promotion on the part of the Viking chieftain Erik the Red to lure fellow Norsemen to this barren land, but a time once existed when the name was quite apt, many millions of years ago. But the timing of onset and the size of the initial Greenland Ice Sheet remain enigmatic. As early as 38 or even 46 million years ago, a few glaciers, descending from mountain valleys, disgorged debris-laden icebergs into the ocean. By 3 million years ago, a small ice sheet probably encased portions of eastern Greenland.

Yet as recently as 2.7 million years ago, Summit, site of a permanent research station at the highest point near the center of the Greenland Ice Sheet, was still tundra. At that time, much of Greenland may have resembled northern Alaska, where a variegated carpet of low shrubs, dwarf willow and birch, sedges, mosses, lichens, grasses, and flowers splash a bright rainbow of colors across the tundra landscape during a brief two-month-long growing season (fig. 7.3).

At Summit, the bottom 13 meters (43 feet) of a 3,000-meter (10,000 foot)-long ice core is laden with silt, organic material, fossil microorganisms, trapped gases, and a water isotope composition that indicates

FIGURE 7.3

Tundra landscape, Neacola Mountains, near Big Valley, Alaska. (U.S. National Park Service/L. Wilcox)

once-warmer conditions. Paul Bierman, head of a team of U.S. geoscientists studying the silt, describes it as an "organic soil that has been frozen to the bottom of the ice for 2.7 million years."[25] The soil contains unusually high concentrations of radioactive ^{10}Be isotopes, created when cosmic rays interact with the atmosphere.[26] Bierman and associates speculate that the soil must have been exposed to cosmic rays for at least 200,000 years and perhaps over a million years—far longer than the longest interglacials of the last few million years. Since the soil obviously formed before its burial under ice, the abundant cosmogenic isotopes demonstrate a long-lived, stable Greenland Ice Sheet, at least at this elevated interior location. Once the ice engulfed central Greenland, it became ensconced there and has remained ever since.

Another group of glaciologists and geologists headed by Joerg Schaefer of the Lamont-Doherty Earth Observatory at Columbia University reached a quite different conclusion. They measured the cosmogenic ^{10}Be and ^{26}Al isotopes, this time in pieces of granitic bedrock from the base of the very same core investigated by Bierman's team. They, too, found a long-lasting ice-free period of 280,000 years followed by at least 1.1 million years of ice cover. But this is just one of several possible scenarios consistent with the data. A number of shorter ice-free episodes lasting several thousand years, presumably during major interglacials, also fit the observed data. Since the elevated interior location of the Summit ice core would be one of the last areas to melt in Greenland, leaving only small ice cap remnants, this implies that 90 percent of the ice sheet would have disappeared. Richard Alley, a glaciologist from Pennsylvania State University and a coauthor in the Schaefer study, does not claim that therefore "tomorrow Greenland falls into the ocean. But the message is, if we keep heating up the world like we're doing, we're committing to a lot of sea-level rise."[27]

Bierman and his colleagues more recently investigated the history of these cosmogenic isotopes in offshore marine sand grains eroded from eastern Greenland's interior by tidewater glaciers. They made note of a major ice sheet expansion 2.5 million years ago. But consistently low ^{10}Be concentrations over the last million years imply the persistence of a large, stable ice sheet in eastern Greenland through even the longest, most intense interglacials. They maintain that the new evidence reinforces their earlier conclusion of a stable, long-lived Greenland Ice Sheet. However, other evidence points to greater mobility, at least along the flanks of the ice sheet.

Spruce and pine forests evidently thrived in southern Greenland during especially warm interglacials, such as those that occurred around 400,000 and 125,000 years ago. The former interglacial stands out because of its prolonged, near-30,000-year duration (395,000 to 424,000 years ago). Abundant spruce and pine pollen grains recovered from offshore marine sediments point to widespread boreal (northern) forests and therefore a substantially smaller southern Greenland Ice Sheet.[28] Sea levels then were also higher than present. The 125,000-year interglacial also enjoyed milder climates and higher-than-present sea levels, as described in the following section.

The Ebb and Flow of the Ice Sheets

Even after Greenland turned white, the Earth continued to cool. A transitional period, roughly 1.2–0.9 million years ago, demarcated the onset of the major ice ages that have dominated the last million years. Much of what we know about these sweeping climatic oscillations comes from the heart of Antarctica, buried deep within the ice.

The ice core archives store a lengthy history of changing Earth climates. They tell a tale of at least eight major ice ages that punctuated the last million years (fig. 7.4; box 7.2). During each successive ice age peak, massive sheets of ice rolled across most of Canada into the northern United States and across Scandinavia, the British Isles, and northwestern Russia. A heavy mantle of ice blanketed all the lofty mountain ranges of the world. Each *glacial-interglacial cycle* lasted roughly 100,000 years, interspersed with shorter episodes of diminishing temperatures that hit rock bottom when ice sheets had expanded to their maximum extent, with a concomitant drop in sea level (note the sawtoothed-shaped curves in fig. 7.4). The cycle terminated abruptly, geologically speaking, with a rapid rise in temperature and sea level and fall in ice volume. Intervening thaws and high–sea level stands typically lasted only 10,000 to 30,000 years—just a small fraction of the entire glacial-interglacial cycle.

The repetitive, sawtoothed pattern of the ice cores is mirrored in ocean sediments. Oxygen isotope ratios of marine foraminifera (box 7.2) tell a tale of cyclically varying ocean temperatures, ice volume, and sea level. The asymmetric curves imply that the entry into each ice age—the ocean

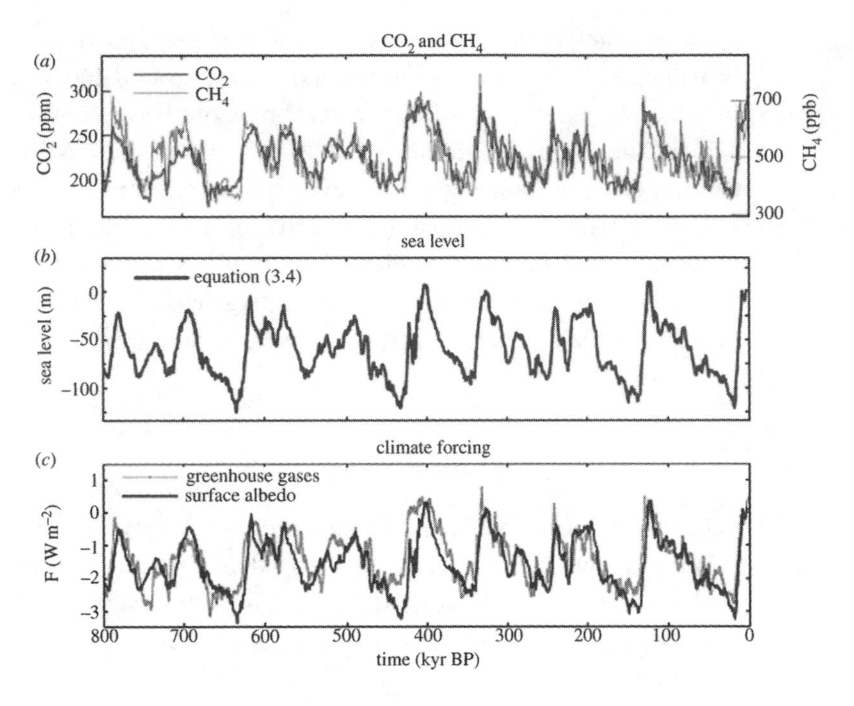

FIGURE 7.4

Sea level, temperature, and CO_2 over the last eight glacial-interglacial cycles. Note the relatively rapid rise in climate warming, sea level, and greenhouse gases following an ice age (*top*) as compared to the more episodic, gradual return to full ice age conditions (*bottom*), creating a sawtoothed-shape curve. (Hansen et al. 2013)

cooling, ice sheet buildup, and associated fall in sea level—was much slower than the exit, characterized by rapidly rising temperatures and sea level.

These major ice sheet and sea level oscillations also match changes in global temperature and greenhouse gases remarkably well.[29] The close associations illustrate the ability of CO_2 to amplify the weak 100,000-year eccentricity signal into one of recurrent glaciations or deglaciations. Nevertheless, as the weakest of the three major orbital cycles (box 7.1), the predominance of the 100,000-year cycle within the last million years and the rapidity of the retreats have long remained puzzling—until recently.

Prior to the mid-Pleistocene transitional period, the ebb and flow of glaciers had marched largely to the beat of the 41,000-year obliquity cycle, with weaker precessional influence. Signs of 41,000-year cycles preserved in the oxygen isotope ratios of tiny deep-sea fossils correspond to alternating colder and warmer periods (box 7.2). Unexplained is how these late Pliocene–early Quaternary obliquity-dominated cycles transformed into a 100,000-year, eccentricity-driven beat within the last million or so years. These cycles grew progressively longer and stronger by 900,000 to 650,000 years ago. Cold periods got even colder, suggesting that Northern Hemisphere ice sheets, in particular, thickened and spread out farther south than before.

A possible solution to the mystery, according to one hypothesis, lies in a positive reinforcement between the eccentricity and precession cycles, enabled by the vast extent of the North American continent and the southernmost reach of the spreading Laurentide Ice Sheet in North America.[30] As mentioned in chapters 4 and 5, the weight of a thick ice sheet depresses the land beneath it, creating a sort of bowl. Liberated of its load when the ice retreats, the formerly glaciated land rebounds. However, the actual uplift lags behind the receding ice. Due to the temporary sag, more ice lies at lower elevations within the ablation zone where melting occurs. This accelerates the pullback. The ice sheet, near the southern margin at its maximum extent, becomes sensitive to even minor climate perturbations that could initiate a retreat (or advance). Obliquity, in this scenario, merely enhances (or lessens) the summer melting that thrusts the ice sheet into a full deglacial (or glacial) mode.

During the transitional period, the temperature contrast between warm and cold periods was smaller than in later cycles.[31] Following the transition, warm periods grew warmer, while glacial periods changed little.[32] Perhaps by then Northern Hemisphere glaciations had accumulated a critical mass of ice that could bypass the obliquity signal to begin the next warming. Maybe only after several obliquity cycles passed could the feedback mechanisms (e.g., the glacial isostatic and ablation feedbacks, among others) strengthen enough to force the ice sheet into the next deglaciation. This apparently recurred on average every 100,000 years. If so, the 100,000-year beat may be merely fortuitous—only weakly related to the eccentricity cycle.

The *Last Interglacial* culminated around 125,000 years ago. Starting 116,000 years ago, the door to the *last ice age* began to open. As in previous cycles, the renewed cold initiated a series of temperature, greenhouse gas, and sea level oscillations that plunged successively lower and lower to a minimum during the *Last Glacial Maximum* around 23,000–19,000 years ago. With Northern Hemisphere summer insolation at a minimum, the northern ice sheets expanded across North America and Europe and sea level quickly dropped. By 20,000 years ago, the orbital cycles had locked into phase for another deglaciation. Ice sheets pulled back and sea levels rose. The last great meltdown was under way.

We scrutinize several of these events more carefully to better understand how ice sheets may affect sea level in a warming world. Two interglacials, at roughly 400,000 and 125,000 years ago, stand out because of higher-than-present sea levels. In addition, we investigate the great meltdown that ended the last ice age.

BOX 7.2: READING THE PAST

Ice doesn't just keep a record of dust, volcanic ashes and the other subtle changes in atmospheric chemistry that together give us our weather. It has an extra property that belongs to no other history book on Earth. Living as it does on a perpetual knife edge between solid and liquid, strong and weak, ice is sensitive enough to capture tiny pockets of air, and strong enough to keep them.

—Gabrielle Walker, *Antarctica: An Intimate Portrait of a Mysterious Continent (2012)*

The Past as a Key to the Future

Key questions facing us are these: How much and how fast will the Greenland and Antarctic ice sheets melt? How high will sea levels rise?

Current global climate and ice sheet models, while greatly improved over earlier versions, may miss important aspects of ice sheet motion that determine how fast its frigid load can be shed. Models still do not fully account for the marine ice sheet instability or the role of ice shelf thinning described in the last chapter.

Past climate analogs therefore offer an alternative means for assessing eventual outcomes, although exact comparisons may not be feasible. This longer record, extending far beyond the 100- to 150-year-long period of instrumental observations, offers a broader perspective on anticipated changes. Plausible bounds on impending ice sheet fluctuations can be set by examining past variations based on natural "archives" contained in microfossils, corals, or marine sediment chemistry and ice cores, and by studying computer climate models. A glimpse backward therefore enables us to envision oncoming ice melt scenarios and thereby to adopt more effective adaptation strategies. Here we explore two major sources of information on long-vanished periods.

Ice Core Time Capsules

The evolution of ancient climates is pieced together from diverse clues left behind by fossils preserved in land and ocean sediments, with other hints buried deep in ice and marine sediment cores. The ice cores hold an amazingly detailed account of past climate history, virtually layer by layer, year by year. The Greenland ice cores reach past the Last Interglacial to 129,000 years ago, whereas the European Project for Ice Coring in Antarctica (EPICA) Dome C ice core in Antarctica yields an 800,000-year history, and a new core from the Allan Hills stretches past 1 million years.*

The ice core time capsules preserve a lengthy account of changing temperatures, atmospheric gas compositions, dust levels, and continental ice volume, among other relics of long-vanished climates. Isotopic thermometers, which exploit variations in oxygen and hydrogen isotope composition, demarcate significant milestones in paleoclimate history. Oxygen exists as several isotopes that differ only in atomic weight. The most abundant isotope, ^{16}O, constitutes 99.8 percent of the total, with ^{18}O a mere 0.2 percent.

Hydrogen occurs as two isotopes: ordinary hydrogen (^{1}H, 99.98 percent) and deuterium (^{2}H or D, 0.015 percent). Variations in ^{18}O-to-^{16}O ratios are expressed in parts per thousand (per mil, ‰). $^{18}O/^{16}O$ ratios can be readily measured in a wide array of materials, including glacial ice, shells of marine foraminifera, corals, cave deposits, and trees, to name just a few. Hydrogen exists in water, ice, and all living systems.

A number of natural processes can alter these basic proportions. Among the more important are evaporation or condensation of rainwater, chemical reactions, and preferential uptake of the lighter isotope by living organisms. In general, the lighter ^{16}O and ^{1}H isotopes are more mobile and react faster chemically than ^{18}O or D. When water falls as rain, the lighter isotopes remain in the cloud, whereas the heavier isotopes (^{18}O and D) precipitate out as rainwater. As air masses migrate toward the poles, or landward, successive evaporation-condensation cycles progressively enrich rainwater in the lighter isotopes. Thus snow falling at the poles, which eventually builds up into an ice sheet, is isotopically lighter than the source ocean water. Conversely, the ocean grows increasingly heavier (i.e., enriched in ^{18}O relative to ^{16}O) as water is transferred to growing ice caps.

The oxygen and hydrogen isotope ratios in polar ice vary according to the local temperatures where the precipitation occurs. Other variables, such as moisture sources, topography, and season, also affect these ratios. Despite some differences, similar isotope ratio patterns from different ice cores suggest a common temperature signal. Therefore, they can serve as valid paleothermometers. For example, each 1°C (1.8°F) rise in temperature increases the O isotope ratio in Greenland ice by ~0.7‰.

The ice cores also capture tiny pockets of the atmosphere. Tiny air bubbles trapped in the ice encapsulate atmospheric greenhouse gases such as carbon dioxide, methane, and nitrous oxide. These greenhouse gases oscillated in sync with temperatures over the 800,000-year history of the EPICA ice core, rising during interglacials and falling during glacials.† Both greenhouse gases, CO_2 and CH_4, increased as temperatures rose and declined as temperatures fell (fig. 7.4). But never during the 800,000-year history of the ice cores did they approach today's high levels.

Dustiness in ice varied inversely with temperatures. Widespread aridity gripped many parts of the Earth during frigid glacial periods. Vegetation cover shrank and deserts expanded. Stronger winds also lifted more dust off the ground. Wind-borne dust like that of the 1930s Dust Bowl swept across large swaths of North America, Northern Europe, and Central Asia. The eolian dust accumulated in extensive deposits of buff-colored, fine silt called *loess*. Some of this fine dust ultimately fell on the Greenland Ice Sheet. With the return of warmer and wetter interglacials, more abundant rainfall restored the vegetation; the winds subsided. Polar ice then cleared up and became almost dust-free.

Thin sulfuric acid aerosol (sulfate) and ash layers in the ice provide evidence of past volcanic eruptions. Some especially explosive eruptions that left an imprint in Greenland ice cores include Tambora, Indonesia (1815); Krakatau, Indonesia (1883); Katmai, Alaska (1912); El Chichón, Mexico (1982); and Mount Pinatubo, Philippines (1991). The largest eruptions, by injecting copious amounts of sulfate aerosols into the stratosphere (upper atmosphere), partly dimmed incoming sunlight and reduced temperatures in following years. For example, the "year without a summer" followed the Tambora eruption in 1815. Brilliant sunsets painted the sky red that year, but crops also failed and famine struck. Volcanic eruptions that maximize the global distribution of aerosols and appear in the polar ice cores tend to occur at lower latitudes.

Marine Archives

The oxygen isotopes embedded in tiny marine creatures—foraminifera[‡]—register sea level changes that mirror changes stored in Greenland and Antarctic ice core archives. The oxygen isotope ratios of these creatures mirror those of the water they inhabit, which in turn responds primarily to temperature and ocean volume variations. (These ratios also vary with the particular species studied and ocean salinity.) In general, $^{18}O/^{16}O$ ratios become progressively heavier at lower temperatures. They also grow heavier during glacials, when large volumes of ocean water are locked up in ice sheets, and conversely, lighter during interglacials, when the ice sheets contract and ocean volume increases. Thus, heavier $^{18}O/^{16}O$ ratios in foraminifera correspond to glacial periods and lower sea levels, and lighter ratios to interglacials and higher sea levels.

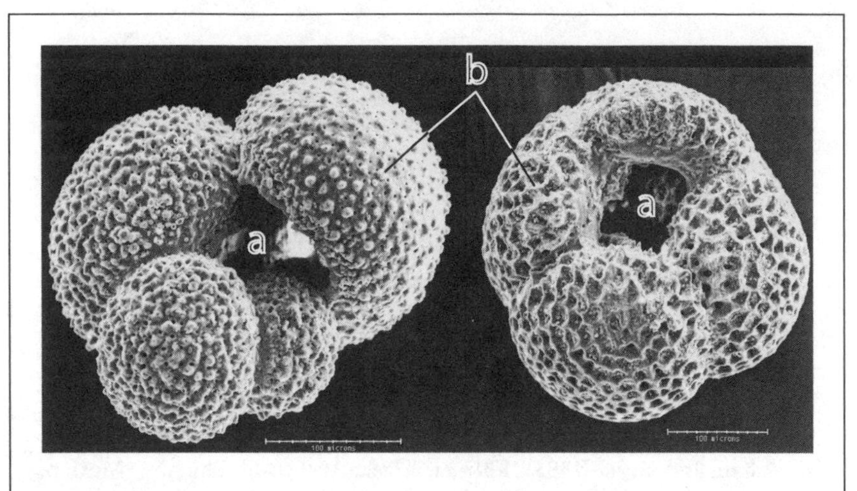

BOX FIGURE 7.2

Planktonic (free-floating) foraminifera *Globigerina bulloides* (*left*) and *Dentoglobigerina altispira* (*right*). (H. J. Dowsett, "Foraminifera," in *Encyclopedia of Paleoclimatology & Ancient Environments*, ed. V. Gornitz [Dordrecht: Springer, 2009], 338, fig. F2. Reproduced by permission of Springer Nature)

However, the $^{18}O/^{16}O$ ratio embeds ambiguity, because while it mainly reflects ice volume (and hence, sea level), it also depends on water temperature. Ice volume changes account for roughly two-thirds of the glacial-interglacial isotope difference, ocean temperatures the remainder. Resolution of this ambiguity requires an independent measure of ocean temperature.

One such measure is the magnesium-to-calcium (Mg/Ca) ratio in marine carbonate shells. Magnesium readily replaces calcium as foraminifera build their calcite shells. The extent of this atomic substitution depends primarily on water temperature. A higher Mg/Ca ratio generally implies a higher temperature. Empirical Mg/Ca-temperature relationships based on living forams have been applied to their fossil ancestors. These relationships depend on the particular species examined. The Mg/Ca method enjoys an important advantage in that both the chemical and isotopic composition can be measured on the same sample.[§]

Other useful seawater paleothermometers include alkenones and TEX_{86}. Alkenones belong to a class of long-chain organic molecules produced by tiny, free-floating marine algae, such as coccolithophores, which are widespread throughout the world's oceans and often produce extensive algal blooms. These algae produce alkenones with different proportions of double bonds between carbon atoms (i.e., more unsaturated carbon bonds), depending on the water temperature in which they grow. This provides a measure of past sea surface temperatures.

Biomarkers, like TEX_{86} found in the free-floating marine organism Crenarchaeota, are also sensitive indicators of past sea surface temperatures.** The TEX_{86} paleothermometer is based on the relative proportion of fatty cell membrane biomolecules within the organism, which readjusts the composition of these molecules in response to changing sea surface temperatures. In spite of local variations in surface currents and temperatures, average worldwide sea surface temperatures respond primarily to major global climate changes. Thus, TEX_{86} serves as an independent means of separating past ocean temperature changes from variations in ice volume or sea level.

* NEEM (2013); Brook (2008); Higgins (2015).

† Brook (2008); Foster and Rohling (2013); Grant et al. (2012).

‡ Foraminifera (or simply, forams) are tiny, single-celled, chiefly marine organisms that can either be free-floating or dwell on the ocean floor or in brackish water. Most species have shells made of calcium carbonate, partitioned into chambers that are added during growth.

§ Higginson (2009).

** The TEX_{86} molecule is a type of biomolecule naturally occurring in fats, waxes, and certain vitamins (e.g., vitamins A, D, E, and K). TEX_{86} contains 86 carbon atoms. It has been found to be empirically correlated with sea surface temperature.

HOW STABLE ARE THE ICE SHEETS?

Looking Through the Mists of Time

As we have seen, ice sheets have repeatedly waxed and waned over the geologic past. The nascent Antarctic ice sheet, 34 to 32 million years ago, grew and shrank repeatedly until it attained near-modern dimensions.[33]

Mid-Pliocene marine archives also record recurrent West Antarctic Ice Sheet collapses.[34] Even the presumably stable East Antarctic Ice Sheet may have retreated during warmer episodes, at times when sediment detritus from the continental interior and fossil diatom deposits indicate ice-free, open water in Wilkes Basin.[35] Large global sea level oscillations paralleled these major ice sheet excursions.

Approaching the present, we zoom in on three periods within the last half million years when the ice relaxed its gelid grip and the ice sheets rolled back substantially: 400,000 years ago, around 125,000 years ago, and 20,000 years ago. Do these drastic past meltdowns foreshadow an increasingly aquatic future?

The 400,000-Year Interglacial

This warm interlude, from approximately 424,000 to 395,000 years ago, lasted nearly 30,000 years. Global temperatures resembled those of the present, although polar temperatures were several degrees higher. Atmospheric CO_2 closely matched preindustrial values of ~280 ppm.[36]

Several geologists reportedly found remnants of 400,000-year-old shorelines 20 meters (66 feet) above present sea level on the tectonically stable islands of Bermuda and the Bahamas. However, prior researchers had neglected to consider glacial isostatic adjustments. The periphery of the North American Laurentide Ice Sheet had bulged upward, but subsequently subsided once the ice melted. Taking this subsidence into account, geoscientists Maureen Raymo and Jerry Mitrovica from Columbia and Harvard Universities, respectively, calculated a more likely, lower sea level rise of 6–13 meters (20–43 feet). They concluded that "both the Greenland and the West Antarctic Ice Sheet collapsed during the protracted warm period while changes in the volume of the East Antarctic Ice Sheet were relatively minor."[37]

Meanwhile, another research team, headed by Alberto Reyes of the University of Wisconsin–Madison, focused their attention on offshore marine sediments of this age from southern Greenland.[38] To them, the absence of glacially eroded sediments in the south, although present farther north and east, suggested a complete deglaciation of southern Greenland—enough to raise global sea level by 4.5 to 6 meters (14.8–19.7 feet).

Abundant spruce pollen grains within the sediments offered further proof of ice-free, forested areas, as noted in "When Greenland Turned White", earlier in this chapter. The unusually prolonged interglacial and higher Arctic temperatures may have spurred the meltdown. A maximal global sea level rise of 13 meters (42.6 feet) implies crumbling of the entire West Antarctic Ice Sheet, with some minor East Antarctic breakup as well.[39]

The Last Interglacial

The penultimate thaw between the ice ages, known variously as the Eemian in Europe and the Sangamonian in North America, lasted from approximately 129,000 to 116,000 years ago. Northern summer sunlight intensified between 130,000 and 127,000 years ago, and average polar temperatures climbed several degrees higher than today, without altering global temperatures significantly. Approaching 10 meters (33 feet) of present sea level, the ocean sped upward at double today's rates. At its peak, sea level then stood 6 to 9 meters (20–30 feet) higher than present levels.[40]

Uplifted shorelines and coral reefs in western Australia suggest the occurrence of two sea level peaks, the first +3.5 meters (11.5 feet) above present levels between ~127,000 and 119,000 years ago, and a second, higher peak up to 9 meters (29.5 feet) above present levels 118,000 years ago.[41]

Oxygen isotope ratios in Red Sea foraminifera also flag two sea level highs—at ~123,000 years ago and 121,500 years ago—followed by a sharp drop-off after 119,000 years ago.[42] At its zenith, sea level may have zipped upward as fast as 2.5 meters (8.2 feet) per century. A Greenland-sized ice sheet could have disappeared within four centuries at an average sea level rise rate of ~1.6 meters (5.2 feet) per century. How much of Greenland actually disappeared remains uncertain.

A new study casts doubt whether a sea level drop separated the apparent two high peaks of this interglacial.[43] After an exhaustive review of the evidence, the authors find no conclusive signs of ice-sheet regrowth and sea level drop during this period. Nor can they account for climate processes that would explain the resurgence of glaciation in the midst of an interglacial. They therefore urge caution in interpreting the proposed high rates of sea level rise following the alleged low sea level stand. Alder and fern vegetation covered southern Greenland between 127,000 and

120,000 years ago, as during an earlier interglacial. Abundant pollen and spores from these plants, as well as glacially eroded silt, were deposited offshore. These signs of vegetation point to ice-free conditions inland. Nevertheless, the reduced ice cover on Greenland raised sea level by only 1.6 to 2.2 meters (5–7 feet).[44]

An international team of scientists headed by Dorthe Dahl-Jensen from the Niels Bohr Institute in Copenhagen drilled the first ice core down to the base of the Eemian during the North Greenland Eemian Ice Drilling (NEEM) project. Although the bottommost layers of the 2,500-meter-long ice core were severely crumpled, a key section spanning the period from 129,000 to 116,000 years ago yielded a wealth of information. An entire layer of Eemian-age ice had melted and refrozen, producing features similar to the widespread melt layer that covered most of the ice sheet during the great thaw of July 2012 (chap. 5). The ice core oxygen isotope paleothermometer registered temperatures 8°C (14°F) higher than present. But despite this unexpectedly high warmth, the Greenland Ice Sheet thinned by only 400 meters (1,300 feet), raising sea level a mere 2 meters (6.6 feet). "The good news," says Dahl-Jensen, "is that Greenland is not as sensitive to climate warming as we thought. The bad news is that if Greenland's ice sheet did not disappear during the Eemian, Antarctica must have been responsible for a significant part of the sea level rise."[45]

Aircraft flying over Greenland between 1993 and 2013 used ice-penetrating radar to delineate the interior structure of the ice sheet. This information enabled scientists to map the distribution of ice layers of different ages. A surprising result is how little Eemian or older ice still exists.[46] Old ice survives only in a fairly small patch of east-central Greenland and a larger area of northwestern Greenland. What became of the very old ice? Two possibilities exist. One is that large portions of the ice sheet containing the oldest ice disappeared during the major interglacials, such as the one 400,000 years ago, or during the Eemian. Subsequent snowfalls during glacial periods helped to rebuild the ice sheet. Most of what we see today has accumulated since the Last Interglacial. The other alternative is that very old ice at the base melts due to the high pressures and geothermal heat and escapes via subglacial streams. Fresh snowfalls make up for losses at the base, in a kind of steady state.

Although multiple lines of evidence point to a reduced Eemian Greenland Ice Sheet, how much smaller remains uncertain. Many studies suggest that Greenland added only 2 to 4 meters (6.5–13 feet) to sea level.[47] Since this falls far short of the total estimated sea level rise, the West Antarctic Ice Sheet presumably made up most of the balance, with some East Antarctic contributions in the highest scenarios. To add to the uncertainty, the recent study alluded to several paragraphs above finds no signs of major sea level fluctuations during the Eemian interglacial.

Since the Earth could reach Last Interglacial temperatures later in this century, will sea level climb to Eemian heights again? Before answering this important question in the next chapter, we zoom in on the last great meltdown.

The Last Great Ice Meltdown

At the height of the last ice age 25,000 to 20,000 years ago, North America lay under a miles-thick ice sheet as far south as present-day South Dakota, Iowa, Illinois, Ohio, and Long Island. In Europe, the ice reached present-day Berlin, Warsaw, and the British Isles, and covered the Alps. New York City, too, was deeply buried in ice. Although the ice has long since vanished, the trained eye can easily spot signs of its passage. The long-gone glaciers left their calling cards throughout Central Park, a leafy oasis in the midst of bustling Manhattan. Among these are well-polished rock outcroppings, worn smooth by the overriding ice sheets and jagged on the "downstream" side; grooves or scratches gouging the rock surfaces; and exotic boulders transported by the glaciers from across the Hudson River or from upstate New York and left behind when the ice melted (see fig. 4.4). A ridge of jumbled sand and gravel slashes across Staten Island, Brooklyn, central Queens, and the middle of Long Island. This ridge represents the terminal moraine—the debris pile that marks the southernmost advance of the ice (see chap. 4).

By 20,000 years ago, as northern summer sunlight gradually strengthened, the glaciers had embarked on their northward journey. Sea level rose slowly at first. The pace quickened by 16,000 years ago, as more and more ice melted. By 14,600 years ago, the meltdown had gathered full speed and sea level rose rapidly as the trickle swelled into major torrents.

After 7,000 years ago, the meltdown neared its end and sea level gradually approached its present level.

Multiple abrupt outbursts, termed *meltwater pulses*, punctuated the overall rise between 15,000 and 7,500 years ago (fig. 7.5; table 7.1). During the earliest spike ~19,000 years ago (meltwater pulse $1A_0$), ocean levels may have sprung upward by 10–15 meters (33–49 feet) within a few hundred years.

The largest upward spurt, meltwater pulse 1A (MWP-1A), occurred between 14,600 and 14,000 years ago, during the milder Bølling-Allerød period, when Greenland's temperatures climbed by 4.5°C to 9.9°C (8°F to 18°F). At its fastest, sea level leaped upward by 16 to 19 meters (54–62 feet) within 300 years, at rates of up to 46 millimeters (1.8 inches) per year (table 7.1).

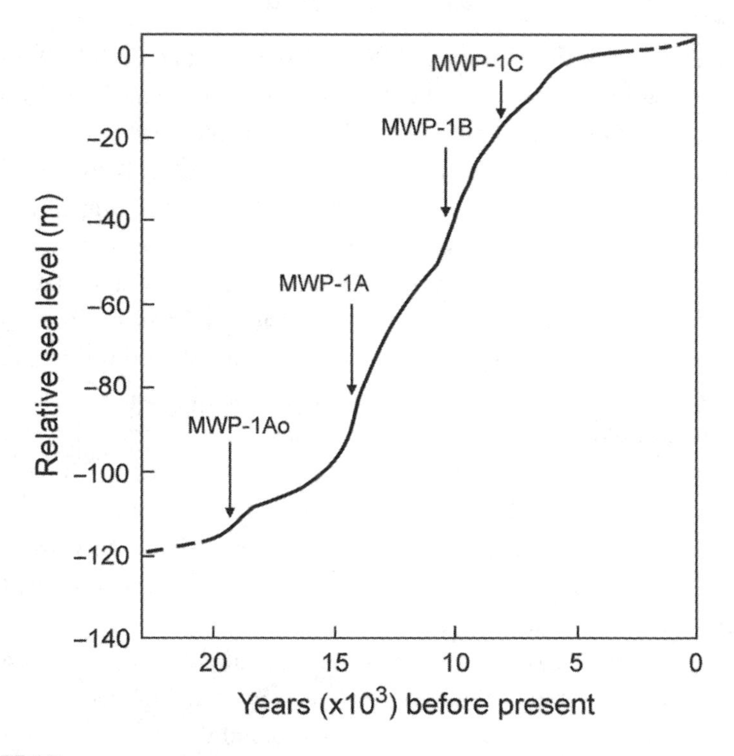

FIGURE 7.5

Generalized curve of sea level rise after the last ice age. (After fig. 5.1b in Gornitz [2013])

TABLE 7.1 Meltwater Pulses During Periods of Rapid Sea Level Rise

MELTWATER PULSE 1A$_0$

TIMING (YR)	RATE OF SLR (MM/YR)	INCREASE IN SEA LEVEL (M)	REFERENCES
~19,000	—	10–15	Yokoyama et al. (2000)
~19,000	≥20	10 m in ≤500 years	Clark et al. (2004)
19,600–18,800	~12.5 (average)	10	Hanebuth et al. (2009)

MELTWATER PULSE 1A

TIMING (YR BP)	RATE OF SLR (MM/YR)	INCREASE IN LEVEL (M)	REFERENCES
14,300–12,800			Stanford et al. (2011)
13,800 (peak)	~26 (max.)	—	Stanford et al. (2011)
14,650–14,300	46	16 m in ~350 years	Deschamps et al. (2012)
14,600–14,300	—	16–19	Carlson and Clark (2012)
~14,500	—	~8.6–14.6	Liu et al. (2016)

MELTWATER PULSE 1B

TIMING (YR BP)	RATE OF SLR (MM/YR)	INCREASE IN LEVEL (M)	REFERENCES
11,500–9,500			Stanford et al. (2011)
10,900; 9,500 (peak)	19; 25 (max.)	—	Stanford et al. (2011)
~10,500	26	~28	Fairbanks (1989)
11,000	~25	~13	Bard et al. (1990)
11,100–11,400	~40	~15 m (Barbados only)	Bard et al. (2010); other sites: none to small SLR acceleration

MELTWATER PULSE 1C

TIMING (YR BP)	RATE OF SLR (MM/YR)	INCREASE IN LEVEL (M)	REFERENCES
8,100–8,300	—	0.9–2.2 (range)	Carlson and Clark (2012)
8,200–8,300	—	0.8–2.2	Li et al. (2012)
7,600–8,200	~12	~6 m in ≤500 yr	Cronin et al. (2007)
~7,600	~10	~4.5 m	Yu et al. (2007)

Source: Revised and updated from Gornitz (2013), table 5.1.

Because of its vast size, the breakup of the Laurentide Ice Sheet in North America was generally assumed to dominate meltwater MWP-1A, with lesser inputs from northern Eurasia. (Greenland contributed little, given its relatively small ice volume.) Vestiges of the glaciers' passage show that Northern Hemisphere sources accounted for at most half of the total meltwater, which leaves the door open for substantial Antarctic inputs.[48] Does the geologic evidence substantiate a significant role for Antarctica during MWP-1A?

The history of ice regression in Antarctica is less clear-cut than that in the Northern Hemisphere, owing to its remoteness, harsh climate, and limited data. In the Mac. Robertson region of East Antarctica, most sediments left by retreating ice sheets are younger than 12,000 years old, although the retreat may have begun as early as 14,000 years ago, during MWP-1A.[49] On the other hand, melting icebergs dumped embedded debris seaward of the Ronne-Filchner Ice Shelf in East Antarctica at least eight times between 20,000 and 9,000 years ago. These events likely marked early stages in the overall regional ice retreat. The most pronounced episode began around 15,000 years ago and peaked between 14,000 and 14,400 years ago, which coincides with MWP-1A.[50] However, the other events were probably too small to result in meltwater pulses. Meanwhile, on the West Antarctic Ice Sheet, Pine Island Glacier retreated to within 100 kilometers (60 miles) of its present grounding line before 10,000 years ago, although exactly when is uncertain.[51] Curiously, ice in the Ross Sea region of Antarctica reached a maximum between 18,700 and 12,800 years ago, followed much later by retreat. This rules out this area as a candidate source for MWP-1A.[52]

Can geophysics shed further light on this contentious history? Different meltwater sources (e.g., the Laurentide, Antarctic, and Barents Sea ice sheets) leave distinctive "fingerprints," i.e., they produce different geographic patterns of sea level rise depending on how much ice melts and where. Water flows away from a dwindling ice sheet because the gravitational pull between it and the ocean weakens. In addition, the ocean level drops near ice-free land that rebounds from the decreased isostatic loading, even as the ocean gains more water overall (see chaps. 4 and 6). The calculated regional sea level signatures that best match observations suggest a substantial Antarctic input for meltwater pulse 1A.[53] However, remaining uncertainties still cloud the extent of Antarctica's role in the meltwater pulse.[54]

The last two meltwater pulses (MWP-1B, MWP-1C) were much weaker than MWP-1A (table 7.1). Meltwater pulse 1B—between 11,500 and 11,000 years ago—materialized upon the return of warmth after a brief relapse into the Younger Dryas neo–ice age between 12,800 and 11,700 years ago. Although fossil corals from Barbados register a distinct sea level jump, elsewhere this event appears as just a minor speedup in an overall rising trend.

Sea level sped up several times again between ~9,000 and 7,600 years ago. Meltwater pulse 1C (MWP-1C) may be related to an episode around 8,200 years ago when colder temperatures gripped Greenland, and European summers chilled by several degrees (table 7.1). The cold may have been triggered by the final catastrophic drainage of giant glacial lakes along the southern edge of the Laurentide Ice Sheet and their discharge into Hudson Bay as the last ice sheet remnants disintegrated.[55] Although the enormous influx of icy freshwater to the world's oceans likely cooled them (and regional climates), it merely elevated sea level a few meters if spread out evenly over the ocean.

The meltwater pulses that punctuated the great meltdown were closely linked to outbursts of glacial meltwater from rapidly crumbling ice sheets and glacial lakes, although specific sources have often been difficult to pinpoint. Since the mammoth Laurentide Ice Sheet that once buried much of North America no longer exists, abruptly disintegrating ice masses may not suddenly unleash massive meltwater pulses—at least not immediately. However, we cannot rule out such an outpouring from Greenland and West Antarctica if global temperatures continue to ramp upward.

This could indeed come to pass in the far future. Having retraced the prolonged, winding path of planetary transformation from greenhouse to icehouse, we turn next to the prospect of reversing course and returning to a greenhouse world—one in which only ragged remnants of today's cryosphere may survive. Though this radical notion may seem far-fetched today, the next chapter examines why this fanciful vision could become our everyday reality at some future date.

8

RETURN TO THE GREENHOUSE

Ninety-nine percent[1] of the planet's freshwater ice is locked up in the Antarctic and Greenland ice caps. Now, a growing number of studies are raising the possibility that as those ice sheets melt, sea levels could rise by six feet this century, and far higher in the next, flooding many of the world's populated areas . . .

Last month [April 2016] in Greenland, more than a tenth of the ice sheet's surface was melting in the unseasonably warm spring sun, smashing 2010's record for a thaw so early in the year. In the Antarctic, warm water licking at the base of the continent's western ice sheet is, in effect, dissolving the cork that holds back the flow of the glaciers into the sea; ice is now seeping like wine from a toppled bottle.

—Nicola Jones (2016)

Our journey into the mists of time opened a window into possible Earth futures. Even so, anticipated conditions may well lie beyond past bounds. The Eemian gave us a foretaste of far higher sea levels, but one would need to revisit the Pliocene to experience the full impact of today's carbon dioxide (CO_2) level (now over 400 ppm), or perhaps even ultimately the hothouse Eocene. Where are the ice sheets headed? How long will it take at today's and tomorrow's ramped-up warming rates before the remaining large ice masses ultimately collapse and unleash renewed meltwater pulses? Will the coming meltdown unfold abruptly, in stark contrast to the last deglaciation, which spanned 8,000 to 10,000 years? MWP-1A followed quickly on the heels of the Bølling-Allerød warming, 14,700–14,000 years ago, as did MWP-1B soon after the end of the cold Younger Dryas, 11,500 years ago (see fig. 7.5, table 7.1). Yet many centuries of prior warming and gradual ice deterioration had primed the ice sheets for these sudden outbursts. Another important distinction is that the Eemian and Holocene interglacials were set off by increasing northern solar insolation, as predicted by the Milankovitch theory. However, increasing anthropogenic greenhouse gas emissions are largely responsible for the current warming trend. Therefore, the hemispheric and seasonal distributions of higher temperatures may differ from those of the past.

More importantly, our excursion into the distant past has demonstrated that large ice sheets may not be as invariant as once assumed. Sea level rose or fell as glaciers and ice sheets repeatedly retrenched and expanded in response to changes in climate and atmospheric CO_2, both then and now (e.g., fig. 7.4; also chaps. 4–6). These close linkages will still persist in the future. As Gabrielle Walker starkly reminds us, "Carbon dioxide has risen and fallen with the seasons, with the ice ages, with the different climate patterns. But in all that time it has never been within striking distance of the amount we have today. Throughout the entire EPICA record [800,000 years], the highest value of CO_2 was about 290 parts for every million parts of air. Now we are at nearly 400 and rising. . . . The deepest voids of Dome C hold a warning that we would do very well to heed."[2]

Not only does today's CO_2 level far exceed those of the past million years; we must turn the clock back 3 million years to match today's high level—to a time when sea level may have been as much as 20 meters (66 feet) higher! The Earth may take a long time to reestablish a new atmospheric equilibrium. Once in the atmosphere, the extra CO_2 is notoriously difficult to eliminate. It stays in the atmosphere for millennia, unlike other greenhouse gases such as methane and nitrous oxide, which have much shorter atmospheric lifetimes and decline within decades.[3] Meanwhile, the heat builds up. Will rising temperatures push the ice sheets into an irreversible downward death spiral?

Already, the Arctic is warming twice as fast as the rest of the planet. Arctic summer sea ice has shown a steady decline over the past 30 years. Any Antarctic sea ice gains have been surpassed by Arctic losses. In February 2017, the temperature at the North Pole briefly spiked 30°C above normal to 0°C (32°F). Both poles also experienced record sea ice lows.[4] Greenland, and recently Antarctica as well, have also been losing ice, as documented in chapters 5 and 6. The explosive disintegration of Larsen B in 2002 illustrates what happens when the integrity of an ice shelf is compromised and it unleashes previously constrained ice streams. Is Larsen B therefore a forerunner of things to come? Could Larsen C be on a similar trajectory, after a giant iceberg split off a rift in the ice shelf on July 12, 2017?[5] In West Antarctica, large icebergs calved off the Pine Island Glacier in 2013, 2015, and 2017. The calving events were preceded by rifts encroaching ever landward (chap. 6). Seongsu Jeong of Ohio State University finds these recent rifts "anomalous," because the fissures originated at the center of the ice shelf and propagated toward the edges. They also opened up from the bottom of the ice, at the grounding line. The ice shelf may have been weakened by thinning caused by incursions of warm marine water. As Nicola Jones starkly reminds us, the West Antarctic ice shelves—the "cork" that keeps the ice in check—are slowly dissolving, spilling more ice into the ocean "like wine from a toppled bottle." Will rising tides be far behind?

THE COMING SUPERINTERGLACIAL

As a matter of fact, summertime sunlight at 65°N is getting a little thin lately. It's approaching the trigger value now . . . if the climate system misses the glacial express this time, the next opportunity will be 50 thousand years from now.

—David Archer, paleoclimatologist (2009)

Most interglacials last around 10,000 years. A few, like the one 400,000 years ago, last around 30,000 years. The orbital alignments for the next glaciation (see box 7.1) could come in as little as a few thousand years or in

50,000 years. But human activities may stall the next glaciation for another half a million years. David Archer, a paleoclimatologist at the Department of Geophysical Sciences of the University of Chicago, and his colleague Andrey Ganopolski reached this startling conclusion after running a climate model that included atmospheres, oceans, and ice sheets that grow, flow, and melt. They simulated changes in the Earth's orbital configurations and atmospheric CO_2 levels. At a preindustrial CO_2 concentration of 280 ppm, they easily initiated an ice age under favorable orbital alignments (see "Recipe for Starting an Ice Age," box 7.1). But the higher the CO_2, the harder and harder it is to regrow ice. On a business-as-usual pathway, we avoid another glaciation for at least 50,000 years, possibly 130,000 years. Worse, if we burn all our coal reserves, the next ice buildup is postponed for much, much longer. "The Earth," they say, "could remain in an interglacial state until the end of not only our current period of circular orbit, but the next circular time, 400 millennia from now."[6]

This is no great excuse for staving off a coming ice age. As the next chapter will show, the potential dangers of warming are already under way. A scary implication of the Archer-Ganopolski study is that "by releasing CO_2, humankind has the capacity to overpower the climate impact of Earth's orbit, taking the reins of the climate system that has operated on Earth for millions of years."

Current observations suggest that both surface melting and ice discharge will dominate future Greenland Ice Sheet attrition, whereas Antarctica will sustain most of its losses via ice outflow into the ocean. Surface melting will likely be negligible, because Antarctica will remain extremely cold in spite of a warmer climate. Increases in interior snowfall will largely cancel losses from surface melting. However, the ice sheet's margins will sustain greater losses that may exceed potential gains inland.

To foresee more clearly where we are headed, climate scientists turn to global climate models (GCMs) that mathematically replicate the major natural and atmospheric processes that govern our climate, based on fundamental principles of physics. These models provide a scientific basis for forecasting anticipated atmospheric greenhouse gas levels. Most combined atmospheric and oceanographic global models (AOGCMs) focus on changes in air and ocean temperatures, circulation patterns, and ocean thermal expansion. Ice losses are usually determined separately. These losses include changes in both surface mass balance (snowfall minus ice melt and runoff) and ice dynamics (calving or outflow of solid ice past the grounding line). Surface mass balance is generally calculated from downscaled global or, preferably, regional-scale climate model projections of temperatures and precipitation (see chaps. 4 and 5).

Dynamic ice flow models additionally estimate future changes in iceberg calving rates and discharge of ice past the grounding line. The latest dynamic ice models also capture significant elements of ice sheet and ice shelf processes, including retreat of grounding lines, changes in ice shelf buttressing, hydrofracturing, ice cliff failure, calving, and basal melting. Even though these models may be much more sophisticated than earlier versions, they may still miss several important aspects of ice loss. For example, they may oversimplify important processes, such as mixing of ocean water with freshwater runoff in the fjord, or ice sliding over obstacles on its bed. Still, they portend troubling news from Antarctica. In contrast to earlier simulations that lacked consideration of some of these ice processes, the newer findings reveal that by 2100, Antarctic ice losses will account for a larger percentage of global sea level rise than previously assumed, and this share will expand in the future.

In a recent study, Robert DeConto of the University of Massachusetts and David Pollard of Pennsylvania State University applied an advanced ice sheet–ice shelf model to Antarctica that mimics the marine ice sheet instability (MISI) and ice cliff fracturing instabilities.[7] They found that at high greenhouse gas emission rates, ice shelves could begin to break apart by midcentury. Antarctica may contribute 1 meter (3.2 feet) of sea level rise by century's end, and over 15 meters (49 feet) by 2500. At such rates, the West Antarctic Ice Sheet (WAIS) would disappear within a few hundred years.

Another, more cautiously worded study finds such high estimates "implausible" given our current understanding of the physical processes involved.[8] Instead, Antarctica would contribute at most around a third of a meter (1 foot) by 2100 and 72 centimeters (2.4 feet) by 2200. But this analysis does not include the consequences of surface melt ponding and ice cliff failure, which could augment these figures.

Semiempirical models adopt a quite different approach to predictions of ice sheet (and ocean) evolution.[9] These models assume that past correlations between observed mean global sea level rise and temperature will continue to hold in the future. Since these models do not simulate underlying individual processes, they provide no further information on glacier or ice sheet contributions. This also limits their ability to account for those processes that, while relatively insignificant today (for example, growing ice sheet instabilities), are expected to play an increasingly important role. The overall balance of processes that causes ice losses may thus change over time. This failure to account for evolving ice sheet behavior is a serious shortcoming of semiempirical methods.

Table 8.1 summarizes the contributions from glaciers and ice sheets to sea level rise by the year 2100. Although most studies find that combined ice mass losses will yield less than a meter of global sea level rise by 2100, adding thermal expansion of the ocean and other processes affecting sea level brings the overall total to nearly a meter or more. The DeConto-Pollard model, which depicts ice sheet/ice shelf behavior more realistically, presents a high-end scenario. But no end-of-century projection should induce complacency. As we shall soon see, the story does not end in 2100.

Runaway Ice Sheets?

> *The future evolution of the Antarctic Ice Sheet represents the largest uncertainty in sea-level projections of this and upcoming centuries . . . In our simulations . . . the region disequilibrates after 60 years of currently observed melt rates. Thereafter, the marine ice-sheet instability fully unfolds and is not halted by topographic features . . . Our simulations suggest that if a destabilization of [the] Amundsen Sea sector has indeed been initiated, Antarctica will irrevocably contribute at least 3 m to global sea-level rise during the coming centuries to millennia.*
>
> —Johannes Feldmann and Anders Levermann (2015)

Repeated past West Antarctic Ice Sheet fluctuations, as well as recent trends and model projections (see the previous section), raise deep concerns over an impending WAIS instability. Thwaites Glacier, along with

TABLE 8.1 Cryosphere Contributions to Sea Level Rise (m) by the End of the Twenty-First Century

	SURFACE MASS BALANCE	DYNAMICS	TOTAL	REFERENCE
GLACIERS				
	—	—	0.10–0.16[a]	Intergovernmental Panel on Climate Change (2013a)
	—	—	0.12–0.18[b]	Kopp et al. (2014)
	—	—	0.08–0.11[c]	Mengel et al. (2016)
GREENLAND ICE SHEET				
	0.03–0.07	0.04–0.05	0.07–0.12	Intergovernmental Panel on Climate Change (2013a)
	0.04–0.09	—	—	Fettweis et al. (2013)
	—	—	0.06–0.14	Kopp et al. (2014)
	—	—	0.04–0.10[d]	Fürst et al. (2015)
	0.07–0.27	0.05–0.07	0.12–0.34	Mengel et al. (2016)
ANTARCTIC ICE SHEET				
	−0.02–−0.04	0.07–0.07	0.03–0.05	Intergovernmental Panel on Climate Change (2013a)
			0.04–0.06	Kopp et al. (2014)
	−0.02–−0.03	0.06–0.13	0.04–0.10	Mengel et al. (2016)
	—	—	0.01 (0.10); 0.10 (0.39)[e]	Golledge et al. (2015)
			0.11–1.05[f]	DeConto and Pollard (2016)
All ice	—	—	0.37–0.56[g]	Meehl et al. (2012)

[a] Median range across four RCP scenarios (RCP2.6, RCP4.5, RCP6.0, and RCP8.5) between 1986–2005 and 2081–2100.
[b] Median range across three RCP scenarios (RCP2.6, RCP4.5, and RCP8.5) between 2000 and 2100.
[c] Median range across three RCP scenarios (RCP2.6, RCP4.5, and RCP8.5) between 1986–2005 and 2100.
[d] Mean of combined climate models across four RCP scenarios (RCP2.6, RCP4.5, RCP6.0, and RCP8.5) between 2000 and 2100.
[e] Low (high) estimates for four RCP scenarios (RCP2.6, RCP4.5, RCP6.0, and RCP8.5) between 2000 and 2100.
[f] Mean range across four RCP scenarios (RCP2.6, RCP4.5, RCP6.0, and RCP8.5) between 2000 and 2100.
[g] Median range across three RCP scenarios (RCP2.6, RCP4.5, and RCP8.5) between 1986–2005 and 2100.

neighboring Pine Island Glacier and other nearby glaciers, has recently accelerated and thinned, suggesting to some observers a potential collapse in progress.[10] Thwaites has greater access to deep marine-based interior basins than do its neighbors. One study suggests that as little as 2°C (3.6°F) of air and 0.5°C (0.9°F) of ocean warming could destroy nearly all ice shelves and trigger a MISI[11] (see also fig. 6.6). Once initiated, the process would become self-sustaining. Ice losses would eventually drain much of the marine-based portions of WAIS, raising sea level by up to 3 meters (9.8 feet) within 300 years, under high greenhouse gas emissions (e.g., RCP8.5, appendix A). Worse yet, Feldmann and Levermann, in another study, suggest that the whole Amundsen Sea sector could destabilize in only 60 years at current ice shelf melting rates. The instability would spill into deep inland valleys and cross into adjacent catchments that feed into presently stable ice shelves (chap. 6). Most of the remaining WAIS would disappear centuries to millennia hence.[12]

The Greenland Ice Sheet may face its own MISI, because of deeply buried subglacial valleys with potential outlets to the sea (recall Greenland's hidden "Grand Canyon," chap. 5). The Northeast Greenland Ice Stream (NEGIS) drainage system, parts of which lie on a reversed slope, cuts deep into the heart of Greenland. It remained stable until the early 2000s, when the Zachariae Ice Stream, a branch of NEGIS, began to retreat.[13] Several other glaciers, including fast-retreating Jakobshavn Isbrae, Helheim, and Kangerlussuaq, sit on reverse slopes as well.

Greenland's tidewater glaciers react quickly to slight ocean and air warming (chap. 5). Increasing surface meltwater works its way down to the glacier bed, where more freshwater not only courses to the sea, but erodes the ice sheet from below. Furthermore, milder water entering a fjord eats away at the glacier terminus, driving the grounding line farther inland (fig. 8.1). This also undercuts ice along the glacier's edges, thereby deepening the fjord cavity into which more seawater can pour. Calving rates increase as the ice edge is undermined. Could this combination of processes also put the Greenland Ice Sheet on the verge of an unstoppable rush to the sea?

Before we wring our hands at the inevitability of runaway ice sheets, we should recall two important negative feedbacks that may slow down

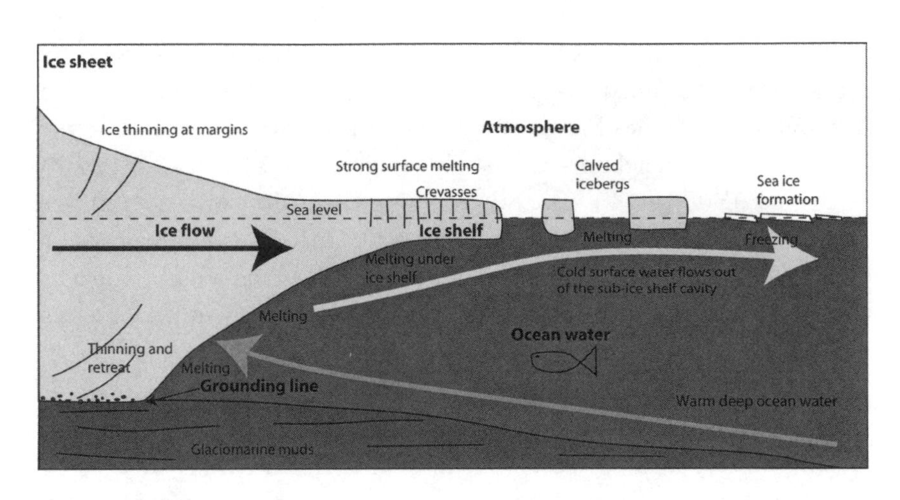

FIGURE 8.1

Schematic diagram showing how the Greenland Ice Sheet and its attached ice tongues and shelves thin from above and below. This weakens the floating ice, which then develops crevasses and rifts, ultimately breaking off icebergs. Discharge across the grounding line increases. (Adapted from Bethan Davies, "Grounding Lines," AntarcticGlaciers.org, http://www.antarcticglaciers.org/glaciers-and-climate/ice-ocean-interactions/grounding-lines/)

or even halt the marine-based instability, at least for a while. These feedbacks involve the way the Earth responds to a lessened ice load and weaker gravity.[14] The land springs back under a thinner ice sheet (glacial isostatic rebound), and local sea level drops. At the same time, the smaller ice sheet exerts less of a gravitational pull on the ocean, so that water flows away into the distance, which further reduces local sea level rise. A lower sea level hinders entry of warmer marine water into subglacial fjord cavities or the base of ice shelves, thereby slowing down ice shelf and ice tongue deterioration. Moreover, an initially reverse slope could steepen seaward due to glacial rebound. Ice shelf disintegration within a narrow embayment can also hinder retreat, because the resulting loose ice mush could provide some temporary buttressing support. Even so, these negative feedbacks may not be enough to prevent a marine ice sheet instability from developing in the face of unabated, mounting CO_2 levels.

Tipping Points and Climate Commitments

The global average temperature of the Earth might be 3°C warmer in the year 2100 than it was in 1950. That doesn't sound like much . . . On the other hand, the climate changes that civilized humanity has witnessed have all been 1°C or less . . . but this is nothing compared with the forecast in 2100 . . .Given the long atmospheric lifetime of fossil fuel CO_2 *. . . it is clear that there is lots of time available to melt ice. The sea level rise one thousand years from now will be much higher than what the coming century will bring. The forecast for 2100 is only the tip of the iceberg. So to speak.*

—David Archer, paleoclimatologist

A few degrees of warming may be all that is needed to destabilize the Greenland Ice Sheet and ultimately return Greenland to a nearly ice-free state. The tipping point may be as low as 1.6°C (2.7°F).[15] Many climate change scenarios foresee much higher temperatures by the end of the century. Deglaciation will not occur overnight, however. It could take as long as 50,000 years for a 2°C regional summer warming. If temperatures soar by 8°C, 20 percent of the ice sheet will melt after 500 years and all of it will melt after two millennia. At the opposite end of the Earth, a half a degree Celsius warmer ocean and several degrees warmer air temperatures would eliminate nearly all ice shelves.[16] While ocean warming and ice shelf breakup cause the greatest initial losses, increasing air temperatures over centuries thin the margins and "soften" the ice sheet, which then accelerates. Once past the initial threshold, Antarctica commits to a substantial sea level rise, which stems largely from the demise of the buttressing ice shelves. A model that features shelf-weakening processes such as hydrofracturing and ice cliff failure finds that Antarctica could potentially drive up sea level by as much as 1 meter (3.3 feet) by the end of the century and over 15 meters (49 feet) in 500 years, at high emission rates.[17] The latter scenario implies significant inputs from East Antarctica. This scenario of a ramped-up ice sheet instability may be somewhat exaggerated. In many other projections, the West Antarctic collapse stretches out longer, over many millennia.

A multiplicity of interacting atmospheric, oceanographic, and geologic variables can influence the future course of ice sheets, which generate large uncertainties in future forecasts. Nevertheless, both past geologic evidence and climate projections demonstrate that given enough warming for long enough, drastic scenarios of ice sheet meltdowns are not beyond the realm of the possible. Whether the ocean rises to such extreme heights within 100, 500, or 1,000 years matters less than the fact that once the big-time meltdown begins, the die will have been cast. The process is irreversible. There is no turning back.

The scientific basis for the close link between changing atmospheric CO_2 levels and temperature, alluded to earlier in this chapter and in previous chapters, has been known for over a century. In 1896, the Nobel Prize–winning chemist Svante A. Arrhenius (1859–1927) helped to establish the foundation for what later became known as the "greenhouse effect"—the heat-trapping ability of atmospheric CO_2 and other greenhouse gases, such as methane (CH_4) and nitrous oxide (N_2O). Motivated by a desire to explain the origin of the ice ages and building upon earlier work by Joseph Fourier, Claude Pouillet, and John Tyndall, Arrhenius applied his knowledge of physical chemistry to calculate the effect of increasing CO_2 on Earth's surface temperatures. His calculations led to a "rule" simply stating that increasing carbonic acid (i.e., CO_2 and H_2O) geometrically would increase temperature nearly arithmetically. Thus, doubling CO_2 would result in a 4–5°C (7.2–9°F) rise in temperature; a fourfold increase in CO_2 would raise it 8°C (14.4°F). This supposedly works both ways: i.e., halving CO_2 would reduce temperatures by 4°C (7.2°F), eventually leading to a new ice age. Arrhenius also realized that the then-modest increases in CO_2 from industrial activities would eventually lead to a warmer climate. But he actually saw this as beneficial, especially in northern climates (such as in Sweden, where he lived) where more crops could be grown to feed "rapidly propagating mankind."[18] However, as we shall soon see, it's not so easy to get rid of the excess CO_2 our activities are generating.

Although progress in atmospheric science, recent observations, and geologic evidence of past climates have vastly expanded our understanding of the greenhouse effect, Arrhenius's simple calculations, given the limited knowledge of his day, were amazingly close to present estimates. For example, climatologists refer to the temperature rise for a doubling of atmospheric CO_2 as the *equilibrium climate sensitivity*. This value is currently believed to lie in a range between 1.5° and 4.5°C (2.7° and 8.1°F), close to Arrhenius's estimate of ~4° to 5°C.[19]

As Arrhenius suspected, our industrial activities, including fossil fuel combustion, cement manufacture, and deforestation, are adding CO_2 to the atmosphere above and beyond what is produced by volcanoes or naturally present atmospheric water vapor (another greenhouse gas). As illustrated in chapter 1, CO_2 has been trending steadily upward since the 1950s (fig. 1.7). The CO_2 level of ~407 parts per million as of December 2017 has far surpassed the preindustrial value of ~280 parts per million.[20] Both temperature and sea level have already responded to this sharp increase. Earth's global temperature in 2016 was the highest ever since the late nineteenth century.[21] Overall, temperatures have gone up 1.1°C (2°F) since the late nineteenth century. Twentieth-century global sea level rose at a rate of 1.2 to 1.9 millimeters (0.05–0.07 inch) per year between 1900 and the 1990s. Since 1993, satellite altimeters have registered an average yearly sea level rise of ~3 millimeters (0.12 inch) per year (chap. 1).

A growing share of the total sea level rise is coming from glaciers and ice sheets (see chaps. 4–6 and tables 4.1, 5.1, and 6.1). Given the expected warming that lies ahead, glaciers and ice sheets will continue to shed even more ice, as suggested in table 8.1 (see also appendix A and "The Coming Superinterglacial," earlier in this chapter).

Most scientists and a growing number of the general public (aside from a few influential climate change skeptics) now concede that the Earth has been warming and that recent changes, especially since the 1960s, bear a considerable human imprint. Yet many point to the large climate oscillations of the geologic past, described in chapter 7, and maintain that the recent warming is merely the latest example of a natural cycle. So why worry? Furthermore, most changes to date are barely perceptible against the wild day-to-day swings in temperature, rainfall, snowstorms, and other extreme weather events.[22] Nevertheless, those living in the Arctic, the mountains, and the major deltas and on coastal barrier islands already notice the changes, as the next chapter will show. Still, most of us find it difficult to think beyond a few years, or at most a human lifetime or that of one's children and grandchildren. Businesspeople plan for this year's or next year's profits, politicians look toward the next election cycle, and urban planners and developers look to the lifespan of a building, roadway, or bridge. Thus, a sharp disconnect separates ordinary human experience from the much longer time spans over which climate change evolves. Therefore, steps to reduce our CO_2 emissions receive a much lower priority than more immediate concerns. However, decisions made now or within the next few decades will ultimately affect not just the fate of the ice sheets, but the fate of our planet as we know it.

Even if we were to stop further CO_2 emissions by midcentury, the emissions we have already generated and will likely add within the next few decades will remain in the atmosphere, where they will linger for millennia before slowly diminishing. Methane lasts roughly 10 years in the atmosphere and nitrous oxide survives a century, while CO_2 interacts with the atmosphere, ocean, and biosphere in complex ways that operate on different timescales. The ocean consumes about a third of the CO_2 released by fossil fuel burning, while the rest winds up in the atmosphere and biosphere. The longevity of atmospheric CO_2 commits us to higher temperatures and sea level long after greenhouse gas emissions have stabilized or are reduced. In a sharp warning, a multidisciplinary, international team of Earth scientists headed by Peter U. Clark proclaim that "20–50 percent of the airborne fraction of anthropogenic CO_2 emissions released within the next 100 years remains in the atmosphere at the year 3000, that 60 to 70 percent of the maximum surface temperature anomaly and nearly 100 percent of the sea-level rise from any given emission scenario remains after 10,000 years, and that the ultimate return to pre-industrial concentrations will not occur for hundreds of thousands of years."[23] If that weren't bad enough, Clark et al. continue: "If CO_2 emissions continue unchecked, the CO_2 released during this century will commit Earth and its residents to an entirely new climate regime"—one in which, they further point out, Greenland would deglaciate completely within 2,500 to 6,000 years and Antarctica would lose the equivalent of 24 to 45 meters (79–148 feet) of sea level rise within 10,000 years. This long-term change occurs because the extra heat already absorbed near the ocean surface requires many centuries to penetrate to great depths. As it does, an ever-increasing volume of ocean water continues to expand and raise sea level. The prolonged warming over timescales comparable to that since the last deglaciation also enables widespread ice sheet melting. Clark et al. furthermore point out that the projected warming "will also reshape the geography and ecology of the world." Clark and his team emphasize that the only way to prevent such a hothouse meltdown is to aim for net-zero carbon emissions—and fairly soon. Our window of opportunity is rapidly closing—possibly within decades.

How do we begin to curb our carbon appetite? One major step forward was the Paris Agreement within the United Nations Framework Convention on Climate Change (UNFCCC), which was adopted by consensus in December of 2015. The Paris Agreement, originally signed by 132 nations, including the United States, took effect in November of 2016.[24] Subsequently, President Donald Trump withdrew the United States from the Paris Climate Accord on June 1, 2017. Meanwhile, Syria, the other remaining holdout nation, signed the accord on November 7, 2017, following Nicaragua's decision to join on October 24, 2017.

The accord ambitiously aims to keep global temperatures from rising by more than 2°C (3.6°F) by 2100, and ideally by only 1.5°C (2.7°F), with the ultimate goal of achieving zero carbon emissions by later this century in order to avert "dangerous anthropogenic interference with the climate system."[25] Participating countries agree to set Nationally Determined Contributions (NDCs) to limit their carbon emissions. While the agreement establishes an important historic precedent, emission targets are self-defined and voluntary, the accord is legally nonbinding, and no provisions exist for enforcement. Furthermore, as recent events have shown, the political climate can change almost as suddenly as the weather, and despite nations' best intentions, actual outcomes may fall short of expectations. Still, hopefully, momentum is slowly building toward a change in course, even in the United States.

Prevention, or at least a dampening of the worst impacts of a potentially large-scale, irreversible deglaciation, requires attacking its ultimate source: the increasing anthropogenic production of atmospheric greenhouse gases. A full treatment of this thorny subject lies beyond the scope of this book and is covered more extensively elsewhere, but here we briefly sketch a few basic steps that can be taken.[26] A broad range of options exist for reducing CO_2 emissions using currently available technology. These include measures such as promoting energy efficiency and conservation in transportation, housing, and energy. In spite of recent declarations of the intention to revive the coal industry, U.S. energy consumption has been moving toward greater utilization of natural gas. Natural gas, or CH_4, does not present a perfect solution, because it too is a fossil fuel, which upon combustion yields CO_2. However, it is far more energy efficient than either coal or oil and also burns more cleanly. Renewable energy derives from natural sources such as the sun, wind, tides, biomass (plant matter and organic wastes), and geothermal energy. Although still a fraction of total energy consumption, renewables' share is growing and has potential for further expansion. Another option is nuclear energy, a route taken by France, Japan, and, to some extent, the United States. While nuclear power emits no CO_2, concerns remain over reactor operating safety after such catastrophic incidents as, for example, the Fukushima, Chernobyl, and Three Mile Island accidents. Another serious consideration is long-term (i.e., 10,000-year) storage of nuclear wastes in a seismic-free, leakproof repository.[27]

Geoengineering, or large-scale anthropogenic interventions in the environment to counteract effects of greenhouse gas emissions, represents an as-yet-untested and potentially risky option. One idea involves injecting tons of sulfur dioxide into the upper atmosphere from high-flying jets to cool the planet. Another plan envisions a string of shiny orbital screens or mirrors to reflect sunlight back into space. However, these "solutions" merely mask the fundamental problem. At best, they work as long as we maintain our atmospheric injections of sulfur or reflecting screens. These fixes may affect cloud formation or plant growth in as-yet-unforeseen ways. They may also fail to prevent other consequences of global warming such as ocean acidification and changes in atmospheric and ocean circulation.

More environmentally friendly approaches involve working with nature. Reforestation soaks up excess atmospheric CO_2 while providing wildlife habitat. Ways of creating biological sinks of CO_2 include expansion of parkland and urban green spaces, and planting rooftop gardens ("green roofs"). Dune rebuilding and coastal ecosystem restoration (e.g., salt marshes, oyster reefs) help mitigate impacts of wave action and storm surges in the face of rising sea levels.

The coming global energy transformation cuts across all economic sectors, but transitioning away from fossil fuels still faces significant technological, economic, and political hurdles. Will our currently available carbon-free technologies suffice to meet growing energy demands? What are the economic costs of the transition, and how can these be met while minimizing dislocations? How best to convince a skeptical public of the urgency of the situation and need for decarbonization? A massive research and development program akin to the Manhattan Project or moon landing may be required to successfully meet this challenge. Meanwhile, existing alternative energy technologies need to be implemented on a more widespread scale than hitherto. Economic incentives to replace fossil fuels with other energy systems should be encouraged. International-scale cooperation, exemplified by treaties such as the Paris Agreement, may be important starting points, but such agreements need stronger enforcement provisions to become truly effective. This may prove challenging, given the disparate agendas and rivalries among, and even within, nations. Meanwhile, a growing number of urban centers, particularly those along the world's coasts, that face the brunt of climatic impacts have been organizing to take steps to adapt to existing and anticipated changes.[28]

Why Care?

What does the fading ice empire mean for the people of the Earth? Why should we care? Many people would welcome a warmer climate, and indeed, many "snowbirds" from northern lands head south each winter to enjoy the sun, warmth, and summer sports. The next chapter investigates in depth why losing ice really matters. The vanishing cryosphere ultimately affects us all, not only those inhabiting the most directly impacted regions—the Arctic and high mountains—but the rest of the world far afield, as Jane Lubchenco, a former NOAA official, reminds us. While the ramifications of a thawing cryosphere extend in multiple directions, we will focus on several that impact large regions or populations. We begin by exploring the epicenter of greatest change: how the big melt affects the Arctic—its people, environment, and economy; how the diminishing summer sea ice opens up new trade routes, travel opportunities, and access to mineral resources; and also how the changing climate alters the landscape as impermanent permafrost thaws, coastlines crumble, and ecosystems slowly adjust to the new norm. We next head for the mountains to determine how reduced winter snowfall will affect water resources for downstream urban and rural populations, and how rapidly shrinking glaciers may make winter sports a fading memory and mountainside living more hazardous.

As alluded to in this and preceding chapters, sea level rise will become one of the most serious consequences of melting glaciers and ice sheets. Glaciers have but little to contribute to the oceans—a half meter (1.6 feet) of rise at most. Greenland has more to give. But when the giant gorilla—Antarctica—reawakens, it may ultimately reconfigure the shorelines of the world and transform the geography of the planet.

The rising waters will imperil coastal populations, megacities and smaller settlements, and deltas and coastal wetlands worldwide. Coastal residents will face more frequent storm flooding. Low-lying lands will be submerged and eventually disappear underwater. Salty and brackish water will creep farther and farther upstream and contaminate rivers and coastal aquifers. Many cities may need to follow the example of Venice and Amsterdam and convert streets to canals, or abandon entire neighborhoods. The ultimate option—a retreat from the coast—may soon become a necessity for some Alaskan villagers, as the next chapter will show. As David Archer reminds us, our actions today could postpone the next interglacial for nearly half a million years—a period that otherwise would likely experience at least another four glacial-interglacial cycles. Welcome to the superinterglacial and hothouse Earth!

9

THE IMPORTANCE OF ICE

We need to save the Arctic not because of the polar bears or because it is the most beautiful place in the world, but because our very survival depends on it.

—Lewis Gordon Pugh, swimmer and environmental champion

A CRY FOR THE CRYOSPHERE

Why cry for the cryosphere? Of what significance is the vanishing ice? The answer is that it plays a vital role in planetary processes. Reverberations of ice losses will propagate to the far corners of the Earth and will strongly affect countless populations far removed from frozen terrains, as this chapter illustrates. The mushrooming shift of land ice to sea will completely transform landscapes and seascapes. We stand, sensitively poised, at the gateway to a new climate regime echoing distant warm geological periods—the Last Interglacial, ~125,000 years ago, or perhaps the mid-Pliocene, 3.6–3.4 million years ago (see, e.g., chap. 7). Only remnants of the ice sheets that now blanket Greenland and West Antarctica would survive; even East Antarctica may shrink a little. Much higher waters would engulf the world's shores, drowning many of the largest coastal cities. The longevity of anthropogenic atmospheric carbon dioxide promises a hotter, more aquatic, and ice-poor future in a coming superinterglacial (chap. 8).

The Arctic lies at the epicenter of global warming, heating up twice as fast as the rest of the planet. The years 2014–2018 were the five warmest since 1880,[1] but truly exceptional heat gripped the Arctic in 2016 and 2017, when in February the North Pole briefly enjoyed temperatures 30°C higher than normal. Winter maximum sea ice reached its lowest extent in recorded history.[2] Instead of there being an unbroken ice sheet as when sea ice normally reaches its maximum extent, open water between ice floes reflected an eerie glow on low-lying clouds. Mark Serreze, director of the National Snow and Ice Center, remarked that "we're starting melt season on a very bad footing." He added: "We knew the Arctic would be the place we'd see the effects of climate change first, but what's happened over the last couple of years has rattled the science community to its core. Things are happening so fast, we're having trouble keeping up with it. We've never seen anything like this before."[3] These still-exceptional Arctic weather events may soon become the new norm.

Perceptible changes already sweep across the Arctic and high mountains. The next section vividly depicts the plight of the Arctic polar, as its favorite ice floe perches melt away.[4] The warming has also altered caribou and reindeer migration routes, while herd populations diminish.[5] Changing ecology, more unpredictable weather, and lessened travel safety under changing ice and weather conditions pose hardships to human food security and health. Stronger storm waves and delayed Arctic Ocean winter freeze-up nibble at over half of the Arctic Ocean coast lined by soft, ice-bonded sediments (see chap. 4). Winter storms threaten at least 30 Native Alaskan villages (see also "Fading Lifeways" and "The First Coastal Refugees," later in this chapter).[6] Thawing permafrost destabilizes the Arctic ground, disrupting economic activity. The thaw shortens periods of passable land-based transportation routes, limiting construction and industry (chap. 4). However, simultaneously, new opportunities open up—shortened shipping routes, growing tourism, and the promise of a rich bonanza of as-yet-untapped natural resources.

The vanishing cryosphere matters not just because of changing Arctic landscapes and fading indigenous lifeways. Waning glaciers in the high mountains of Asia, the Andes, and the Sierra Nevada in the United States lessen dependability of water resources, with consequences to agriculture, hydroelectric power generation, flood control, and tourism.

Tropical mountain glaciers are fading fast. Soon, the snows of Kilimanjaro in Tanzania may exist only in Hemingway's book, while only classic paintings and fading postcards preserve memories of once-mighty snow-clad peaks. Far beyond the Arctic looms the worldwide threat of rising seas, putting populations in low-lying coastal areas at risk. Coastal storm flooding will become more frequent, beaches will wash away, and saltwater will encroach inland. Threatened small island states in the Pacific and Indian Oceans already prepare to evacuate and relocate.[7] Therefore, what happens in the Arctic does not stay in the Arctic. Instead, it offers a preview of the Earth's climatic future. The coming meltdown truly matters to all!

TRANSFORMING THE ARCTIC

Plight of the Polar Bear

Pack ice is . . . the platform on which polar bears travel, hunt, and rest. Bearded seals and ringed seals haul out on its floes to catch spring sun. Pack ice is the staging ground for human hunters as well.

Seasonal ice is a villager's highway, a hunter's path in spring to the ice edge where bowhead and beluga whales, walruses, and bearded seals can be found.

—Gretel Ehrlich, *In the Empire of Ice: Encounters in a Changing Landscape* (2010)

Polar bears, close cousins of the brown bear, live on the shores surrounding the Arctic Ocean and on its sea ice.[8] These are large animals, the male weighing 350–700 kilograms (772–1,543 pounds), the female half that size. Although born on land, polar bears spend most of their time on sea ice. They tend to seek out spots where ice meets open water, such as leads or polynyas,[9] where they can hunt their favorite food—ringed and bearded seals. After locating a seal breathing hole, the bear waits patiently, hours at a time, until a seal appears. Reaching into the hole with its forepaw, the bear drags the seal onto the ice and quickly dispatches it. Even when not actively hunting, bears like to rest on ice.

FIGURE 9.1

Plight of the polar bear. (V. Gornitz)

As the Arctic climate warms, the polar bears' preferred domain—sea ice—steadily shrinks (fig. 9.1). September 2016 tied with 2007 for the second-lowest summer sea ice minimum.[10] (September 2012 ranks as the lowest to date.) Meanwhile, Arctic maximum winter sea ice extent continued on a record low level streak for the fourth straight year, after 39 years of satellite observations.[11] Habitat loss leads to malnutrition and even starvation of the bears. Although they are excellent swimmers, the much-reduced ice cover forces them to swim longer distances and deplete their energy reserves, which may result in drowning. Most sea ice is now thinner, weaker seasonal or first-year ice, upon which it is more difficult for bears to hunt. As a result, they appear to be spending more time ashore, especially in the fall, awaiting winter freeze-up. The growing numbers of displaced bears in populated areas bring menace, but also create a big tourist draw.[12] Numerous bears congregate around places like Churchill on Hudson Bay, Canada, especially around Halloween. Children celebrating the holiday are warned not to wear bear costumes, for fear of attracting the beasts. Grizzly bears from the south, encouraged by the changing climate, expand into polar bear territory. In rare cases, grizzlies have even hybridized with polar bears, perhaps heralding the birth of a new subspecies.[13]

Polar bears are not the only species that will regret the loss of ice. Walruses and seals also enjoy stretching out on these icy platforms. Although also good swimmers, walruses hunt in relatively shallow waters to catch their seafood prey. But as the summer ice edge recedes northward into deeper water beyond the continental shelf, the animals' swimming abilities reach their limits. Like the polar bears, walruses begin to spend more time ashore.[14] Beaches grow densely covered with thousands of bellowing and writhing creatures. What effect will a landward shift have on their social and reproductive life? To find out, scientists partner with traditional hunters to closely observe how the large sea mammals adjust their migratory patterns to changing sea ice characteristics. But it's not only about bears or walruses. "We are all ice-dependent species," says Julienne Stoeve, a sea ice specialist at University College, London, as this chapter demonstrates.[15]

Fading Lifeways

Author Gretel Ehrlich points out that the Inuit (Eskimo) people and the wildlife upon which many still rely "co-evolved with ice"—i.e., became well adapted to the frigid Arctic climate. Sheila Watt-Cloutier, an Inuit spokeswoman, poignantly pleads the case for the right of her people to pursue their traditional hunting and fishing activities. She points out that their hunting culture thrives on the snow and ice, or, as she puts it, the "right to be cold."[16] But as we have seen, a major transformation is reshaping the Arctic. The lengthened open-water season drives Inuit hunters farther offshore to locate polar bears, and also seals or walruses, needed for meat and furs, which still provide many with a livelihood.

The older generation clearly notices the marked climate changes within their lifetimes. Herb Anungazuk from Wales, Alaska, observes: "We had good ice most of the time from December until the third week of June. Now, by mid-April or May the ice goes out, and we have years when there is almost none at all."[17] He also notes how melting ice has changed the salinity of the sea, which in turn affects phytoplankton and fish, migrating birds that feed on fish, and the large marine mammals such as whales, walrus, and seals. Animals are modifying their customary migration patterns, which makes it harder for the hunters to predict their whereabouts.

Anungazuk recounts how "the walruses always came in about June 10 but now they come earlier. . . . The drift ice is almost gone and the pack ice goes out beyond the continental shelf where it's too deep for them to dive for food." Another Alaskan native laments that "these days there is no more multiyear ice. No old ice at all. That's what we depended on for water because all the salt had percolated out. We started losing it maybe 10 or 15 years ago. That's how long ago things like wind and ice began to go fast."

Greenlanders too keenly feel the loss of ice. Jens Danielsen, a northern Greenland hunter, recalls: "We used to hunt from September through June on the sea ice. Now we don't know when the ice will be strong enough to hold us.[18] We had multiyear ice that never melted. Now it is all *hikuliak*, new ice. . . . In the last two or three years there were big storms and high waves in November. That was new. We never had storms like that break up the ice before. . . . It's hard on hunting. There are now so few days of getting food. This year the ice didn't come until December, later than we've ever known. . . . Ten years ago the ice was six feet thick. When the ice was thick, nothing bothered us when it was cold like this. Now the ice is so thin, just a little wind wave breaks it. It has been like this for the last three years."

As the subsistence hunting lifestyle gradually fades, many younger Inuit migrate to the larger towns, where they seek better jobs and new opportunities (see "The New North," later in this chapter). But with the new life come difficulties in adjusting to the modern pace, and social problems develop. Still, many continue to hunt on the remaining ice, even if less frequently than before. As Jens puts it: "Our lives are based on ice."

Changing Landscapes

Soils, plants, and animals quickly react to changes in ice or snowfall. While most of us welcome an early spring, this may not be the case for songbirds that fly north at the end of winter.[19] The birds anticipate a feast on the Arctic's abundant, rich food resources. New tundra growth and insects appear shortly after the snow begins to melt, which occurs much earlier than it used to. However, the migratory songbirds stick to their customary punctual spring arrival schedule, cued by the lengthening hours of sunlight, which remain fixed. Will the birds adapt fast enough to narrow a growing mismatch with local vegetation and insect cycles, so as to maintain healthy breeding and feeding habits? Elsewhere, grazing animals such as reindeer and caribou struggle to break through a glazed ice-snow crust formed after more frequent winter or early-spring rainfall. Sami herders must realign their herds' migratory needs to the altered weather patterns.

As the Arctic warms, brown and grizzly bears, red fox, fish from lower latitudes, and other southern species advance into polar bear, Arctic fox, and beluga whale territory. As their ecological ranges overlap and the animals intermingle, they may also begin to interbreed, a possibility that concerns wildlife ecologists.[20] A hybrid polar bear–brown bear was recently killed by a hunter in the High Arctic. The mixed-breed bear had a white fur body, but brown paws. Its head was also shaped more like that of a grizzly, yet it was equipped with a polar bear's longer claws. Marine animals such as seals, walruses, and sea lions could potentially hybridize more readily, since they share the same number of chromosomes. But the hybrids may not be as well adapted to Arctic living as the native species; such an adaptation would require acclimatization over multiple generations. Furthermore, even native species may have trouble adapting fast enough to a rapidly changing environment. Ecologists also worry that the newcomers may carry diseases for which Arctic sea mammals lack immunity and that could wipe out entire communities. Opening of ice-free shipping corridors could also introduce invasive species and pests.[21]

Someday, the northern shores of Canada and Greenland may harbor the last refuge for sea ice–dependent mammals. Robert Newton, an oceanographer at the Lamont-Doherty Earth Observatory of Columbia University, says: "The Siberian coastlines are the ice factory and the Canadian Arctic Archipelago is the ice graveyard."[22] Even so, there may be limits to how many polar bears the region could accommodate, and there are questions as to whether endangered bears from elsewhere should be transported there.

Aside from its effect on wildlife conservation, the waning sea ice may steer midlatitude weather patterns in new directions (see chap. 2). Some scientists see linkages among a warming Arctic, summer sea ice meltdown, and a sluggish, wavier jet stream that delivers more extreme weather to the midlatitudes. One such weather pattern is the wintertime "warm Arctic–cold Eurasia"—characterized by a reduced Arctic-midlatitude air pressure difference that leads to colder, snowier midlatitude winters.[23] Polar warming also decreases the north-south temperature difference, causing the weakened jet stream to meander like a mature river nearing a delta. As the meanders carry cold air south and warm air north, frigid air outbreaks spread across Eurasia and the eastern United States. The sluggishness also sets up more "blocking" patterns in which cold air outbreaks persist longer. However, an emerging understanding of these climate connections embraces a more complex network of multiple cause-and-effect pathways whose exact linkages have yet to be established.[24]

Environmental changes are also afoot farther south. An intense forest fire ripped through Fort McMurray (pop. 61,000) in Alberta, Canada, in May 2016. The blaze scorched much of the city and forced a mass evacuation. At its worst, the thick smoke turned day into night and reduced visibility to a few feet. Although this particular event can be blamed on unusually dry weather conditions and strong winds, boreal forest wildfires are on the rise in recent decades.[25] Boreal (northern) forests, populated by coniferous trees such as pines, spruces, and larches, sprawl across Russia, Canada, Alaska, and Scandinavia. Because winter snow thaws earlier, more exposed dried leaves, dead needles, and drying peatlands supply ample tinder for lightning-triggered wildfires. In North America, four times the area burned in the 2010s than in previous decades. If current trends persist, conflagrations such as the one at Fort McMurray may become more commonplace. Some of that soot eventually reaches Greenland, adding to the load that darkens the ice sheet and triggers more melting (see chap. 5).

To offset the effects of disappearing sea ice, some scientists even propose large-scale geoengineering schemes to chill the north. One such scheme would employ wind-powered pumps to convey warmer water to the surface in winter. As it spreads out, the water would thicken the sea ice upon freezing. Aside from the huge expense and untested practicality, such vast tinkering with the Earth's systems may produce unpleasant, unforeseen consequences or yield counterproductive results. For example, more heat may simply head north to compensate for the induced changes in ocean and atmospheric circulation.

But the picture is not altogether bleak. The next section investigates how the new north holds much promise for travel, improved access to natural resources, and expansion of opportunities. The big thaw may unfreeze a fresh new world.

THE NEW NORTH

> Some say that the Northern Lights are the glare of the Arctic ice and snow;
> And some that it's electricity, and nobody seems to know.
> But I'll tell you now—and if I lie, may my lips be stricken dumb—
> It's a mine, a mine of the precious stuff that men call radium.
> It's a million dollars a pound, they say, and there's tons and tons in sight.
>
> —Robert W. Service, "The Ballad of the Northern Lights" (1907)

The picture of a rapidly changing Arctic has an upside as well. Increasingly navigable waters will open up new trade routes, shorten travel times, cut fuel costs, and open up the Arctic to development of its bountiful mineral resources. The Arctic stores vast mineral wealth in hydrocarbons, much of which lies in shallow offshore marine deposits.[26] Most of the undiscovered gas and many other mineral resources lie in the Russian Arctic. On the other hand, in a fragile Arctic environment, improperly managed development of mineral resources could create potential risks to the sensitive tundra and its wildlife. Because of the short growing season and slow growth rates, recovery from damages to sensitive ecosystems may take many years. But the lure of finding hidden treasure is strong, and many bold explorers and adventurers have risked life and limb for a chance at instant wealth. This urge holds as true today as it did back then.

The Lure of Northern Treasure

> *GOLD! GOLD! GOLD! Sixty-Eight Rich Men on the Steamer Portland. STACKS OF YELLOW METAL! Some Have $5,000, Many Have More, and a Few Bring Out $100,000 Each.*
>
> —*Seattle Post-Intelligencer*, Klondike Edition, July 17, 1897

The discovery of gold in 1896 along tributaries of the Klondike River in the Yukon of northwestern Canada set off a wild stampede that at its peak in

1897 attracted an estimated 100,000 eager prospectors.[27] Only somewhat over a third actually reached the Klondike goldfields because of the harsh climate and rough physical environment. Miners were forced to haul a year's supply of food and other needed gear up a steep slope of the Chilkoot Pass (elevation 1,067 meters; 3,057 feet)—one of the most dangerous routes to the goldfields. In winter, people climbed slowly, single file, up steps carved into the slippery ice. Several trips were needed to haul all of the miners' supplies to the summit. By 1897, construction of a tramway had eased the passage for wealthier miners. But the gold boom was short-lived, and only a handful of the miners ever struck it rich. In 1898–1899, larger quantities of gold were discovered in Nome, Alaska. By then, the Klondike gold rush was history.

The dream of striking it rich still endures. Diamonds in Canada? Once believed as unlikely as finding the proverbial pot of gold at the end of the rainbow, but sometimes luck prevails! As told in chapter 4, two persistent prospectors, after many fruitless years of search, finally hit the "mother lode"—diamond-bearing kimberlite pipes near Lac de Gras. Ranked as the world's fourth-largest diamond producer, in 2015 Canada produced 11.6 million carats of diamonds worth $1.7 billion (in U.S. dollars).[28] The Ekati Diamond Mine, 320 kilometers (200 miles) north of Yellowknife, Northwest Territories, and 160 kilometers (100 miles) south of the Arctic Circle, first opened in 1998, soon followed by the Diavik Diamond Mine in 2003, and several others later.

Development of the Diavik Diamond Mine, about 300 kilometers (186 miles) northeast of the nearest town, Yellowknife, posed unique challenges because of its remoteness and harsh Arctic environment.[29] Heavy equipment and supplies can be trucked in for only two months in winter, when the ice grows thick enough over permafrost and lakes to support the weight of trucks. Extraction of the precious ore, located in a kimberlite pipe covered by the Lac de Gras, presented another special engineering challenge. The overlying lake had to be diked and drained, and special devices had to be installed to keep the permafrost frozen. Other difficulties included working in extreme cold and darkness, and providing for careful disposal of debris to minimize contamination. But the efforts have paid off. Diamond production from Diavik alone yielded 91 million carats of high-quality gems from 2003 to 2014, worth an average of $135 to $175 per carat. When the mine eventually closes, the land will be restored as nearly as possible to its original condition.

In Russia, the Mir (or Mirny) diamond pipe in Sakha Republic, Siberia—the world's second-largest excavation after the Bingham Canyon copper mine in Utah—operated for nearly half a century between 1957 and 2004. The severe Arctic winters necessitated blasting through rock-solid permafrost with strong jets and dynamite to get into the ore-bearing pipe. In its heyday, the mine produced over 10 million carats of diamonds a year, a fifth of which were of gem quality. Production from this single mine helped boost the Soviet Union into a major economic power. A second Russian diamond mine, the Udachnaya ("lucky") pipe in Yakutia, Siberia, initially a surface operation starting in 1971, has tunneled underground since 2014.

The Popigai diamond deposit on the Taymyr Peninsula of northern Siberia is located within a 100-kilometer (62 mile)-wide crater gouged out by an asteroid collision 35 million years ago. The intense temperatures and pressures from the blast almost instantly transformed any initial carbon present in the target rocks into diamond.[30] Although inferior as gems, their superior cutting and polishing properties make them highly desirable for industrial applications. However, the remote location and lack of infrastructure limit the full exploitation of Popigai's potential.

Gold and precious gems are not the only mineral treasures enticing prospectors north. The Arctic promises a vast, untapped cornucopia of mineral resources ranging from fossil fuels to metals and other critical, strategic elements. The thawing Arctic is making this hidden bonanza more accessible.

Reaping a Bonanza

On August 2, 2007, two *Mir* submersibles descended over 4 kilometers (2.5 miles) and planted a titanium Russian flag on the seabed beneath the North Pole. This feat was no mere gesture, but a deliberate move intended to bolster Russia's claim to the underwater Lomonosov Ridge as an extension of its continental shelf. If the claim is proven, Russia would gain exclusive rights to its mineral resources. But Denmark and Canada, also sharing Arctic coastlines, counter the Russians with competing claims of their own. This still-unresolved territorial dispute highlights the growing interest in reaping the Arctic's potential bonanza.

The United States Geological Survey estimates that around 30 percent of the world's undiscovered natural gas and 13 percent of its undiscovered oil lie north of the Arctic Circle, largely offshore under less than 500 meters (1,640 feet) of water.[31] The Alaska Platform section of the continental shelf may contain over 31 percent of yet-to-be-discovered Arctic oil, with other promising areas in northern Canada, Russia, and Greenland. Estimates of Russia's undiscovered oil range between 66 and 142 billion tons of oil equivalent (a unit of energy equivalent to that obtained by burning a ton of oil). Russian energy resources constitute 52 percent of the Arctic total, while Norway holds 12 percent.[32] Arctic hydrocarbon resources are important to Russia: 91 percent of its natural gas production and 80 percent of its explored reserves of natural gas lie there.[33]

Among the northernmost mines in the world, Sveagruva sits on the island of Spitsbergen in the Norwegian-held Svalbard archipelago. One of the largest underground coal mines in Europe, it has produced 1 to 2 million metric tons of coal annually in recent years.[34] In 2015, $64 million worth was shipped to Denmark, Sweden, the Netherlands, Germany, and the United Kingdom. Coal mining is controversial in Norway, which relies heavily on hydroelectric power. However, coal mining on Svalbard has had a long history. Because it is a mainstay of the local economy, islanders fear job losses if mining should end. New jobs in tourism or research would not make up the difference.

Fossil fuels comprise just one of many Arctic mineral resources. Many important metal deposits lie buried in the permafrost. As of 2006, Russia operated 25 active mines in the Arctic.[35] Although most are nickel-copper mines, tin, phosphate, aluminum ore, uranium, and other ores are also extracted. The largest mine was the Norilsk Nickel plant at Pechenga, northern Kola Peninsula, which produced nickel, copper, and cobalt. Notorious for its heavy pollution that created a barren zone surrounding the mine, the Norilsk plant was forced to close in 2008. After merging with two other plants, Norilsk Nickel became one of the world's largest mining groups, now a major producer of nickel, palladium, platinum, and copper.

Elsewhere in the Arctic, the Red Dog zinc and lead mine,[36] located in a remote corner of northwest Alaska, was named for the dog of a bush pilot and prospector and the red-stained creeks he discovered while flying. Geologists later confirmed extensive mineralization, and mining commenced in 1989. Red Dog is one of the world's largest producers of zinc and holds the largest zinc reserves. It also yields significant amounts of lead and silver.

The Mary River Mine, operated by the Baffinland Iron Mine Corporation, on Baffin Island, Nunavut, Canada, began operations in 2015. Expansion plans in 2016 called for a railway that would transport ore to the docks and open up shipping for ten months of the year, extending into ice-covered wintertime periods.[37] Subsequent plans call for building a second railway to a proposed port at Steensby Inlet, to the south. As of October 2018, the Nunavut Impact Review Board is reviewing the latest Baffinland Iron Mine Corporation's expansion plans and expects to complete its assessment by June 2019.[38]

The longer open-water season and receding ice opens the door to development of Greenland's rich mineral potential. High-grade deposits of iron, zinc, ruby, zirconium, rare earth elements, uranium, and other metals are scattered over the island. With the promise of an economic boom, Greenlanders question: Who will benefit? As the traditional hunting and fishing lifestyle slowly fades, mining is increasingly seen as a stepping-stone to economic growth, self-sufficiency, and eventual independence for the island. Thus, many Greenlanders favor mining for the new jobs it offers and as a gateway to full independence. However, possible negative impacts from the proposed Kvanefjeld uranium and rare earth element (REE) mine near Narsaq in southern Greenland in a still-pristine environment, and from a nascent tourist industry, have aroused controversy. "In Greenland, we are nature people. We live from the nature," said Julie Rademacher, a former member of Danish Parliament and advisor to a Greenlandic labor union. "We don't want to damage what's important to us, and to our children and future generations. We want mines, but not for [any] price."[39] Issues between Greenland and Denmark over the latter's strict regulations of radioactive materials have also yet to be resolved.

Beyond mining, the warming of the Arctic encourages many new undertakings that would not have been dreamt of a few decades ago, as the next section shows.

Promise and Peril

On August 21, 2016, a giant luxury cruise ship, the *Crystal Serenity*, sailed into Nome, Alaska, carrying over 1,000 passengers and a crew of nearly 600.[40] The ship made an unprecedented 32-day journey from Seward, Alaska, through the fabled Northwest Passage, with several stops in the Arctic along the way, before heading to New York City, its final destination, that September. As recently as a decade or so ago, only ice-breaking ships were capable of traversing the passage, and until that voyage, only a few smaller cruise ships had ventured through. This crossing, the first for a cruise ship of its size, promised a new era in northern tourism as the Arctic enjoys more ice-free summers. The *Crystal Serenity*'s second voyage in August of 2017 may have been its last. The Crystal Cruises company may substitute a smaller, specially built, Polar Code–compliant megayacht, the *Crystal Endeavor*, which holds 100 guest suites. Meanwhile, the Norwegians are planning a 2019 cruise from Alaska to Greenland with a large, hybrid-electric cruise ship, the MS *Roald Amundsen*.[41]

Is the region ready for mass tourism? The sudden influx of large numbers of tourists may overwhelm the small Arctic towns. Furthermore, existing emergency facilities may be incapable of handling unanticipated accidents or extreme weather events. Environmentalists voice concerns over possible oil pollution or ecological disturbance. Crystal Cruises, for its part, promises to adhere to high environmental safety standards. However, since their steps are voluntary, would other cruise ship companies adopt the same strict measures?

Cruise ships are not alone. In 1969, Humble Oil's icebreaking supertanker SS *Manhattan* undertook the journey. Locked in ice repeatedly, the *Manhattan* had to be freed by an accompanying Canadian icebreaker. The first cargo ship, escorted by icebreakers, made it through successfully in September 2013. Today, although still hazardous, the Northwest Passage has become more navigable over longer periods. It sharply cuts travel distance between East Asia and Europe via the Suez or Panama Canal, and also shortens the route from Asia to eastern North America. Similarly, the Northern Sea Route (Northeast Passage), via the Russian Arctic, reduces the trip between European and East Asian markets. These shipping shortcuts save ship owners considerable sums. Although some scientists envision an ice-free Arctic summer by the 2040s,[42] expansion of trans-Arctic shipping also depends on other factors: duration of the shipping season (still fairly short), the shortage of ice-ready vessels, the price of ship fuel, and unpredictable ice conditions.

These short-term impediments do not deter nations from staking claims, exemplified by the symbolic Russian flag planting on the ocean floor under the North Pole. However, the status of the Northwest Passage still awaits resolution. Canada maintains that it lies within its territorial waters, whereas the United States insists that it remain an international waterway. Meanwhile, China casts an eager eye on Greenland's mineral potential. Although China's attempt to purchase an unused naval base was thwarted, a subsidiary of a Chinese company bought a 12.5 percent stake in Greenland Minerals and Energy (GME), which holds a license to develop the Kvanefjeld uranium and REE (rare earth elements) deposit. Chinese have also expressed interest in developing a large iron deposit near Nuuk, Greenland's capital.[43] The Chinese activities may be part of a wider goal of investment and eventual control over strategic resources.

The opening of the Arctic to development brings environmental hazards as well as economic benefits. The fragility of the tundra ecosystem has been examined in chapter 3. Scars on the landscape from mining, road building, or other large-scale construction may take years to heal because of the short growing season and partially thawed permafrost. The harsh Arctic climate and long periods of winter darkness make cleanup of potential oil spills more difficult than farther south. Alien species may increasingly invade ice-free Arctic waters. Smaller Arctic invaders, such as small crabs, barnacles, and copepods, may hitch rides on the hulls of ships or in ballast tanks,[44] or simply swim or fly north. More astounding was the journey of a gray whale spotted off the coast of Israel in 2010. More showed up in 2013 off the coast of Namibia in southern Africa! These Pacific Ocean whales presumably crossed an ice-free Northwest Passage into the Atlantic Ocean.[45]

The consequences of vanishing ice are not confined to the Arctic. As documented in chapter 4, the glaciers of the world's high mountain ranges are also in retreat. What do the darkening peaks mean for the people who live or play in the narrow valleys and adjacent lowlands? The next section examines some of the ways the changing icescapes impact their lives.

DARKENING PEAKS

You can think of glaciers as hydrological Prozac—they smooth out the highs and lows.

—Jeffrey McKenzie, hydrogeologist

The worldwide retreat of glaciers described in chapter 4 is a highly visible and easily grasped manifestation of global warming. In the simplest terms, temperature rises; ice melts; glaciers withdraw upslope.[46] As high mountain peaks shed their icy capes, numerous consequences unfold farther downstream. An earlier spring thaw follows a lessened winter snowfall. Stream runoff peaks sooner, in spring and early summer, but slows to a trickle during the hottest summer months. This altered peak river runoff timing cuts water supplies to urban centers and agricultural irrigation when they are most needed. It also introduces greater uncertainty in hydroelectric power generation. As glaciers retreat, mountain village life grows increasingly hazardous, as slope instability mounts and glacial lake outburst floods become more frequent. Whether in the Alps, the Andes, or the Rockies, changes in snow and ice cover will impact the tourist economy, affecting winter skiing and summertime mountain climbing, and transforming the serene beauty that draws countless visitors to the lofty summits.

Water Woes

Peruvian farmers anxiously scan the Andean peaks and see the snow line receding higher and higher. They worry what will happen when the snow disappears entirely. Where will their water come from? The Andes hold nearly all of the world's tropical glaciers, and they are going fast. Glaciers along the Cordillera Blanca in the central Andes, in Peru, shrank by 33 percent between 1980 and 2006.[47] In neighboring Bolivia to the south, along the Cordillera Real, glacier surface area has declined by 48 percent. The Chacaltaya glacier, once the site of Bolivia's only ski resort, completely disappeared in 2009.

Recent Andean climate trends give the farmers good cause for concern. Andean snow cover declined by 2 to 5 days per year and the snow line altitude climbed by 10–30 meters (33–98 feet) per year south of 29°S in Chile between 2000 and 2016.[48] The largest changes occurred on the east side of the Andes, particularly during Southern Hemisphere winter. These changes were linked to declining precipitation and increasing temperatures. However, these trends are also strongly influenced by regional variations and sources of decadal climate oscillations such as El Niño/Southern Oscillation (ENSO).

Diminishing Andean glaciers jeopardize water resources, agriculture, and hydroelectricity in a South American region with a population of 77 million.[49] Many major Andean cities lie above 2,500 meters (8,200 feet) and depend to varying degrees on glacial meltwater. Glaciers provide cities such as Quito, Ecuador; La Paz, Bolivia; and Lima, Peru, with a significant portion of their drinking water supplies. Meltwater also supplies drinking water to numerous rural communities, not to mention water for irrigation. A large percentage of electricity generation in Colombia, Peru, Ecuador, and Bolivia comes from hydropower. Reduced glacier runoff would seriously impact the hydropower industry and diminish energy output.

Dependable water supplies from the "Third Pole"—the high mountains of Central Asia—affect a quarter of the population of South Asia. As pointed out in chapter 4, most of the major rivers that drain into Southwest, South, Southeast, and East Asia originate in the Himalayas or high Central Asian Plateau. Water from the Indus River alone feeds some 237 million people. Rivers streaming down the southern flanks of the Himalayas help irrigate the wheat-growing "breadbasket" regions of northern India—Uttar Pradesh, Punjab, Madhya Pradesh, and Haryana states—in addition to the water supplied by the summer monsoon.[50] Mountain water runoff also contributes to hydropower generation that meets a significant fraction of the region's electricity demand.[51]

Glacier meltwater helps many Himalayan rivers maintain a steady flow, which is especially crucial for watersheds in the drier western portions of the mountain range as well as the high Central Asian Plateau. Glacier-fed rivers, which act as buffers against drought, partially mitigate high water stress during dry years in the Indus, Aral, Tarim, and other drainage basins of Central Asia.[52] Although generally less critical for the wetter eastern monsoon-fed Himalayan drainage basins, glacier meltwater also alleviates shortages in years of low monsoon rainfall. Upstream runoff is essential in ensuring storage and release of water farther downstream when it is most needed for agriculture. In the Indus basin, for example, two major dams (the Tarbela Dam on the Indus River and Mangla Dam on the Jhelum River) regulate a major irrigation network upon which millions of people depend. Both rivers lie in the upper Indus basin and are fed mostly by meltwater.[53] Any projected future glacier ice losses in these and several other Himalayan drainage basins may substantially curtail runoff, only partially compensated for by monsoonal rainfall, and thereby adversely affect downstream crop yields.[54]

In the future, as Himalayan and other mountain glaciers recede, river runoff may temporarily increase, partially offsetting potential drought scarcities. However, their ability to buffer droughts will decline in the long-term, as available ice-melt and hence, runoff diminishes.[55]

Runoff from some regions, such as the Karakoram Mountain range, will probably remain steady, since local meteorological conditions favor abundant winter snowfall and ice build-up (see "Bucking the Trend," in chap. 4). While some studies foresee more abundant monsoon rainfall that would partially compensate for glacier runoff shortfalls, increased monsoon variability could instead diminish reliability of water resources.

A recent study found that in roughly half of 56 large, glacier-fed drainage basins examined worldwide, annual glacier runoff will reach a maximum "peak water" as more and more of a glacier melts back, after which runoff will steadily diminish.[56] (Many of the remaining basins have reached this point already.) Annual glacier runoff is expected rise until approximately midcentury for river basins fed by high Asian glaciers, after which a decline will set in. However, the exact timing of this switch-over will vary considerably over the region, depending on glacier size and local differences in summer monsoonal precipitation. Furthermore, the largest decreases in glacier runoff relative to the entire basin runoff are projected for the Central Asian basins (particularly the Indus and Aral Sea drainages) in September–October and for the South American basins in February–March (note the reversed seasons in the Southern Hemisphere).[57]

In many western U.S. states, mountains stand as islands of moisture surrounded by arid desert "seas." But the snowpack shrank by an average of 10 to 20 percent between the 1980s and 2000s.[58] The outlook may worsen as future temperatures rise. Projected snowpack declines of 30 to 60 percent may occur as early as the 2030s.[59]

In California, over 60 percent of the state's water needs derive from the Sierra Nevada snowpack, on average.[60] Water from snow-fed rivers and the state's two major water-diversion networks provide much of the water for more than 25 million people and more than 1.46 million hectares (3.6 million acres) of farmland.[61] California recently recovered from an exceptionally severe multiyear drought, in part worsened by the historically dwindling Sierra Nevada snow cover. While the heavy 2017 winter snowfall brought much-needed relief to the state, higher future temperatures will limit snowfall to higher elevations; also, more precipitation will fall as rain instead of snow. The higher spring runoff is accompanied by greater flood risks, but also reduced summer river flow to fulfill agricultural, industrial, and municipal demands.

Mountain Hazards

White blankets cover the surface of the Rhône Glacier to slow down the melting ice in a desperate attempt to save a popular Swiss tourist attraction—an ice grotto that is recarved each year. [62] In the late nineteenth century, this glacier nearly reached the tiny village of Gletsch; since then it has retreated 1.42 kilometers (0.88 miles) up the steep mountainside.[63] Proposed attempts to rescue another dying Swiss glacier—the Morteratsch Glacier—by blowing artificial snow on it would take an enormous number of snow machines for just one glacier—a costly and as-yet-untested method. Swiss glaciers lost nearly 28 percent of their area just between 1973 and 2010.[64]

These losses are beginning to unglue the high mountains. In 2006, a mountain guide in Grindelwald, a Swiss tourist resort at the foot of the majestic Jungfrau, Mönch, and Eiger peaks, witnessed an "immense cloud" of dust generated when a huge chunk of the Eiger collapsed.[65] The mountain slopes surrounding the Lower Grindelwald Glacier evidence signs of slope instability. The 2006 landslide released 2 million cubic meters (900,000 cubic yards) of rock at the foot of the Eiger. The debris formed a rock dam, augmented by subsequent landslides, at the entrance to a narrow glacial gorge.[66] In Tibet, massive, catastrophic avalanches cascaded down the slopes of two adjacent glaciers in July and again in September of 2016.[67] The collapse was triggered by an unusually high amount of basal meltwater that accumulated over weak bedrock material during a particularly warm and wet summer.

Rockfalls and avalanches are normal occurrences in the high mountains, but as glaciers dwindle, removal of ice and permafrost that once secured portions of surrounding mountains threatens to unleash more rockslides and avalanches. As glaciers retreat upslope and expose bare rock or loose glacial sediments in the high mountains, more frequent landslides and other debris flows are likely to occur.[68] Thawing of high-altitude permafrost may also enhance slope instability, leading to rockfalls, rockslides, and rock avalanches in the Alps, British Columbia, and elsewhere. Monte Rosa in the Italian Alps has been scoured by ice and rock avalanches since the 1990s, peaking with a massive event in 2005. Other Alpine slope collapses have occurred on Mont Blanc (1997), Punta Thurwieser (2004), and Dents du Midi and Dents Blanches (2006).[69] Behind this picture of emerging mountain instability lies a complex intermeshing of multiple geomorphological processes. For example, slope failures can unload heavy sediment or rock piles that dam streams and clog downstream flow, in turn magnifying flood hazards.

At the mouth of the Lower Grindelwald Glacier, a stream rushes down the narrow gorge, above which lies a large meltwater lake. This lake began to grow, and by spring 2008 had broken through a weak glacial moraine and the 2006 rocky dam in a sudden-outburst flood. This event prompted the construction of a water-diversion tunnel in 2009–2010 to prevent further flooding downstream.[70] Before the diversion, the summer floodwaters had regularly carried large ice blocks down the gorge in an event that sounded like tumbling "battle tanks," according to a local hotel owner in the town of Grindelwald.

Glacial lake outburst floods (GLOFs) present an increasing risk from melting glaciers as moraine-, ice-, or landslide-formed dams suddenly give way, discharging huge volumes of water such as the heavy flow that briefly menaced Grindelwald. Because of higher temperatures in

Switzerland over the last several decades, new glacial lakes have appeared where none had been seen before. GLOFs are most prevalent in southern Switzerland during summer months. While Alpine glacial lakes are generally small, their proximity to mountain villages and downstream towns poses a risk from outburst floods.

In recent years, a number of destructive Himalayan GLOFs have wiped out bridges, homes, farmland, and even small hydropower plants in Nepal and Bhutan.[71] At least 20 out of more than 2,000 glacial lakes in Nepal are "potentially dangerous." Neighboring Bhutan has suffered damage from a growing number of sudden GLOFs within the last few decades. Moraine-dammed glacial lakes, enlarged by rapid retreat of glaciers, can drain rapidly when the moraine collapses under pressure from the excess water buildup. Lake water may also overtop the dam after heavy rains, avalanches, or rockslides. Earthquakes, not uncommon in the tectonically active Himalayas, may also provoke a GLOF.

Himalayan GLOFs present one of the most serious natural hazards facing the rapidly growing mountain population. Knowledge of where and when these events have occurred is important in land use planning, dam siting, and construction decisions. However, until now, because of the remoteness and difficulty of access, the number of GLOFs and their geographic distribution have been poorly documented. A recent study analyzing Landsat imagery from 1988 to 2016 expanded the inventory of known Himalayan GLOFs by 91 percent, even though the study area covered only 10 percent of the high Himalayas.[72]

Other geohazards associated with thawing ice include subglacial floods, or jökuhlaups (see glossary), and icequakes triggered by glacier movement, calving, and glacial isostatic rebound.[73] Iceland's Vatnajökull ice cap covers some active volcanoes, such as Grimsvötn and Kverkfjöll. As the ice cap thins a half meter per year, the lithosphere responds by rebounding upward a few centimeters per year. Pressures in the asthenosphere—the birthplace of magma—slowly drop, making it easier for magma to melt and trigger future volcanic eruptions. Of greater concern is a remote possibility of volcanism or geothermal activity under the West Antarctic Ice Sheet, which could flare up if the ice sheet thins in the future and reduces the overburden pressure.[74] A subglacial eruption could potentially help mobilize and destabilize the ice sheet via basal melting, possibly unleashing a jökuhlaup, as in Iceland. On the other hand, buried volcanoes could present topographic obstructions that act as brakes on ice flow.

Even larger numbers of people worldwide will be concerned over the increasing specter of rising sea levels. The next section examines some of the ramifications.

ENCROACHING OCEANS

But it is the horizontal incursion of the rising sea that is most apparent.

—Henry Pollack, *A World Without Ice* (2009)

Ice into Water

Water and ice are two faces of the same substance—one liquid, flowing, mobile, the other solid, largely inert, yet slowly creeping downslope under the relentless pull of gravity. Water is in constant motion around the world. As it evaporates from oceans and land, it rises into the atmosphere. As it reaches saturation, the water vapor condenses into clouds of tiny water droplets or ice crystals, then releases its burden as rain or snow. From sea to land and back to sea via streams and rivers; from water to ice and back to water with changing temperatures—such is the never-ending dance of the hydrological cycle, the continuous journey of H_2O around the globe.

Ever since the end of the last ice age, the Earth has enjoyed a lengthy period of relative stability and warmth, during which time sea level has varied by no more than a few meters. A rough balance has prevailed in the cycling of water between land and sea—ice to water and back. This balance is now being upset as more ice melts each year than refreezes, returning excess meltwater back to the ocean. Furthermore, warming ocean water expands upward. Since 1993, sea level has been climbing by around 3 millimeters (0.12 inch) per year (chap. 1). A high-end "business-as-usual" emissions trajectory could send sea level up by 0.45–0.82 meters (1.5–2.7 feet) by late this century.[75] The cryosphere's share amounts to 0.04–0.53 meters (0.13–1.7 feet). An even more dire forecast that assumes continued high carbon-emission rates revs up the marine ice sheet instability (MISI) in West Antarctica, in which ice cliffs crumble and ice shelves break apart by midcentury.[76] Half a century later in this scenario, Antarctica's ice losses alone would raise sea level by 1 meter (3.2 feet), and by over 15 meters (49 feet) by 2500, which would wipe out the West Antarctic Ice Sheet (WAIS)! Although this is a highly unlikely scenario within the designated time frame, it delineates a possible pathway to a much more aquatic future.

Until now, this book has documented the slow liquefaction of the cryosphere. The ocean is its ultimate repository. But the trends listed in tables 4.1, 5.1, 6.1, and 8.1 offer no clues about how the advancing edge of the sea may alter the lives of the tens of millions of people who live along the world's shorelines. This issue is briefly addressed in the next section.

TOWARD A MORE AQUATIC FUTURE

> *The truth is that the First Pulse was a profound shock, as how could it not be, raising sea level by ten feet in ten years. That was enough to disrupt coastlines everywhere, also to grossly inconvenience all the major shipping ports around the world . . . People stopped burning carbon much faster than they thought they could before the First Pulse.*
>
> *Too late, of course. The global warming initiated before the First Pulse was baked in by then and could not be stopped by anything the postpulse people could do . . . the relevant heat was already deep in the oceans*
>
> —Kim Stanley Robinson, *New York 2140* (2017)

Loud, piercing sirens disrupt the tranquility of Venice whenever "exceptionally high water" (*acqua alta* in Italian) reaches or exceeds 140 centimeters (55 inches) above the city's 1897 reference datum. The locals quickly place movable raised walkways across the plazas, shopkeepers and hotel owners sandbag their entrances, and people walk in hip-high rubber boots. The number of "very intense" 110-centimeter tides (enough to trigger the alarm system), which occurred on average once every 2 years in the early twentieth century, jumped to roughly four times per year by century's end.[77]

One of the earliest manifestations of a rising sea is the increasing number of days of "nuisance flooding," or flooding in coastal neighborhoods at times of higher-than-average tides or minor storms. To be fair, a higher local sea level cannot be blamed entirely on changing climate. In Venice, overpumping of groundwater as well as natural geological subsidence led to the mounting incidence of piazza flooding.

The frequency of nuisance flooding since the 1950s has grown in the United States as well.[78] Two-thirds of these flood days can be attributed to higher sea levels caused by mounting temperatures alone.[79] Nationwide, the number of such flood days increased by over 80 percent for the period 1985–2014 relative to 1955–1984. Although minor in extent, such flooding creates a growing public inconvenience due to more frequent street, driveway, and basement flooding. Repeated exposure to saltwater incursions may eventually degrade or damage buildings, private homes, and infrastructure in the flood-prone zone.

> *At sea level, raised to its current height just forty years before, the tenuous brittle fragile rebuilding efforts of humanity and all other living species were particularly vulnerable to superstorms in the new categories established, sometimes called class 7, or force 11.*
>
> —Kim Stanley Robinson, *New York 2140* (2017)

Tropical storms such as cyclones, typhoons, or hurricanes generate strong winds and heavy rainfall. In addition, these powerful meteorological disturbances create an elevated storm surge that results in coastal flooding. The storm surge is an increase in ocean water level above that of the normal astronomical tide, caused by wind pushing water toward the coast and the reduced atmospheric pressure.

Regions most vulnerable to the ravages of tropical cyclones or hurricanes include Southeast Asia, especially near major river deltas; the southern and southeastern United States; and the northeast coast of Australia.

Bangladesh faces a particularly high risk due to its low-elevation coast, a shallow continental shelf, and storm tracks that tend to recurve near the apex of the Bay of Bengal. An intense cyclone in November 1970 killed several hundred thousand people; another 11,000 perished in a May 1985 storm. Since then, the death toll from more recent storms has dropped significantly, thanks to the coastal storm warning system and evacuation centers instituted by Bangladesh. In neighboring Myanmar, which lacks these facilities, Cyclone Nargis killed over 138,000 people in 2008. In the United States, Hurricane Katrina flooded 80 percent of the city of New Orleans and killed ~1,200 people on August 29, 2005, causing $106 billion (2010 dollars) in damages.[80] The low-lying Gulf Coast, Florida, and the Carolinas are at high risk from hurricanes, and even the Northeast is not immune.

Nontropical storms also provoke serious coastal flooding. The heavy rainfall accompanying these storms amplifies the impacts of coastal flooding, inasmuch as these storms encompass a large region and often persist over several tidal cycles. Superstorm Sandy, a hybrid hurricane/nontropical storm that struck New York City in late October 2012, generated the highest recorded water level (3.38 meters [11.1 feet]) in nearly 300 years at the Battery in southernmost Manhattan.[81] Its intensity was magnified by a powerful combination of forces including strong easterly winds, coastal geometry, and a peak storm surge aligning with high tide and a full moon. Storm damages totaled an estimated $19 billion.[82] The local historic sea level rise (0.46 meters [1.5 feet], the highest since 1856[83]) exacerbated the impacts of the flooding. Superstorm Sandy dramatically underscores existing coastal flooding hazards in low-elevation neighborhoods of a major urban center—a hazard only magnified by rising seas.

The vulnerability of the world's shorelines to coastal storm floods will grow with rising sea levels. Many of the world's most populous coastal cities will confront serious inundation risks: New York City, Los Angeles, London, Tokyo, Shanghai, Hong Kong, Mumbai, and Bangkok, among others. Today's 100-year flood, with a chance of occurring at least once per century, will occur more frequently—perhaps as often on average as once per decade, depending on locality, even without significant changes in storminess. In the year 2000, an estimated 189 million people lived in the 100-year flood zone worldwide, according to one study.[84] The exposed population grows to 411 million by the year 2060, for an assumed sea level rise of 21 centimeters (8.3 inches). This study finds that Southeast Asia will face the highest future exposure to coastal flooding.

In one high-end scenario that foresees up to 1.8 meters (5.9 feet) of sea level rise by 2100, an estimated 13.1 million people in the United States would be at risk. Over two-thirds of these people would reside in the southeastern United States. Florida alone would account for nearly half, pointing to very regionally concentrated impacts.[85]

Sea level rise would also lead to intrusion of saltwater upstream and into coastal aquifers. This in turn could contaminate city drinking water supplies and farmland soils. The low topography and unique geology of the city of Miami, Florida, create hazards from the encroaching sea: episodic hurricanes, storm surge flooding, beach erosion, saltwater intrusion, and increasing incidence of nuisance flooding.[86] Overpumping of groundwater from coastal aquifers accelerates the process. The near-surface, highly porous, permeable limestone bedrock—the Biscayne Aquifer—is the city's major water resource. Saltwater infiltrates into the existing aquifers during heavy storms, or at times of low rainfall. Salinity control and water management have therefore become major priorities in Miami in light of the threat of increasing saltwater infiltration.

Shoreline erosion is worsened by towering, storm-driven waves and higher water levels, superimposed over a higher sea level. Thawing permafrost-cemented sedimentary coasts of the Arctic already experience high erosion rates (chap. 4, "Crumbling Arctic Coasts"). Farther south, coastal erosion also chews away at the land and degrades the natural environment.

Most U.S. East and Gulf Coast beaches are eroding, although periodic beach nourishment projects mask most losses. Seen from space, much of the delta of the mighty Mississippi River resembles a giant green Swiss cheese—riddled with water-filled holes lined by dying, waterlogged trees (fig. 9.2). The region suffers considerable land loss and coastal erosion from both natural and human-induced causes. Geological land subsidence magnifies local sea level rise, aggravated further by overpumping of groundwater and hydrocarbons and dredging of canals. Upstream dams have also trapped sediments that otherwise would have

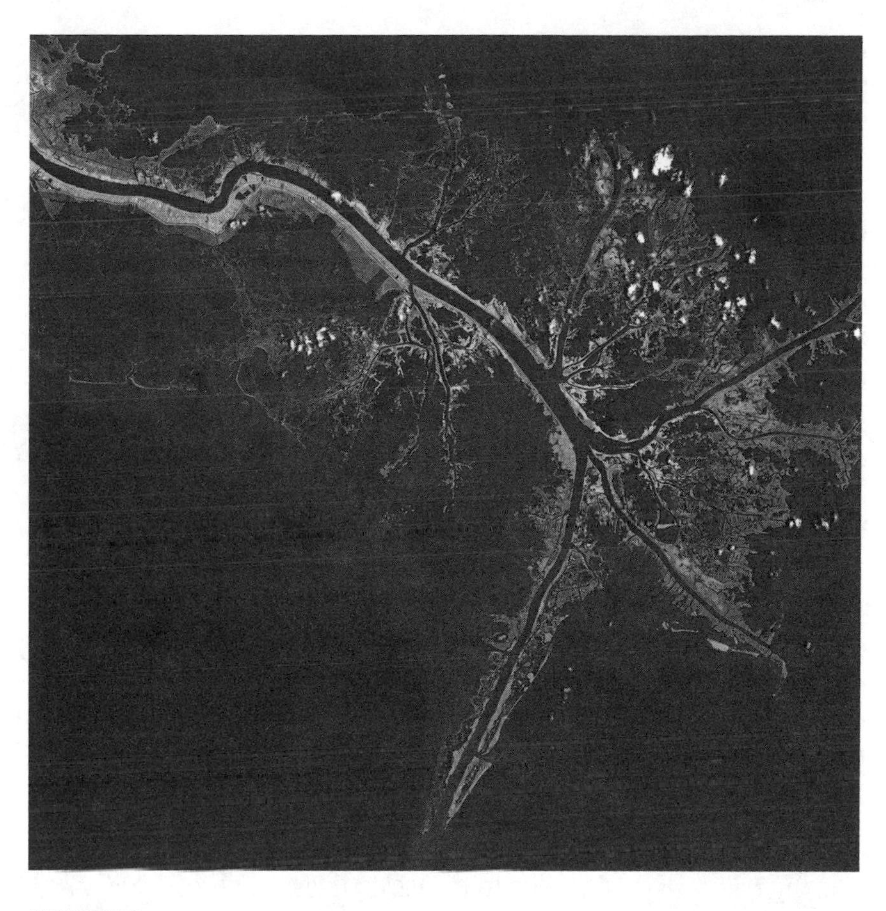

FIGURE 9.2

Mississippi Delta from space. (NASA Goddard Space Flight Center, ASTER Terra satellite image, May 24, 2001)

reached the delta and replenished erosion losses. As a result, erosion rates along the Louisiana coast average 8.2 meters (26.6 feet) per year, while elsewhere along the U.S. Gulf Coast erosion removes an average of ~3 meters (~10 feet) per year. Sixty-one percent of the measured shoreline is retreating.[87]

Chesapeake Bay, the largest estuary in the United States, was a river valley submerged by the advancing sea some 9,000 to 8,000 years ago, at the tail end of the last ice age. The estuary gradually assumed its modern form between 7,000 and 6,000 years ago, when the pace of sea level rise slowed considerably. However, as elsewhere, this pace has accelerated since the late nineteenth century into the twenty-first century.

A number of small islands in Chesapeake Bay, some occupied since the time of European settlement, have shrunk considerably in area or have almost totally disappeared due to historic sea level rise.[88] Sharps Island, one such island, covered 360 hectares (889 acres) in 1660. By 1848, only 175 hectares (432 acres) remained, and by 1900 the island occupied only 36 hectares (89 acres). Today, only a partially submerged, ruined lighthouse survives. The rising sea continues to eat away at remaining land on still-inhabited Smith and Tangier Islands, communities dependent on fishing and tourism. With a population of 700, Tangier Island lies a mere 1 to 2 meters (3.3–6.6 feet) above sea level. Between 500 and 800 meters (1,600–2,600 feet) have eroded since the 1850s, particularly along its western shoreline. Sea level in Norfolk and Hampton Roads, Virginia, near the mouth of Chesapeake Bay, climbs faster than elsewhere along the U.S. East Coast, amplified there by ground subsidence due to excess groundwater extraction. Neighborhood street flooding has become commonplace. These examples from Miami, the Mississippi Delta, and Chesapeake Bay give just a small foretaste of what lies in store for other low-lying coastal regions on our way to a more watery future.

Adapting to the Rising Tides

Three major strategies exist to prepare for steadily encroaching tides. These include defending the coast, accommodation, and, ultimately, retreat.[89] The first line of defense is to armor the shore with "hard" structures, such as seawalls, bulkheads, boulder ramparts (revetments, riprap), groins, jetties, and breakwaters that fortify the shoreline (fig. 9.3). These engineered structures prevent slumping or erosion of unconsolidated sediments, reduce wave scouring, and limit flood damage. Nevertheless, these ramparts can still be overtopped by extreme storm surges, and if poorly designed or inappropriately sited, they may instead intensify erosion. Dikes, tidal gates, and storm surge barriers protect against extreme floods or permanent inundation, but as sea level rises, these defenses will require periodic reinforcement and elevation.

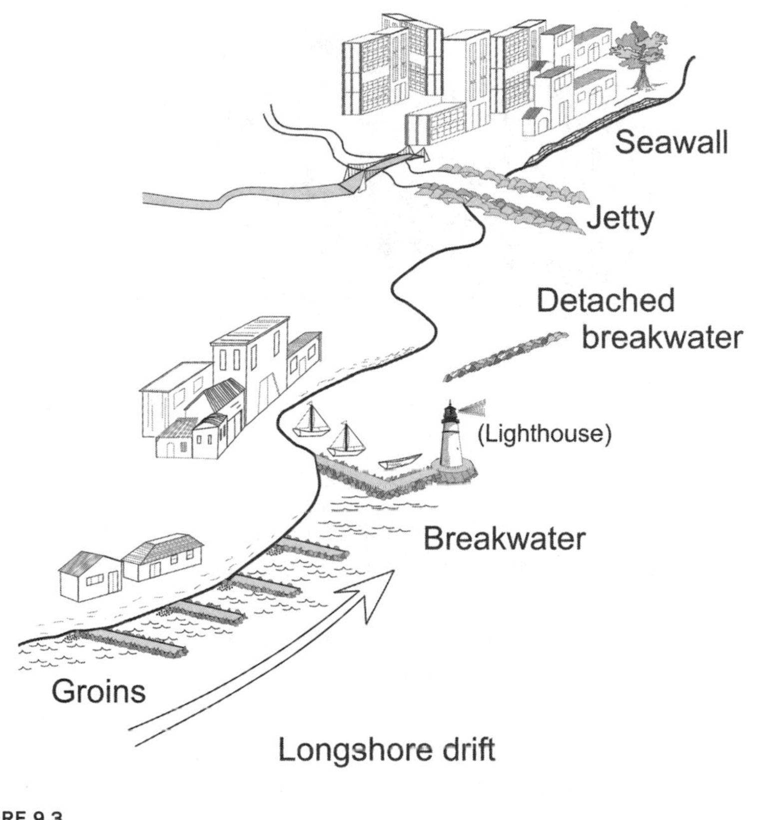

FIGURE 9.3

Schematic diagram of "hard" coastal defenses. (After fig. 8.11 in Gornitz [2013])

Alternatively, "soft" defense responses include beach nourishment and rehabilitation of dunes and coastal wetlands. Wide beaches and salt marshes act as buffers against storm surges and high waves and also provide habitat for coastal wildlife and recreational activities. Brooklyn Bridge Park, located along a former industrial and shipping waterfront on the East River in New York City, exemplifies a nature-based coastal protection system. The park's shoreline is fortified by a mix of replanted salt-tolerant native plants, bulkheads, riprap, and rocks that stabilize the water's edge and create new habitat (fig. 9.4). Refurbished piers also serve multipurpose recreational uses.

FIGURE 9.4

Example of a soft-edge shoreline, Brooklyn Bridge Park, New York City. (Department of City Planning, City of New York, 2011)

Accommodation means learning to live with the water. Stricter building codes ensure stronger, more flood-resilient structures. Buildings subjected to frequent flooding can be elevated above the current or projected 100-year flood zone, or constructed on stilts or pilings. The ground floor can be converted to other, nonresidential uses such as business, parking, or community centers. Planting of more street trees or expanding parks increases drainage and water infiltration into the ground, which reduces street flooding.[90] Floating neighborhoods can be built, where buildings rise and fall with the tides. People can live on houseboats or barges, as along the canals of Rotterdam, Seattle, Sausalito, and Bangkok. The Dutch have innovated multiuse flood defenses in which dikes or levees combine surge protection with housing, parking, parks, and commercial activities. Underground garages store excess water at times of high river or ocean levels.[91] Multipurpose levees from Dordrecht, the Netherlands; Hamburg, Germany; and Toyko, Japan, serve as models for other coastal cities.[92]

The Venetian Solution

In Kim Stanley Robinson's recent and not-so-far-fetched science fiction novel *New York 2140*, two major meltwater pulses have inundated New York City under 50 feet of water. Buildings in Lower Manhattan south of 40th Street are submerged up to the second or third floor, and as in

Venice, people commute in small "vaporetti" along the major transportation arteries, now transformed into canals.[93] The tall skyscrapers still stand, mostly occupied by coastal refugees from drowned lands, while older, less well-constructed buildings periodically "melt" into the polluted water, and superskyscrapers cluster along the higher elevations of uptown. In fact, uptown still thrives, with untold overseas wealth streaming in.

Too much money, real estate, and infrastructure have been invested in the city for people to just walk away. They will dig in and remain, no matter what. Construction of a huge seawall enclosing all of New York City's 837 kilometers (520 miles) of shoreline, or tidal barriers from Sandy Hook, New Jersey, to the Rockaways, as some have proposed, will not protect people living beyond these coastal defenses, nor will they function beyond a certain elevation of sea level. A partial solution would be to "venetianize" New York and other major coastal megacities, turning streets into canals and moving buildings' electrical and heating systems and living quarters to higher floors.

The innovative ideas expressed in Guy Nordensen's *On the Water: Palisade Bay*[94] offer a futuristic vision of a city that lives with water. Construction of an archipelago of artificial islands and reefs in a future New York–New Jersey Upper Bay would dampen powerful storm surges and allow growth of new estuarine habitats. A rigid waterfront would become a "broad, porous, 'fingered' coastline" of tidal marshes, parks, and piers used for recreation and residences. Some of these concepts have already been implemented. A nearly continuous series of parks with pathways for pedestrians and bike lanes lines the waterfront of Manhattan's West Side, and former shipping piers hold sports fields and picnic tables. On Manhattan's East Side, construction is under way on a section of the Dryline (formerly the BIG U)—a protective ribbon of berms and waterfront parks designed to shield Lower Manhattan neighborhoods from future floods and higher sea levels, while also expanding outdoor recreation and cultural spaces.[95]

Coastal defenses or accommodation work well in the short term, but can generate a false sense of security that promotes increased development and may eventually become indefensible under more ominous, multimeter sea level–rise scenarios. Nevertheless, densely populated large port cities will still defend whatever they can, at whatever cost.

Smaller or less prosperous communities may have fewer choices. However, farsighted coastal management and sensible land use practices, such as limited development in high-risk, flood-prone areas, may avert some adverse consequences. Expanded parklands, widened beaches, restored wetlands, and preserved natural habitats can function as buffer zones. Other approaches involve creation of erosion setbacks, or easements, and *buyout programs* to reimburse shorefront landowners for property abandoned in high-risk zones.

The First Coastal Refugees

> *It is beautiful—like white, sandy beaches. The sea is very clear. If you wanted a holiday, that is a place you will want to go . . . We can no longer tell when the strong winds are coming. The climate is changing and changing fast.*
>
> —Ursula Rakova, a Carteret Islander

The tiny village of Shishmaref, Alaska, a traditional Iñupiaq Eskimo settlement on a barrier island that subsists on hunting and fishing, may soon wash into the Chukchi Sea (fig. 9.5).[96] As in many other parts of the Arctic, a much-reduced sea ice cover, which otherwise would buffer storm surges, exposes the shore to relentlessly pounding waves that eat away at its shoreline. Coastal erosion has intensified, in spite of seawalls and rock revetments that keep the sea at bay. Some homes have already washed into the sea. A slim majority of the 600 villagers have recently voted to relocate their village to two potential mainland sites. But opponents of the move cite lack of barge access for delivery of much-needed fuel and food supplies and high relocation costs. While villagers could move to larger towns such as Nome, Kotzebue, or Anchorage, they would then risk loss of their unique cultural heritage—language, hunting skills, traditional songs, dances, and arts. (Shishmaref whalebone and walrus ivory carvings are highly collectible.) At least 31 Alaskan towns face similar difficult decisions. Of these, Shaktoolik ranks among the top four "at imminent risk of destruction."[97] Nevertheless, in spite of insufficient funds to armor their village against an angry sea, the residents have decided to "stay and defend"—at least for now.

FIGURE 9.5

View of Shishmaref, Alaska—Iñupiaq village threatened by coastal erosion. ("Shishmaref- Erin (53)," August 5, 2014, by Bering Land Bridge National Preserve/ Wikimedia Commons, https://commons.wikimedia.org/wiki/File:Shishmaref-_Erin_(53)_edit_(15653586503).jpg)

In the Mississippi Delta, the Isle de Jean Charles in Louisiana's Terrebonne Parish is slowly slipping into the Gulf of Mexico, forcing the small Biloxi-Chitimacha-Choctaw Native American tribe to flee to higher ground farther north. Within the last 60 years, the island has lost 98 percent of its land to the rising seas and storm surges.[98] The island, about 11 miles long by 5 miles wide in 1963, has shrunk to a thin strip two miles long by a quarter of a mile wide. The new community will be built with money allocated by the U.S. Department of Housing and Urban Development as part of a natural disaster resiliency program. Houses will be elevated at least 10 feet to be safe from hurricane storm surges. In spite of strong emotional ties to the tiny island sliver, many tribal members acknowledge the move's necessity, yet express concerns over loss of tribal identity, fate of the remaining land, and whether islanders will like their new location.

The Carteret Islands, 86 kilometers (53 miles) northeast of Bougainville Province in Papua New Guinea in the South Pacific, reach a maximum elevation of 1.5 meters (5 feet) above sea level. Storm surges and high waves increasingly sweep over the islands, imperiling their food crops. Islanders have already obtained 85 hectares of land on Bougainville, a small, neighboring island. So far, only 100 people out of 2,700 have relocated. Funding for construction of new homes has been hard to find. Yet conditions on Carteret are deteriorating, and Ursula Rakova, who heads the local organization Tulele Peisa that oversees relocation efforts, hopes that the islanders will be able to resettle within the next 10–20 years.[99]

The Coming Tidal Wave of Coastal Refugees

Sea level rise may eventually force a tidal wave of tens to close to 200 million coastal refugees worldwide to relocate by 2100, if the present coastal zone becomes uninhabitable. In the United States, a projected high-end 1.8-meter (5.9 feet) sea level rise by 2100 would displace over 2.5 million residents from South Florida alone, with another 500,000 fleeing Louisiana and 50,000 leaving the metropolitan New York region.[100] Neighboring New Jersey and Virginia, farther south, would also lose population. Many of these emigrants would head for Texas, Georgia, North Carolina, and, not surprisingly, the continental interior. In other parts of the world, especially in Southeast Asia, millions living in the deltas of the Ganges-Brahmaputra, Chao Phraya, Yangtze, Yellow, and Mekong Rivers, and also the Nile and Niger Rivers in Africa, would be compelled to seek higher ground in already-overcrowded lands.

Retreat from the shore may be involuntary and compelled by untenable conditions, or by a managed relocation/realignment program in which land is voluntarily abandoned and structures relocated farther inland. Managed realignment in the United Kingdom and Germany has been implemented at dozens of sites to create more intertidal habitat and mitigate flood risk.[101] So far, very few if any people have been displaced. In the United States, the Federal Emergency Management Agency (FEMA) Hazard Mitigation Grant Program finances property buyouts following hurricanes or floods. The program purchased 36,707 properties between 1993 and 2011, relocating an estimated 93,000 people.

Although small-scale managed relocation efforts have begun elsewhere, these plans often meet stiff resistance because of strong community and geographic ties, contested land access rights, and economic uncertainties in the new location. A successful managed-retreat program works when risks become intolerable, affected populations and governments reach mutual agreement on action, costs vs. benefits can be justified, and society at large also benefits. In parts of the world already on the brink, relocation becomes a necessity. As we have seen, the bayous of Louisiana and the Chesapeake Bay islands are drowning. Destructive waves imperil Alaskan coastal villages, as well as a number of low-lying Pacific islands. And traditional ways of life disappear as the first wave of coastal refugees prepares to move to higher ground. The transition will not be easy for them, and even less so for the millions more expected in the coming centuries.

HEADING TOWARD WATERWORLD?

> *Is it the fate of the world to lose its ice? If an ice-free world comes to pass, future generations will gaze over vast areas of the planetary surface that have not seen the light of day or felt the warmth of sunshine for thousands or even millions of years. They will see the drab, gray rock beneath Greenland and Antarctica slowly rebound . . . But these same generations will also watch low-lying areas of the continents being flooded by the sea—areas that have not been submerged beneath the ocean since the Pliocene, or the Paleocene, or the Cretaceous, or perhaps ever.*
>
> —Henry Pollack, *A World Without Ice* (2010)

David Archer, in his book *The Long Thaw: How Humans Are Changing the Next 100,000 Years of Earth's Climate*, informs readers that the carbon dioxide we are injecting into the atmosphere may delay or even prevent another ice age for tens of thousands of years, or even longer. Perhaps, as Arrhenius speculated, this might even prove beneficial. Siberia or Arctic Canada would enjoy a temperate climate in which industry, agriculture, and new cities could blossom, without worry over heavy winter snowfalls, icy blizzard whiteouts, and impassable roadways. But no matter how balmy a future Arctic might be, the length of polar night—dark, sunless winter days—would not change.[102] Thus, any new vegetation growth or croplands would go dormant during the long Arctic winter. Furthermore, Arctic tundra soils may not be the most suitable for growing needed food crops, nor would new construction find thawing permafrost terrain an ideal substrate (see chap. 4).

Even more troubling consequences await the rest of the world: millions of people fleeing higher waters in need of resettlement, and abandonment of large sections of many coastal cities. If that were not enough, the new climate regime could greatly alter global rainfall patterns, turning some areas into deserts and causing other areas to suffer crop failures.

The planet has experienced greenhouse climates before, with elevated temperatures and much higher sea levels, as illustrated in chapter 7. But the human race had not yet evolved back then. Must we condemn our descendants to live in an environment for which humans are poorly adapted?

Now is the time to awaken from our complacency. This book has painstakingly depicted the dramatic changes rapidly sweeping across the cryosphere. Increasingly, visible signs point to a cryosphere in trouble. Previous chapters have meticulously probed the closely intertwined links among mounting temperatures, a shrinking cryosphere, and higher oceans. This book has aimed not just to open the reader's eyes to this rapid transformation, but also to emphasize its effects on our future well-being. As consequences of ice loss ripple across the globe, countless people far removed from still-frozen realms will feel their impacts. Hopefully, society will finally realize the serious environmental challenges ahead and make the necessary, well-informed choices that will help avert potential disaster. We might be wise to heed the advice of Native American elders who tell their children to think in terms of seven generations hence. The Big Meltdown ignores political boundaries and will touch us all!

> *I can't think of anything more important than the environment we leave to our children and our children's children.*
>
> —Lewis Gordon Pugh

EPILOGUE

> *People tend to focus on the here and now. The problem is that, once global warming is something that most people can feel in the course of their daily lives, it will be too late to prevent much larger, potentially catastrophic changes.*
>
> —Elizabeth Kolbert, *The New Yorker*, April 25, 2005

In the course of this book, we have encountered the far-flung members of the cryosphere—free-floating and attached floating ice, impermanently frozen soils, glaciers, and the two big gorillas—the Greenland and Antarctic ice sheets. To varying degrees, all are beginning to feel the Big Thaw. At present, this thaw is like the first days of spring—tentative, small-scale, and largely imperceptible except to local residents and experts. But little by little, the spring thaw gains strength as the sun rises higher each day and the hours of daylight lengthen. The first flowers—snowdrops, crocuses—poke through the melting snow, buds appear on trees, songbirds return, the last patches of dirty winter snow disappear, river and lake ice melts, and up north, Arctic sea ice begins to break apart. And as spring turns to summer, a lush carpet of green covers the land, birds twitter, crickets chirp, and a balmy warmth fills the air. Meanwhile, the tiny trickle of water from the first snowmelt turns into a raging stream, a cascading torrent, and floating ice quickly breaks apart and melts.

The Arctic Ocean becomes a deep-blue liquid carpet, glaciers recede farther uphill, and flotillas of icebergs sail majestically out to sea. Ships then sail back and forth between the Pacific and Atlantic across a year-round ice-free Arctic Ocean, and thawing permafrost uncovers new mineral treasures. But at the same time, the permafrost turns into a mushy bog, releasing more greenhouse gases.

At midlatitudes, spring and summer are delightful seasons, filled with long, bright, sunny days; warmth; and the promise of a bountiful harvest. But as the planet's temperature soars, summer may turn into a near-endless supersummer that rivals distant geologic greenhouse periods. Some regions may bake in the heat and endure more intense droughts, while others soak under hypermonsoon rainfalls. A new climate regime emerges—one in which the Big Thaw morphs into a Big Meltdown. The retreating glaciers that bare the mountain peaks alter hydrologic regimes, and water unleashed by the vanishing ice redraws the geography of the world's shorelines.

Climate change is no hoax. It's upon us and growing. The year 2016 tied with 2014 as the warmest in over 150 years, topping the records of the last few decades. Nowhere are the signs more evident than in the Arctic. Giant cruise ships now regularly tour the fabled Northwest Passage that once doomed the Franklin expedition, thawing permafrost tilts spruce trees at crazy angles, and shorelines are fast washing away. Greenland glaciers rumble loudly as they disgorge heavy loads of icebergs into the sea. However, the warming is not just the Arctic's problem. In the mountains to the south, glaciers march upslope, avalanches and rockfalls menace villages, flash floods scour valleys, and stream flow becomes more erratic. More ominously, the rising oceans threaten to engulf major deltas and low-lying Pacific coral islands, and may increasingly inundate many coastal cities, especially during heavy storms.

Even the sleeping giant gorilla, Antarctica, stirs and begins to flex its muscles, with potentially dire future consequences for distant lands.

So what? you may think. Hasn't our brief journey to the past revealed vastly different climates than those of today—both hotter and colder? The Earth's climate has always been changing. Why is the current situation any different? Why blame us? What does this have to do with human activities?

Svante A. Arrhenius's predictions of atmospheric CO_2 buildup and warming (in 1896) due to coal burning—then the dominant fossil fuel—yielded results close to present-day observations (chapter 8, Tipping points and climate commitments). Thus, the basic science of the greenhouse effect is well established. Furthermore, ice cores and marine sediments preserve close relationships between CO_2 temperature and sea level variations far into the geologic past. However, the present is quite anomalous. Changes unfold much faster than ever—several millennia of warming are time-compressed into a century or less, making consequences difficult to predict. Even worse, the surplus CO_2 will linger in the atmosphere for a long time—long after further emissions cease. As chapter 8 points out, during the persistent warm spell, more heat will penetrate into the abyssal ocean, more land ice will continue to melt, and sea level will rise even further than before. Our window of opportunity to avoid the Big Meltdown is rapidly closing—possibly within decades.

Potential alternative solutions are multifaceted, although often politically unpopular, occasionally technically challenging, and economically difficult to implement. Covered elsewhere at great length, they lie beyond the scope of this book. Here, just a few parting words will suffice. In a nutshell, avoidance of a hothouse world and major meltdown requires a timely transition to a decarbonized energy future. Regardless of the pathway forward that we choose, the reader should keep in mind that the decisions we make today and in the next several decades will affect future generations for centuries—even millennia—to come.

APPENDIX A

ANTICIPATING FUTURE CLIMATE CHANGE

The Intergovernmental Panel on Climate Change (IPCC), a United Nations–sponsored international organization of leading experts in the natural and social sciences, has assessed the state of the Earth's climate periodically since 1990. Its most recent Fifth Assessment Report (AR5) was published in 2013.[1] This report concluded that "warming of the climate system is unequivocal, and since the 1950s, many of the changes are unprecedented over decades to millennia. The atmosphere and ocean have warmed, the amounts of snow and ice have diminished, sea level has risen, and the concentrations of greenhouse gases have increased. . . . Over the last two decades, the Greenland and Antarctic ice sheets have been losing mass, glaciers have continued to shrink almost worldwide, and Arctic sea ice and Northern Hemisphere spring snow cover have continued to decrease in extent. . . . It is *extremely likely* that human influence has been the dominant cause of the observed warming since the mid-20th century." (*Extremely likely* in IPCC parlance means an outcome with a likelihood of 95–100 percent.)

The twentieth-century rise in global sea level stems primarily from melting of glaciers and ice sheets and thermal expansion of warming ocean water. Ice mass losses already dominate the total rise in sea level and will constitute a growing share of it in the future. The 2013 IPCC sea level rise projections draw upon a suite of climate models that incorporate current trends in various elements of the climate system as well as scenarios of economic development, population growth, and fossil fuel consumption, ranging from "business as usual" to strong mitigation efforts. Future climate scenarios are portrayed in a set of four Representative

Concentration Pathways (RCPs). The RCPs represent trajectories of greenhouse gas emissions, aerosols, and land use/land cover that produce changes in energy at the top of the atmosphere equivalent to greenhouse gas concentrations of 475 parts per million (ppm) (RCP2.6), 630 ppm (RCP4.5), 800 ppm (RCP6.0), and 1,313 ppm (RCP8.5), respectively, by 2100. In conjunction with the RCPs, a suite of coupled atmospheric and oceanographic global climate models (AOGCMs) mathematically simulate physical interactions among the atmosphere, ocean, continents (including permafrost), and sea ice, in order to project future trends in temperature, precipitation, and sea level rise.

While AOGCMs compute the oceanographic contributions to sea level, including thermal expansion, cryospheric contributions are determined separately. Changes in ice mass of glaciers and ice sheets are derived from surface mass balance models,[2] which incorporate AOGCM projections of temperatures and precipitation. Dynamic ice flow models additionally estimate future changes in discharge of ice past the grounding line, and calving rates of icebergs. The former is calculated from estimates of ice velocity and ice thickness at the grounding line, the latter from satellite images (see chaps. 4–6). The IPCC report bases its projections on extrapolated current trends and several studies using dynamic models.

Table A.1 summarizes the main IPCC projections for selected elements of climate change by the end of this century.

TABLE A.1 Climate Change Projections in 2081–2100 Relative to 1986–2005 for the Four IPCC RCPs

SCENARIO				
CLIMATE ELEMENT	**RCP2.6**	**RCP4.5**	**RCP6.0**	**RCP8.5**
Temperature (°C)	1 (0.3 to 1.7)	1.8 (1.1 to 2.6)	2.2 (1.4 to 3.1)	3.7 (2.6 to 4.8)
Arctic sea ice extent (% change)				
September (minimum ice)	−43			−94
February (maximum ice)	−8			−34
Global sea level rise (m)				
Thermal expansion	0.14 (0.10 to 0.18)	0.19 (0.14 to 0.23)	0.19 (0.15 to 0.24)	0.27 (0.21 to 0.33)
Glaciers	0.10 (0.04 to 0.16)	0.12 (0.06 to 0.19)	0.12 (0.06 to 0.19)	0.16 (0.09 to 0.23)
Greenland, SMB	0.03 (0.01 to 0.07)	0.04 (0.01 to 0.09)	0.04 (0.01 to 0.09)	0.07 (0.03 to 0.16)
Greenland, rapid dynamics	0.04 (0.01 to 0.06)	0.04 (0.01 to 0.06)	0.04 (0.01 to 0.06)	0.05 (0.02 to 0.07)
Antarctica, SMB	−0.02 (−0.04 to 0.0)	−0.02 (−0.05 to −0.01)	−0.02 (−0.05 to −0.01)	−0.04 (−0.07 to −0.01)
Antarctica, rapid dynamics	0.07 (−0.01 to 0.16)	0.07 (−0.01 to 0.16)	0.07 (−0.01 to 0.16)	0.07 (−0.01 to 0.16)
Land water storage	0.04 (−0.01 to 0.09)	0.04 (−0.01 to 0.09)	0.04 (−0.01 to 0.09)	0.04 (−0.01 to 0.09)
Total global sea level rise (m)	0.40 (0.26 to 0.55)	0.47 (0.32 to 0.63)	0.48 (0.33 to 0.63)	0.63 (0.45 to 0.82)

Note: First number is mean value; numbers in parentheses are the 5–95 percent ranges of the models' distribution (assumed to be normal, or Gaussian).
Sources: IPCC (2013b), table SPM.2 (temperature); and IPCC (2013a), chap. 13, table 13.5, p. 1182 (sea level rise).

APPENDIX B

EYES IN THE SKY—MONITORING THE CRYOSPHERE FROM ABOVE

Remote sensing provides powerful tools with which to monitor the cryosphere. A vast array of satellite "eyes in the sky" now routinely examine terrestrial targets in various regions of the electromagnetic spectrum (table B1). Instruments onboard these satellites give a hemisphere-wide view and yield a wealth of information on the state of the cryosphere. Among these are seasonal and long-term changes in snow cover and sea ice extent, sea ice and ice sheet thickness, thinning and breakup of ice shelves, wandering icebergs, and surging glaciers. Satellite remote sensing proves immensely useful not just for monitoring the cryosphere, but also for tracking changes in global sea level, ocean circulation, land use changes, crop conditions, and weather patterns, and for discovering new mineral deposits.

Instruments aboard satellites detect radiation either reflected or emitted from the Earth's surface over the visible, infrared, and microwave regions of the electromagnetic spectrum. Others monitor the Earth's gravitational field. Not intended as a comprehensive review of the subject, which lies beyond the scope of this book, appendix B instead provides a basic introduction to remote sensing, mentions some of the most frequently used satellites and instruments for cryospheric studies, and lists some useful sources.

Table B.1 summarizes some of the main cryospheric applications of satellite sensors covering various parts of the electromagnetic spectrum. For more detailed information, please consult the references listed in the Bibliography.

TABLE B.1 Satellite Sensors Used for Remote Sensing of the Cryosphere

	VIS/NIR/IR MULTISPECTRAL	PASSIVE MICROWAVE	SAR	INSAR	RADAR SCATTEROMETRY	RADAR ALTIMETER	LASER ALTIMETER	GRAVIMETER
Satellite/sensor	Landsat-7 ETM+, 8	DMSP-SSM/I	TerraSAR-X TanDEM-X	CryoSat-2 SIRAL	QuikScat	(ERS-1, 2)	(ICESat)	(GRACE) GRACE-FO
	Terra ASTER, MODIS	(SSMR)	CryoSat-2	TerraSAR-X	(ERS-1, 2)	(Envisat)	ICESat-2	IceBridge
	(IKONOS)		Copernicus Sentinel-1, 3	(ERS-1, 2)		CryoSat-2	IceBridge	
	(Quickbird)		RADAR-SAT-(1), 2	RADARSAT-(1), 2		Jason-(1), 2, 3*	Airborne laser altimetry	
	SPOT-6, 7		(ERS-1, 2)					
	WorldView-1–4		(Envisat)					
Icebergs	I	I	M		I			
Sea ice extent	I	M	M		O			
Sea ice thickness			I	M	M	I	M	
Ice shelf retreat	M		M	O				

Snow cover extent	M (esp. MODIS)	M					
Glacier extent	M		O				
Ice sheet extent	I		M	O	O		
Ice motion	M		O	M		O	
Ice sheet/shelf, glacier elevation/ thickness changes	I			O	M	M	
Ice sheet mass changes	O				M	M	M
Ice/snow/firn separation	M		I				

*The Jason-1–3 radar altimeters are chiefly used for determination of sea surface heights and monitoring of ocean circulation. They have been extensively used for determination of recent sea level–rise trends.

Note: M: Major data source; I: Important data source; O: Occasional data source. Satellites in parentheses are no longer active.

VISIBLE/NEAR-INFRARED/THERMAL INFRARED (VIS/NIR/TIR)

Like human eyes, satellite-borne instruments that are sensitive to visible light (~0.4–0.7 micrometers, μm) detect sunlight reflected off surfaces. Because fresh ice and snow reflect more light, they appear much brighter than most other materials, including vegetation, bare soil, rock, and water, and therefore contrast markedly with their surroundings. However, visible sensors operate effectively only in daylight and in clear, cloud-free weather. Satellite sensors that "see" the near-infrared region and beyond have greatly expanded our understanding of glaciers and other cryospheric constituents. Infrared instruments, for example, detect the amount of heat emitted by objects at the Earth's surface, which is a measure of their temperature. Infrared sensors detect features even at night and readily distinguish sea ice from ocean water because of the sharp temperature contrast. For example, while sea ice winter temperatures plunge to −20° to −40°C (−4° to −40°F), open ocean water remains above freezing (otherwise it too would be ice covered). Dark meltwater pools on ice in summer have a low albedo, approaching that of ocean water. Clouds, which also absorb, emit, and reflect infrared radiation, can complicate the detection of sea ice. Therefore, corrections need to be made to remove atmospheric interference.

Since 1972, a series of eight Landsat satellites (NASA/USGS) that span the VIS/NIR/TIR regions of the spectrum have mapped the Earth's surface.[1] Landsats 1–3 carried multispectral scanners (MSS) in four spectral bands over the VIS/NIR, while Landsats 4–7 Thematic Mapper and Enhanced Thematic Mapper Plus (ETM+) expanded the number of bands and spectral range to 10.4–12.5 μm in the TIR. Ground resolution also increased from roughly 80 meters (262 feet) on the earliest Landsats to 15 meters (49 feet) on Landsats 7 ETM+ and 8.[2] Cryospheric applications of Landsat data include monitoring changes in glacier and ice sheet area; tracking features to monitor ice flow velocities; mapping changes in sea ice extent and structure, snow cover, melt ponds, and ice albedo; and differentiating among types of ice.

Some newer satellite imaging systems detect ground features as small as 1.5–6 meters (4.9–20 feet) for both SPOT-6 (2012–present) and SPOT-7 (2014–present), 0.82 meters (2.7 feet) for IKONOS (no longer operational), and 0.65 meters (2 feet) for Quickbird (no longer operational). The most recent satellites, such as WorldViews-1–4, detect objects less than one meter (3.3 feet) across. WorldView-4 (2016–present), flying at an altitude of 617 kilometers (383 miles), resamples each ground pixel every 4.5 days or less. Its multispectral sensor covers the visible spectral range (0.45–0.69 μm); its panchromatic camera spans 0.45–0.80 μm into the NIR.[3]

The NASA-Japan ASTER (Advanced Spaceborne Thermal Emission and Reflectance Radiometer) aboard the Terra (EOS AM-1) and Aqua (EOS PM-1) satellites operates in 14 spectral bands, ranging from 0.52 μm in the visible to 11.65 μm in the TIR.[4] Sweeping across a 60-kilometer (37-mile) strip, the instrument's ground resolution increases from 15–30 meters (49–98 feet) in the VIS/NIR to 90 meters (295 feet) in the TIR. The broader spectral range improves discrimination among fresh snow, various states of ice (fresh, dirty, debris covered, etc.), and water.[5] Furthermore, the ability of thermal infrared bands to detect emitted radiation (i.e., heat, rather than reflected light) at night compensates for their coarser ground resolution. Yet, as with visible and NIR sensors, they lack the ability to see beneath clouds.

The NASA MODIS (Moderate Resolution Imaging Spectroradiometer), another key instrument aboard Terra and Aqua, encompasses the visible-to-infrared region from 0.4 μm to 14.4 μm in 36 spectral bands[6] (see also ASTER, previous paragraph). Circling 705 kilometers (438 miles) above the Earth, they cover the entire Earth every 1–2 days, over 2,330-kilometer (1,447-mile)-wide swaths. Ground resolution ranges from 250 meters (820 feet) in bands 1–2 (0.620–0.876 μm) to 500 meters (1,640 feet) in bands 4–7 (0.545–2.16 μm), and 1 kilometer (0.6 mile) for bands 8–36 (0.405–14.39 μm). MODIS is used mainly to monitor changes in snow cover, depth, and albedo; snow grain size; sea ice extent; and ice temperatures. MODIS has also captured images of major iceberg calving events and can therefore be applied to iceberg tracking as well.

Laser Altimetry

Laser altimeters, such as on NASA's Ice, Cloud, and Land Elevation Satellite (ICESat), which circled the Earth from 2003 to 2009 at an altitude of ~600 kilometers (373 miles) in a near-circular, near-polar orbit (94°N inclination), operates in a manner analogous to that of radar altimeters, but substitutes an intense, narrow laser beam (see "Radar Altimetry," later in this appendix). ICESat could detect elevation changes as small as 1 centimeter per year over a 200-square-kilometer (77 square mile) area. Its successor, ICESat-2, launched in September, 2018, carries the Advanced Topographic Laser System (ATLAS), which emits green laser pulses at 0.53 μm.[7] In place of the original ICESat design, ICESat-2 employs a micropulse, multibeam approach that provides dense cross-track coverage perpendicular to the spacecraft's forward motion. Its main objectives are to measure changes in ice sheet and glacier elevations and their effects on global sea level rise, improve elevation estimates in rugged terrain and on rough surfaces such as ice crevasses, map ice drainage systems and divides, and measure changes in sea ice thickness.

To bridge the gap between ICESat and ICESat-2, laser altimeters on board Operation IceBridge aircraft provided continuity in measurements of changes in polar land and sea ice thickness. Radar instruments, also on board, probe the ice column from surface snow through ice, down to the underlying bedrock[8] (see the next section) A gravimeter and magnetometer are used together help locate the bed where radar is unable to image, for instance, beneath water. IceBridge acquires surface elevation data for ice sheet mass balance determinations, and for selected areas of rapid sea ice changes. It also produces related data on ice sheet bed topography, grounding line positions, and ice/snow thicknesses.

MICROWAVE/RADIO WAVE

The Earth's surface also emits microwave radiation at relatively low energy levels. Unlike infrared radiation, microwave emission depends more on the physical properties of the surface, such as atomic composition or roughness. Because of low energy levels, radiation data must be acquired over a larger area, which reduces the spatial resolution. Instruments sensitive to this radiation are called *passive microwave* detectors. Examples include NASA's Nimbus-7 Scanning Multichannel Microwave Radiometer

(SMMR; 1978–1987), succeeded by Special Sensor Microwave/Imagers (SSM/I) aboard Defense Meteorological Satellite Program (DMSP) satellites (F8, F10, F11, F13, F14, F15, F16, and 17), beginning in 1987.[9] The satellites operate in circular to near-circular sun-synchronous and near-polar orbits 833 kilometers (518 miles) above the Earth over a swath 1,394 kilometers (866 miles) wide, completing an orbital cycle in 102 minutes. Nearly all parts of the globe at latitudes over 58°N or S are covered at least twice daily, except for a narrow latitude band near the poles. Because of their ability to see through clouds day and night, passive microwave sensors such as the SSM/I series have proved especially useful in monitoring sea ice changes over time.

Satellite sensors that beam microwaves toward the Earth's surface and measure the returned emissions are called *active microwave*, or radar, detectors. Radar is widely employed in tracking of aircraft, ships, and speeding automobiles. In addition to tracking moving objects, radar can be used to build up images, much like a photograph, but with radar instead of visible light. These high-resolution images can be acquired regardless of weather conditions, any time of day or night.

Radar Altimetry

Radar instruments aboard satellites are used to map surface elevations on land, sea, and ice. Satellite radar and laser altimeters measure glacier and ice sheet topography, ice motion, and sea ice freeboard (height above ocean surface). Radar altimeters emit microwave pulses and record the time elapsed for the pulse to reach the Earth's surface and return. Since the speed of light is known, this method allows calculation of the exact distance between the spacecraft and surface, after correcting for atmospheric and instrumental effects. The satellite's position along its orbit is carefully tracked via Global Positioning System (GPS) to determine the height relative to the center of mass of the Earth. In practice, the altitude of the satellite is measured with respect to a fixed reference frame—a *reference ellipsoid* (a mathematical shape approximating the Earth's surface, with flattened poles and bulging equator), or a *geoid* (an idealized shape whose surface coincides with mean sea level in the absence of tides, ocean currents, and winds). The difference between the satellite's altitude and distance to the fixed reference frame gives the topographic elevation to within an inch. Repeated observations over multiple satellite orbits over a given time period establish a trend in glacier or ice sheet elevation. Radar altimetry works best on low slopes of less than 1 degree, such as for the smoother interiors of the Greenland and Antarctic ice sheets.

Ocean Surface Topography Mission (OSTM)/Jason-2 and -3 (NASA/NOAA/CNES/EUMETSAT), launched in 2008 and 2016, respectively, succeed the earlier U.S.-French TOPEX/Poseidon (1992–2005) and Jason-1 (2001–2013) missions.[10] Their main instrument is the Poseidon-3 radar altimeter, but they also use the DORIS instrument and GPS for precise orbit location and tracking information. Like their predecessors, Jason-2 and -3 orbit the Earth at an altitude of 1,336 kilometers (830 miles) inclined to the equator at an angle of 66°, flying over the same spot of ocean every 10 days. Primarily designed for ocean studies, these radar altimeters have detected a sea level trend of 3.2–3.3 millimeters per year from 1993 to 2017,[11] up from the twentieth-century average of 1.2–1.9 millimeters per year (chap. 1). The radar altimeter aboard Jason-3 measures variations in sea surface heights as small as 3.3 centimeters (1.3 inches), aiming for 2.5 centimeters (1 inch), to monitor evolving ocean circulation patterns, ENSO (El Niño–Southern Oscillation) events, hurricane intensities, and sea level change.

The European Space Agency's (ESA) CryoSat-2 (2010–present), successor to ERS-1 (1991–2000), ERS-2 (1995–2011), and Envisat (2002–2012), carries a pair of enhanced radar altimeters and synthetic aperture radar (SAR) instruments/interferometric radar altimeters (SIRAL-2) (see the following section). These instruments are used to detect and measure changes in ice sheet elevation and extent, and thus, indirectly, changes in mass. CryoSat-2's radar altimeter acquires accurate sea ice thickness data that track seasonal variations. Over sloping land ice surfaces, CryoSat-2 switches to SAR/interferometry mode to record changes in sheet thickness and motion (see more in the following section).

Other ESA satellites, such as Copernicus Sentinel-6, continue the work of their predecessors using radar altimeter instruments that accurately measure land, ocean, and ice sheet topography.

Synthetic Aperture Radar Satellites

Synthetic aperture radar (SAR) functions as a type of imaging radar. In synthetic aperture radar, microwave pulses are beamed toward the Earth's surface at very short time intervals, and the return time of the echo is measured precisely. Because of the extremely short time interval between pulses (e.g., as short as 50 microseconds for CryoSat; see the following paragraphs), a whole series can be treated as one burst. The cross-track return echoes (in a direction perpendicular to motion) are combined into strips. The satellite's forward motion produces slight frequency shifts (i.e., via the Doppler effect) in the return signal, allowing strips from successive bursts to be superimposed and averaged to reduce noise. This generates an image of the ground terrain. An advantage of radar is that it penetrates through clouds and can "see" under all weather conditions. The radar echoes also provide useful information on surface characteristics. Radar can establish the degree of roughness, and also has the ability to separate sea ice from water, since ice reflects more radar energy than does the surrounding ocean. Thicker, older sea ice appears brighter (more reflective) than young sea ice, which is still full of myriad air bubbles. Thus, SAR becomes a useful tool in separating multiyear from first-year sea ice.

ESA's Earth remote sensing satellites, ERS-1 (1991–2000) and ERS-2 (1995–2011), carried a SAR, radar altimeter, and some additional instruments. Instruments on Envisat (2002–2012) included a radar altimeter, DORIS (Doppler Orbitography and Radiopositioning Integrated by Satellite)—a microwave tracking system to locate the satellite's position precisely—and an enhanced form of SAR, the advanced synthetic aperture radar (ASAR), which enabled continuity with ERS-1 and -2 data. Among its accomplishments, the ASAR tracked changes in sea ice coverage, movement of sea ice, and calving of icebergs.

ESA's CryoSat-2, launched in 2010, holds a pair of duplicate SAR/interferometric radar altimeters (SIRAL-2), which can measure changes in ice thickness to within 1.3 centimeters (0.5 inch). The radar altimeter on board ESA's CryoSat-2 satellite, although derived from earlier instruments, has been enhanced to improve measurements of ice-covered surfaces.[12] The SAR/SIRAL-2 operates as a conventional altimeter over relatively even terrain on the Antarctic ice sheet when in low-resolution mode.[13] At high resolution, it also operates in SAR mode to monitor sea ice and ice shelves. As an interferometric synthetic aperture radar (InSAR), it uses two or more SAR images to map changes in surface elevation or slope, using differences in the phase of waves returned to the satellite or aircraft. This capability enables it to investigate slanted or rough surfaces such as the sea ice/land boundary and improve precision in measuring ice sheet topography and sea ice thickness (height of ice floes above water, to within 1.6 centimeters [0.63 inches] per year).

ESA's Copernicus Sentinel program consists of a series of Earth-observing satellites (Sentinel-1–6 and Sentinel-SP) that examine vegetation, coastal zones and waterways, the ocean, land and sea temperatures, and atmospheric conditions.[14] Sentinel-1 uses an advanced SAR that also operates in interferometric mode. The radar imagery maps sea ice conditions for safe navigation and also differentiates thinner first-year sea ice from thicker, more hazardous, multiyear sea ice. The Sentinel-3 dual-frequency SAR altimeter, based on Jason and CryoSat instruments, will cover the entire Earth's surface in SAR mode and will accurately measure sea ice and ice sheet topography, among other tasks.

TerraSAR-X (2007–present), a joint German Aerospace Center–private operation, carries a high-resolution SAR that generates radar images of up to 1-meter (3.3 feet) resolution and covers most of the planet. A nearly identical follow-on satellite, TanDEM-X, was launched in 2010.[15] Both satellites have been flying in close orbits, separated by only a few hundred meters. Together they form the first configurable synthetic aperture radar/interferometer pair in space. The satellites, imaging the entire Earth's surface in three years, produced a global topographic digital elevation model data set accurate to within 10 meters. Although TanDEM-X was not primarily designed for cryospheric studies, the data have diverse applications in hydrology, geology, climatology, oceanography, and environmental observations.

The Canadian Space Agency's RADARSAT-1 (1995–2013) operated in a near-polar, sun-synchronous orbit at 798 kilometers (496 miles) of altitude. Its successor, RADARSAT-2 (December 2007–present), following the same orbit as RADARSAT-1, is equipped with a SAR sensor that images the Earth at a ground resolution of between 3 and 100 meters (10–330 feet).[16] RADARSAT data are used for sea ice–edge detection, discrimination between first-year and multiyear sea ice, and ice topography, among other applications. RADARSAT-2, in conjunction with several other satellites, mapped the flow of ice across the Antarctic ice sheet and revealed previously unsuspected information about the mechanics of ice sheet motion (see, e.g., "Moving Antarctic Ice," chap. 6).

SATELLITE GRAVIMETRY

Not all remote sensors probe different regions of the electromagnetic spectrum. Gravimeters onboard satellites respond to changes in the Earth's gravitational field caused by subtle changes of mass. The Gravity Recovery and Climate Experiment (GRACE), jointly operated by NASA and the German Aerospace Center, has been orbiting the planet since 2002.[17] The two twin GRACE satellites, circling the Earth 15 times a day, respond to tiny gravitational variations along their orbits. A stronger gravitational pull beneath one satellite accelerates it relative to its twin; thus the distance between the pair increases. As the first one slows down once past the anomaly, its twin accelerates upon reaching that spot. By combining the changing distances with precise positions given by GPS, scientists construct detailed gravity maps of the planet. Both altimetry and gravity measurements need to be corrected for glacial isostatic adjustments (GIA) (chap. 4; glossary). GRACE-2 approached the end of its useful lifetime and was decommissioned in October 2017. NASA and GFZ Potsdam launched a follow-on mission, GRACE-FO, that closely resembles the orbit and design of GRACE.

GRACE has uncovered numerous changes in mass distribution around the planet. Variations in gravity due to mass changes arise naturally from differences in rock densities or topography. Another source of large-scale mass redistributions involves significant water transfers between land and sea from both natural and anthropogenic processes. An example of the former is extremely heavy precipitation, as during the very strong 2010–2012 La Niña event in Australia. Examples of the latter include storage of water in large reservoirs and groundwater mining, both of which alter river runoff and, ultimately, sea level. Of greater interest to cryospheric studies, however, GRACE has provided evidence for recent loss of mass from glaciers and ice sheets (chaps. 4–6).

APPENDIX C

GEOLOGIC TIME SCALE

ERA	EON	PERIOD	EPOCH	YEARS BEFORE PRESENT
Cenozoic		Quaternary	Holocene	11,700
			Pleistocene	2.58 my
		Neogene	Pliocene	5.33 my
			Miocene	23 my
		Paleogene	Oligocene	33.9 my
			Eocene	56 my
			Paleocene	66 my
Mesozoic		Cretaceous		145 my
		Jurassic		201.3 my
		Triassic		251.9 my
Paleozoic		Permian		298.9 my
		Carboniferous		358.9 my
		Devonian		419.2 my
		Silurian		443.8 my
		Ordovician		485.4 my
		Cambrian		541 my
Precambrian	Proterozoic			2,500 my
	Archean			4,000 my
	Hadean (formation of Earth)			~4,600 my

Source: Dates from International Chronostratigraphic Chart, International Commission on Stratigraphy, v2018/08, http://www.stratigraphy.org.

GLOSSARY

ablation zone The zone in which mass losses of glaciers or ice sheets by surface melting, runoff, sublimation, wind scour, and calving exceed accumulation.

accumulation zone The zone in which mass gains of glaciers or ice sheets through precipitation (snowfall) and wind drift exceed losses by ablation.

active layer Near-surface layer above permafrost that freezes in winter and thaws in summer.

albedo The fraction (or percent) of solar radiation (or sun's energy) striking a surface that is reflected by the surface.

Antarctic Circumpolar Current (ACC) A wind-driven surface current that flows west to east around Antarctica.

Anthropocene The period beginning approximately with the Industrial Revolution when humans have increasingly transformed the Earth's surface. (Note that varying start periods have been proposed, including 1945—the beginning of the Atomic Era.)

Arctic haze Reduced atmospheric visibility, particularly in late winter and spring, caused by the presence of aerosols (tiny liquid or solid particles suspended in air) of either natural or anthropogenic origin.

Arctic Oscillation (AO) A seesaw pattern of atmospheric pressure between the Arctic and midlatitudes, similar to the NAM (Northern Annular Mode) and to the **North Atlantic Oscillation (NAO)**. During the positive phase, low polar latitude pressure and high midlatitude pressure steer midlatitude storms north, increasing rainfall in Alaska and Northern Europe. During the negative phase, the pressure pattern reverses and cold air outbreaks tend to grip North America and Europe.

asthenosphere A weak zone within the Earth's upper mantle, ~100 to 200 kilometers (62 to 124 miles) beneath the lithosphere, where rocks deform by plastic flow under pressure. It is the source of lava that erupts at the midocean ridges.

astronomical theory of ice ages The theory that ice ages are caused by periodic variations in the Earth's orbit. Also referred to as the Milankovitch theory, after one of its first proponents. (*See also* **eccentricity**; **obliquity**; **precession**.)

Atlantic Meriodional Overturning Circulation (AMOC) North-south deep-ocean circulation driven by density differences (temperature, salinity), winds, and tides.

basal melting Increased melting at the base of an ice shelf or ice tongue, generally due to warmer subsurface ocean water.

bergy bit A small iceberg between 5 and 14 meters (15 and 46 feet) long and 1–5 meters (3–13 feet) high.

bipolar seesaw An out-of-phase relationship in millennial climate variability between the Northern and Southern Hemispheres. The cause has been attributed to changes in relative strengths of hemispheric ocean deepwater formation and circulation following massive iceberg discharges.

boreal forest Northern forest south of the **tundra** zone, characterized by spruce and larch.

brash ice A mix of ice and icebergs less than 2 meters (6.5 feet) across.

brine rejection The process of expelling salt impurities from sea ice as it recrystallizes and matures.

calving The break-off of ice chunks from the edge of a glacier or a floating ice shelf into the ocean or a lake. The broken ice masses are called icebergs.

Circumpolar Deep Water (CDW) Part of a worldwide system of ocean circulation (or "oceanic conveyor belt") that flows eastward around Antarctica at depths of between 488 and 914 meters (1,600 and 3,000 feet).

cirque A bowl-shaped, steep-sided hollow high on a mountain, carved out by a glacier.

congelation growth (of sea ice) Growth at the base of a newly formed thin sheet of sea ice in which ice crystals grow vertically with their c-axes oriented parallel to the ocean surface.

continental shelf Shallow seaward extension of a continent, with average depths between 130 and 135 meters (430 and 443 feet).

continental slope The steeper slope between the outer edge of the continental shelf and the continental rise, near the base.

crevasse A deep crack in a glacier caused by extensional stresses as ice moves over an uneven surface.

cryosphere The frozen portions of the Earth's surface including glaciers, ice sheets and caps, ice shelves, permafrost, snow, and sea ice.

diatoms Unicellular algae living in the ocean, coastal lagoons, or freshwater lakes. Sensitive to various environmental parameters such as temperature, salinity, and water chemistry, they are used to reconstruct past climates and sea levels.

drumlin Smooth, streamlined, oval hill consisting of sediments deposited beneath an ice sheet and shaped by its flow; it is blunt and steeper on the up-glacier side and tapered in the downstream direction.

East Antarctic Ice Sheet (EAIS) Part of the Antarctic ice sheet that lies east of the Transantarctic Mountains.

eccentricity The degree of deviation of the Earth's elliptical orbit from a circle. It varies over roughly 100,000- and 400,000-year cycles.

El Niño-Southern Oscillation (ENSO) A quasiperiodic oscillation of the ocean-atmosphere system in the tropical Pacific manifesting as changes in air pressure, sea surface temperature, and wind and rainfall patterns, recurring approximately every three to seven years.

equilibrium line altitude (ELA) The altitude at which the annual **accumulation** of snowfall on a glacier is exactly balanced by **ablation** (see above).

esker Long, sinuous ridge consisting of sands and gravels deposited by glacial meltwater at the edge of an ice sheet or in an ice-walled tunnel.

eustatic sea level rise A global change in sea level due to both addition of water from ice losses and thermal expansion of seawater. However, the magnitude of the sea level change may vary from place to place (*see* **relative sea level rise**).

firn Snow that has survived a summer's melt season; a transitional stage in the conversion of snow to ice.

first-year ice Sea ice that forms in autumn and winter and melts in summer.

fjord Long, narrow, U-shaped valley bounded by steep cliffs and carved by glacial erosion but subsequently filled with seawater.

foraminifera (forams) Single-celled organisms that live in the ocean, brackish water, and lakes. Marine forams can be either planktonic (free-floating) or benthic (bottom-dwelling). Many species have shells of calcite (calcium carbonate). Variations in oxygen isotope ratios in the shells of marine forams are used to infer past sea levels and ocean temperatures.

frazil ice Spongy, slushy aggregate of small, needle- or plate-shaped ice crystals suspended in turbulent seawater.

frost heave Ground movement due to freezing of water.

gas hydrate (methane hydrate or clathrate) A form of ice in which cage-like spaces can accommodate gas molecules, usually methane, or, less commonly, carbon dioxide, or low-molecular-weight hydrocarbons. It is stable only under high pressure and low temperature, and occurs beneath permafrost and buried marine sediments.

geothermal gradient The average rate of increase in temperature with depth in the Earth's interior.

glacial erratic A boulder or rock transported by a glacier or ice sheet that has been deposited far from its source.

glacial drift Sediments of all sizes transported and deposited by a glacier.

glacial deposits These include eskers, kettles, kames, drumlins, till, etc.

glacial-interglacial cycle A roughly 100,000-year cycle that includes a glacial period of prolonged cold (glaciation) and advancing ice sheets followed by a warm period (interglacial) during which ice sheets retreat.

glacial isostasy Changes in land elevation resulting from an added or decreased load of ice mass.

glacial isostatic adjustment (GIA) Response of the Earth's lithosphere and upper mantle to changes in ice mass loading/unloading during glacial-interglacial cycles. (Gravitational, rotational, and isostatic changes due to recent mass redistribution between ice sheets and ocean are often referred to as "fingerprinting.")

glacial lake outburst flood (GLOF) A flood generated when a dam (either rock, sediment, or ice) containing a glacial lake fails. A *jökuhlaup* is a similar type of flood caused by a subglacial volcanic eruption (as in Iceland) or, by extension, any subglacial flood.

glacier A large mass of ice on land formed by compaction and recrystallization of snow that slowly slides down the slope of a mountain under the pull of gravity. Glacier types include:

cirque glacier A small glacier that forms within a cirque basin, generally high on the side of a mountain.

hanging glacier A glacier that originates high on the wall of a mountain, but descends only part of the way to the main glacier.

outlet glacier A glacier or ice stream emerging from an ice sheet or ice cap through a valley.

piedmont glacier A fan- or lobe-shaped glacier located at the front of a mountain range. It forms when one or more valley glaciers flow from a confined valley onto a plain, where it expands.

tidewater glacier A glacier that reaches the coast, often with a floating extension (ice tongue or shelf). Also called a *marine-terminating glacier.*

valley glacier A glacier confined for all or most of its length within the walls of a mountain valley. Also called an *Alpine* or *mountain glacier.*

greenhouse effect, anthropogenic Theory that atmospheric greenhouse gases (e.g., carbon dioxide, methane, nitrous oxides) generated by fossil fuel combustion, deforestation, and industrial activity are warming the Earth more than the warming due to naturally occurring greenhouse gases such as water vapor and carbon dioxide. These latter gases heat the Earth by 33°C (59°F) above temperatures that would exist in their absence.

grounding line Boundary between a glacier or ice stream resting on land and an attached ice shelf or tongue floating on water.

growler The smallest-size iceberg—under 5 meters (3 feet) long, ~1 meter (39 inches) high.

hydroisostasy Changes in ocean elevation resulting from the addition or loss of water due to changes in the mass of ice sheets.

hydrological cycle Movement of water between ocean and land reservoirs (lakes, rivers, soil, groundwater, glaciers) by means of evaporation and precipitation as rain, snow, or hail.

ice age Period of ice accumulation near the poles. Ice has covered both poles for at least the last 2.8 million years. The Antarctic ice cores record as many as eight interglacial-glacial cycles within the past million or so years during which the ice sheets retreated and expanded.

iceberg A large chunk of drifting ice, of varying shapes and sizes, that has broken off a glacier or ice sheet.

iceberg calving *See* **calving**.

ice cap A large, dome-shaped ice mass covering underlying topography, generally less than 50,000 square kilometers (19,300 square miles) in area.

ice edge Outer margin of densely packed sea ice.

ice floe Floating slab of ice.

ice front Seaward edge of a tidewater glacier or ice shelf.

ice island Flat iceberg calved off ice shelves (especially around northern Ellesmere Island, Canada).

ice lens Mass of buried ice that grows by drawing in water as the ground freezes; found in permafrost regions.

ice mélange (Fr., a mixture, blend) A densely packed mix of icebergs and sea ice at the terminus of many Greenland glaciers. An ice mélange may impede calving rates.

ice-rafted debris (IRD) Sediment or pebbles transported by icebergs and deposited in the ocean.

ice sheet A large, continental-scale ice mass covering an area of over 50,000 square kilometers (19,300 square miles) that spreads outward from a center of accumulation, usually in all directions. The major ice sheets today are in Greenland and Antarctica.

ice shelf Floating slab of ice attached to the shoreline as a marine extension of an ice sheet or glacier.

ice/snow-albedo feedback Fresh ice and snow are bright and highly reflecting (i.e., have high albedo). A reduced summer sea ice cover (or winter/spring snow cover) exposes more dark surface water (or bare ground) that absorbs more of the sun's incoming energy and raises surface water (or ground) temperatures. This, in turn, triggers more ice/snow melting (*see* **albedo**).

ice stream A faster-flowing portion of an ice sheet. Ice streams behave as glaciers and account for much of the ice shed by ice sheets into the ocean.

ice tongue A narrow, floating extension of a tidewater glacier, usually within a fjord or bay.

Intergovernmental Panel on Climate Change (IPCC) United Nations Environment Programme (UNEP) and World Meteorological Organization (WMO)–sponsored international group of scientists and technical experts convened to assess the current state of climate change and its environmental and socioeconomic impacts. The IPCC has issued reports in 1990, 1995, 2001, 2007, and, most recently, in 2013.

interstadial A warm period within a glacial cycle.

isostasy A state of gravitational equilibrium in which the different elevations of continents and oceans largely reflect differences in thickness and density of crustal blocks. Isostatic changes are further caused by loading or unloading of sediments at river deltas, volcanic lava accumulations, or ice masses (*see* **glacial isostasy**).

kame A stratified sand and gravel deposit in a hollow near the edge of a glacier; it forms small hills, mounds, knobs, hummocks, or ridges.

katabatic wind A cold, dense wind blowing off ice sheets and glaciers.

keel Submerged portion of an iceberg. (The exposed portion of the iceberg is called a *sail*.)

kettle A depression formed by melting of a stagnant block of ice buried in glacial deposits.

landfast ice (fast ice) Floating ice that forms and remains attached to the shore, to grounded icebergs, or between shoals.

Last Glacial Maximum The period near the end of the last ice age when ice sheets and glaciers reached their maximum global extent—roughly between 23,000 and 19,000 years ago.

Last Glacial Termination The period following the last ice age, between around 20,000–19,000 years ago, and 7,000 years ago.

last ice age (or glaciation) The cold period following the end of the Last Interglacial, between ~116,000 and 20,000–19,000 years ago.

Last Interglacial The warm period preceding the last ice age lasting from ~130,000 to 116,000 years ago; also called the *Eemian* (Europe) or *Sangamonian* (United States).

lead Linear opening in sea ice.

lithosphere The rigid, strong crust and upper mantle above the asthenosphere. It varies in thickness from several to 100 kilometers (62 miles) beneath the ocean and 100 to 150 kilometers (62 to 93 miles) beneath continents.

Little Ice Age A colder period from 1400 to 1850 CE with temperatures slightly below the millennial average, during which time many glaciers advanced.

loess Wind-deposited dust and silt in unglaciated areas, especially in the central United States, northern Europe, Russia, and China during the last ice age.

marine ice sheet instability (MISI) A potential ice instability that is set up where the grounding line, resting below sea level, begins to retreat into a region where the subglacial topography plunges steeply landward. (*See* **grounding line**.)

mass balance (of a glacier or ice sheet) The sum of all processes adding to or removing ice from a glacier or ice sheet, including accumulation (gain), ablation, meltwater runoff, calving and ice discharge (losses). When positive, accumulation exceeds mass losses; when negative, ice losses exceed accumulation. (*See also* **ablation**; **accumulation**; **surface mass balance**.)

meltwater pulses Periods of very rapid sea level rise during the Last Glacial Termination due to major discharges of glacial meltwater.

Medieval Warm Period (MWP) A period of relative warmth between roughly 950 and 1100 CE, when mean Northern Hemisphere temperatures were several tenths of a degree Celsius above the 2,000-year hemispheric average. The medieval warmth has not been exceeded until recent decades.

methane hydrate A form of ice in which methane or carbon dioxide is locked within cage-like crystal structures under pressure at subfreezing temperatures; a potential source of natural gas found beneath Arctic permafrost and in marine sediments.

moraine A ridge of rocky debris deposited by a glacier at its leading edge or along the sides of the glacier valley.

lateral moraine A sediment ridge on the glacier's edge formed by the accumulation of rock debris from the valley walls.

medial moraine A sediment ridge located toward the middle of a glacier. It forms when the lateral moraines of two glaciers merge.

recessional moraine A ridge of rock and sediment left by a glacier in successive stages as it retreats up a valley.

terminal moraine The moraine marking the outermost advance of a glacier or ice sheet.

moulin (Fr., mill) A tubular channel through a glacier or ice sheet formed by meltwater.

multiyear ice Sea ice that has survived two or more summer melt seasons. (*See also* **first-year ice**; **perennial ice**.)

nilas ice A thin, elastic crust of ice that freezes at the sea surface. Young ice, 10–30 centimeters (3.9–11.8 inches) thick, represents a transitional stage between nilas and first-year ice (typically 0.30–2 meters [11.8 inches to 6.6 feet] thick), which forms during the winter season.

nor'easter An extratropical cyclone, most active between November and March, that affects the Atlantic Coast of North America.

North Atlantic Deep Water (NADW) A branch of the Atlantic Meridional Overturning Circulation (AMOC) where warm, northward-flowing surface currents become cold, dense, saline water that sinks and helps drive ocean circulation (the latter is sometimes called the *global ocean conveyor system*).

North Atlantic Oscillation (NAO) Climate variability in the North Atlantic region related to the atmospheric pressure difference between Iceland and Lisbon, Portugal (or, alternatively, the Azores). The Arctic Oscillation (AO) and the Northern Annular Mode (NAM) are similar types of climate oscillation.

nunatak Mountain peak or ridge that protrudes above an ice sheet or glacier.

obliquity The angle between the Earth's rotation axis and the perpendicular to the ecliptic (the plane of the Earth's orbit around the sun). It is presently 23.4°, but varies between 22° and 25° over ~41,000 years. This angle affects the contrast between seasons.

ogive An arc-shaped, convex, down-glacier-pointing series of waves, or alternating dark and light banding, on the surface of a glacier.

outwash plain A broad, low-sloped plain composed of sediment deposited by glacial meltwater at the margin of a glacier. Typically, the sediment becomes finer grained with increasing distance from the glacier edge.

oxygen isotope variations Changes in the proportions of ^{18}O and ^{16}O in marine carbonate shells due mainly to changes in ice volume and ocean temperatures; also affected by changes in ocean chemistry or by biological activity.

pack ice An area containing high concentrations of sea ice.

pancake ice A type of sea ice that forms in turbulent water as frazil ice fragments repeatedly collide and adhere into pancake-shaped masses with upturned rims. (*See* **frazil ice**.)

patterned ground Surface features found in landscapes affected by frost action. These include circles, polygonal networks, and stepped or striped ground on sloped surfaces.

perennial ice Sea ice that survives at least one summer season.

periglacial Relating to landforms and processes associated with present-day cold climates; these landforms are also found at the edges of formerly glaciated terrain.

permafrost Frozen ground where temperatures remain below 0°C (32°F) for two consecutive winters and the intervening summer. **Continuous permafrost** underlies over 90 percent of an area; **discontinuous permafrost** underlies between 50 and 90 percent of an area.

pingo Ice-cored, dome-like hill found in permafrost regions.

polynya Area of large and persistent open water within the pack ice.

precession of the equinoxes, axial A wobble in the Earth's rotational axis, which varies over a ~26,000-year period, caused by the force of the moon and sun on the Earth's equatorial bulge.

precession, elliptical The change in timing of perihelion with respect to the seasons at the Earth's closest approach to the sun, which varies over a ~22,000-year cycle. Today, perihelion nearly coincides with the Northern Hemisphere winter solstice; 11,000 years ago it coincided with the Northern Hemisphere summer solstice.

pressure ridge Ridge formed by compression as slabs of sea ice collide.

reference ellipsoid A flattened sphere approximating the Earth's shape that includes the equatorial bulge and polar flattening caused by rotation. It is used as a reference surface for measuring sea level.

relative (local) sea level rise Geographic differences in the rate of sea level rise caused by ongoing glacial isostatic adjustments, tectonic movements, subsurface fluid extraction, or seasonal to interannual oceanographic processes (e.g., ENSO), in addition to climate-related sea level rise.

roche moutonnée Asymmetric rock outcropping sculpted by glacial erosion. It is smoothly streamlined on the side of the oncoming glacier, shattered and jagged on the lee flank.

sea ice Slabs of ice floating on the surface of the sea.

sea ice area Area occupied by sea ice per cell. For a given sector or an entire hemisphere, it is the sum of the area of each cell times its fractional ice concentration.

sea ice concentration The relative amount of a given area covered by sea ice, expressed as a percentage.

sea ice extent Area covered by at least 15 percent ice. (*See also* **sea ice area**; **sea ice concentration**.)

sea level rise equivalent The rise in sea level, in millimeters, due to a given loss in ice mass. Conversion factor: 362.5 GT ice = 1 millimeter of global sea level rise, assumed to be evenly spread over the oceans. Units: one GT (one billion metric tons, or one trillion kilograms) = 1 cubic kilometer of freshwater, or 1.1 cubic kilometer of ice.

steric changes Ocean water density and ocean height changes due to variations in temperature and/or salinity.

storm surge An increase in water level above that of the astronomical tide caused by high winds and low barometric pressure associated with a storm.

sublimation The transformation of ice directly into water vapor without passing through the liquid state.

surface mass balance (of a glacier, ice sheet) (SMB) The net balance between accumulation and surface ice mass loss. When positive, accumulation (snowfall) exceeds mass loss (surface melting plus SUBLIMATION [ice to water vapor] + runoff). When negative, ice loss exceeds accumulation. (*See also* **mass balance**.)

taiga Subarctic forest south of the tundra, dominated by conifers (mostly pines, spruces, and larch) and birch. Synonymous with *boreal (northern) forest*.

talik Unfrozen lens or zone within permafrost.

thermal expansion Increase in ocean water level due to a rise in water temperature.

thermal inertia A measure of the ability of a material to conduct and store heat. It denotes the ability of the material to store heat during the day and reradiate it at night. Materials that have a high heat capacity, such as water, also generally have a high thermal inertia, showing only minor temperature changes during the diurnal cycle.

thermokarst Landforms in permafrost terrain characterized by depressions resulting from thawing of ground ice.

tidewater glacier *See under* **glaciers**.

till A heterogeneous mixture of different-sized materials deposited by a glacier.

tundra A treeless, mossy, shrubby plain in Arctic and sub-Arctic regions underlain by permafrost.

utilidor An above-ground, covered utility conduit used in Arctic communities where permafrost prevents burial of water and sewer pipes.

West Antarctic Ice Sheet (WAIS) Part of the Antarctic ice sheet that lies west of the Transantarctic Mountains. It is potentially unstable because much of the ice sits below sea level and slopes down inland. (*See* **marine ice sheet instability** [MISI].)

yedoma A type of permafrost that is carbon rich (around 2–5 percent carbon) with an ice content of 50 to 90 percent by volume.

NOTES

1. WHITHER THE SNOWS OF YESTERYEAR?

1. Barry and Gan (2011), p. 167; Intergovernmental Panel on Climate Change (2013a), table 4.1.
2. Barry and Gan (2011), p. 3.
3. Oxygen and hydrogen atoms in ordinary ice form tetrahedral bonds 109.5° apart. Stacking tetrahedra into a three-dimensional lattice creates a crystal with overall hexagonal symmetry (as in a snowflake). Ice can exist in at least 15 distinct phases, each with a different crystal structure, over a broad range of very low temperatures and high pressures.
4. "Eskimo," once applied to all indigenous Arctic peoples of eastern Siberia, Alaska, Canada, and Greenland, now refers more narrowly to the Iñupiaq and Yupik people of Alaska, while those in Arctic Canada and Greenland are collectively called "Inuit." The timing of the early migration based on genetics is discussed in Achilli et al. (2013).
5. Ingstad and Ingstad (2001).
6. Wikipedia, "Northwest Passage," https://en.wikipedia.org/wiki/Northwest_Passage (last modified October 6, 2018).
7. "Lost Franklin Expedition Ship Found in the Arctic," CBC News, September 9, 2014, https://www.cbc.ca/news/politics/franklin-expedition-ship-found-in-the-arctic-1.2760311; "Franklin Expedition Ship Found in Arctic ID'd as HMS Erebus," CBC News, October 1, 2014, https://www.cbc.ca/news/politics/franklin-expedition-ship-found-in-arctic-id-d-as-hms-erebus-1.2784268.
8. Paul Watson, "Ship Found in Arctic 168 Years After Doomed Northwest Passage Attempt," *Guardian*, September 12, 2016, http://www.theguardian.com/world/2016/sep/12/hms-terror-wreck-found-arctic-nearly-170-years-northwest-passage-attempt/; Government of Canada, "Parks Canada Media Statement—Validation of Discovery of HMS Terror" (accessed January 1, 2017).
9. Henderson (2009).

10. The North Polar Regions: A Geographical and Navigational Study, "Appendix D, Chronology of North Polar Exploration," http://norpolar.tripod.com/chron.html (accessed July 10, 2013).
11. Piri Reis, an Ottoman admiral and cartographer, compiled a world map in 1513 from various sources. This map, according to some, purportedly shows the coastline of Antarctica as it would appear without ice. This knowledge supposedly came from advanced ancient sources. However, when centered at 0° latitude, 0° longitude on an azimuthal equidistant projection, the map outlines the coastlines of France, Spain, West Africa, and South America down to southern Brazil fairly well, but deviates sharply beyond Uruguay, showing a coastline with little resemblance to Antarctica's. See Steven Dutch, "The Piri Reis Map," http://web.archive.org/web/20171226235246/https://www.uwgb.edu/dutchs/PSEUDOSC/PiriRies.HTM.
12. Two others, a British naval captain, Edward Branfield, and an American sealer, Nathaniel Palmer, also sighted Antarctica within days of Bellingshausen.
13. Quoted in Walker (2013), p. 144.
14. Wikipedia, "Comparison of the Amundsen and Scott Expeditions," http://en.wikipedia.org/wiki/Comparison_of_the_Amundsen_and_Scott_Expeditions (last modified September 3, 2018).
15. Although best known for the voyage of the *Endurance*, Shackleton previously participated in the National Antarctic (*Discovery*) Expedition (1901–1903) and led the Nimrod Expedition (1907–1909) in which he and his party set a new southern record at 88°23′S, 162°E (only 180 kilometers [118 miles] from the South Pole) in January of 1909. His final expedition to Antarctica ended in South Georgia with his death in 1922.
16. Summerhayes (2008).
17. National Research Council (2012).
18. Goddard Institute for Space Studies, National Aeronautics and Space Administration (NASA), "Long-Term Warming Trend Continued in 2017: NASA, NOAA," January 18, 2016, https://www.giss.nasa.gov/research/news/20180118/ (accessed January 19, 2018).
19. Arctic Monitoring and Assessment Programme (2017); "Surface Air Temperature," in *Arctic Report Card 2017*, ed. J. Richter-Menge, J. E. Overland, J. T. Mathis, and E. Osborne, pp. 5–12 (NOAA, 2017), ftp://ftp.oar.noaa.gov/arctic/documents/ArcticReportCard_full_report2017.pdf (accessed February 12, 2018).
20. Arctic Monitoring and Assessment Programme (2017).
21. Comiso (2012); Stroeve et al. (2012).
22. Defined as ocean with an ice concentration of 15 percent or more.
23. Arctic Monitoring and Assessment Programme (2017).
24. National Snow & Ice Data Center, "Arctic Sea Ice 2017: Tapping the Brakes in September," October 5, 2017, http://nsidc.org/arcticseaicenews/2017/10/arctic-sea-ice-2017-tapping-the-brakes-in-september/ (accessed October 6, 2017).
25. National Snow & Ice Data Center, "Arctic Sea Ice Maximum at Second Lowest Level in the Satellite Record," March 23, 2018, http://nsidc.org/arcticseaicenews/2018/03/arctic-sea-ice-maximum-second-lowest/ (accessed April 24, 2018).

26. National Snow & Ice Data Center, "Freezing in the Dark," November 2, 2017, http://nsidc.org/arcticseaicenews/2017/11/ (accessed February 13, 2018); Andrea Thompson, "Sea Ice Hits Record Lows at Both Poles," February 13, 2017, https://www.scientificamerican.com/article/sea-ice-hits-record-lows-at-both-poles/ (accessed February 17, 2017).
27. Waldman (2017).
28. Overduin and Solomon (2011).
29. The Little Ice Age lasted roughly from the thirteenth century to the mid-nineteenth century.
30. Revkin (2008).
31. Van den Broecke et al. (2016); Forsberg et al. (2017); Harig and Simons (2015).
32. The Greenland and Antarctic ice sheets together would raise sea level by 66 meters, if both melted entirely (see table 1.1).
33. Straneo and Heimbach (2013); Joughin et al. (2012).
34. Joughin et al. (2014); Rignot et al. (2014).
35. Stroeve et al. (2012).
36. Viñas (2012).
37. Tedesco et al. (2016).
38. Quinn (2007). The Arctic haze introduces conflicting feedbacks. Although sulfate aerosols scatter sunlight, cooling the surface, black carbon or soot weakly absorbs solar energy. Black carbon or soot disproportionately affects regional climate by creating a haze that spreads over highly reflecting Arctic snow and ice. Sulfate aerosols help increase the number of cloud droplets, whitening clouds and making them more reflecting (cooling effect). However, smaller droplets induce more downwelling infrared emission and, hence, warming.
39. Hadley and Kirchstetter (2012).
40. Alley (2007).
41. The Atlantic Meridional Overturning Circulation (AMOC) includes winds and tides, in addition to salinity- and temperature-driven ocean currents (see also Broecker, 2010).
42. Hu et al. (2011); Yin et al. (2010).
43. Intergovernmental Panel on Climate Change (2013a); Intergovernmental Panel on Climate Change (2013b).
44. Earth System Research Laboratory, National Oceanic and Atmospheric Administration, "Trends in Atmospheric Carbon Dioxide," https://www.esrl.noaa.gov/gmd/ccgg/trends/ (accessed February 14, 2018); CO_2.Earth, "Annual CO_2 Data," http://www.co2.earth/annual-co2/ (accessed June 29, 2018). The upward curve is punctuated by a rhythmic seasonal rise and fall, due largely to Northern Hemisphere vegetation. Photosynthesis consumes carbon dioxide during northern spring and summer; it is released during winter, as the vegetation becomes dormant
45. Intergovernmental Panel on Climate Change (2013a); Intergovernmental Panel on Climate Change (2013b).
46. Greenhouses also retain heat because they protect from the wind and reduce heat losses by conduction and convection.

47. United Nations Environment Programme (2007).
48. Navab (2011).
49. Bradley et al. (2006).
50. Bolch (2017); Pritchard (2017).
51. Belmecheri et al. (2015).
52. Huggel et al. (2012).
53. Horstmann (2004).
54. National Oceanic and Atmospheric Administration, "NOAA: 2017 Was 3rd Warmest Year on Record for the Globe," January 18, 2018, https://www.noaa.gov/news/noaa-2017-was-3rd-warmest-year-on-record-for-globe (accessed January 19, 2018); Goddard Institute for Space Studies, NASA, "Long-Term Warming Trend Continued in 2017."
55. Sea Level Change, NASA, "Key Indicators: Global Mean Sea Level," https://sealevel.nasa.gov/understanding-sea-level/key-indicators/global-mean-sea-level (Sea level rise plot covers January 1993 to October 9, 2017); AVISO+, "Mean Sea Level Rise (January 5, 1993 to November 17, 2017)," https://www.aviso.altimetry.fr/en/data/products/ocean-indicators-products/mean-sea-level.html (posted March 26, 2018).
56. Gornitz (2013).
57. Dieng et al. (2017); Rietbroek et al. (2016).
58. Intergovernmental Panel on Climate Change (2013a); Intergovernmental Panel on Climate Change (2013b). See also appendix A.
59. Neumann (2015). The 1-in-100-year flood is one with a 1 percent chance of occurring in any given year.
60. Hauer et al. (2016).
61. Glacial isostatic rebound (see glossary); McGuire (2012).
62. Gautier et al. (2009).
63. Teck, "United States Operations: Red Dog Operations," https://www.teck.com/operations/united-states/operations/red-dog/.
64. United Nations Framework Convention on Climate Change (2010).
65. Miller et al. (2012).

2. ICE AFLOAT—ICE SHELVES, ICEBERGS, AND SEA ICE

1. Rignot et al. (2013).
2. Barry and Gan (2011), pp. 278–279.
3. National Snow & Ice Data Center, "Larsen B Ice Shelf Collapses in Antarctica," March 18, 2002, http://nsidc.org/news/press/larsen_B/2002.html (updated March 21, 2002). The ice slab was 220 meters (720 feet) thick and ~3,250 square kilometers (1,255 square miles) in area.
4. Rignot et al. (2004); Scambos et al. (2004).
5. Khazendar et al. (2015).

6. NASA press release, May 14, 2015. "Study Shows Antarctica's Larsen B Ice Shelf Nearing Its Final Act." https://www.nasa.gov/press-release/nasa-study-shows-antarctica-s-larsen-b-ice-shelf-nearing-its-final-act/
7. Rebesco et al. (2014).
8. Holland et al. (2015).
9. Hogg and Gudmundsson (2017); Viñas (2017).
10. NASA Earth Observatory, WorldView-2 satellite image, March 23, 2013, http://earthobservatory.nasa.gov/IOTD/view.php?id=81174.
11. Pelto (2015).
12. Paolo et al. (2015). See also chap. 6, "Ice Shelves in Trouble."
13. Quoted in Fox (2012).
14. *Ilulissat*, in Greenlandic; An et al. (2017).
15. David Bresson, "The Science Behind the Iceberg That Sank the Titanic," *Scientific American*, April 14, 2012, https://blogs.scientificamerican.com/history-of-geology/the-science-behind-the-iceberg-that-sank-titanic/; BBC History, "The Iceberg That Sank Titanic—Origins of the Titanic Iceberg," http://www.bbc.co.uk/history/topics/iceberg_sank_titanic.
16. Joughin et al. (2014).
17. Chelsea Harvey, "One of the World's Fastest Melting Glaciers May Have Just Lost Its Biggest Chunk of Ice on Record," *The Washington Post*, August 19, 2015, https://www.washingtonpost.com/news/energy-environment/wp/2015/08/19/one-of-the-worlds-fastest-melting-glaciers-may-have-just-lost-its-biggest-chunk-of-ice-ever/?utm_term=.4afdc3c25606 (accessed October 5, 2017).
18. McGrath et al. (2013)
19. Steig (2016); Bromwich et al. (2014).
20. Cook and Vaughan (2010).
21. Fox (2012). Note that the −9°C refers to the average *annual* temperature threshold, whereas the 0°C refers to the average *summer* temperature line.
22. Davies (2017); Fox (2012); Qiu (2013); Scambos et al. (2004).
23. MacGregor et al. (2012).
24. Alley et al. (2015); Schmidtko et al. (2014).
25. Paolo et al. (2015).
26. Depoorter et al. (2013); Rignot et al. (2013).
27. Rignot et al. (2004); Scambos et al. (2004).
28. Shepherd et al. (2012).
29. Favier et al. (2014).
30. Mercer (1978); Joughin and Alley (2011); Joughin et al. (2012).
31. Straneo and Heimbach (2013); Joughin et al. (2012).
32. Joughin et al. (2012); Joughin et al. (2014).
33. An et al. (2017).
34. Matilsky and Duwadi (2008).

35. Matilsky and Duwadi (2008).
36. Wadhams (2013).
37. Bruneau (2004); United States Coast Guard, International Ice Patrol (IIP), "About International Ice Patrol (IIP)," http://www.navcen.uscg.gov/?pageName=IIPHome; Canadian Coast Guard, Ice Navigation in Canadian Waters, Chapter 3, "Ice Climatology and Environmental Conditions," http://www.ccg-gcc.gc.ca/folios/00913/docs/icenav-ch3-eng.pdf.
38. Marko et al. (2014).
39. United States Coast Guard Navigation Center, https://www.navcen.uscg.gov/?page?Name=IIPHome. (Last updated September 25, 2018).
40. Barry and Gan (2011), p. 291.
41. Barry and Gan (2011), p. 293; Diemand (2001/2008), p. 1262.
42. Wadhams (2013); Barry and Gan (2011), p. 293.
43. Diemand (2001/2008), p. 1262.
44. Very large icebergs may extend into subzero water at depth, due to the lowering of the freezing point with salinity.
45. Diemand (2001/2008), p. 1261.
46. National Snow & Ice Data Center, "2016 Ties with 2007 for Second Lowest Arctic Sea Ice Minimum," September 15, 2016, http://nsidc.org/arcticseaicenews/2016/09/2016-ties-with-2007-for-second-lowest-arctic-sea-ice-minimum/ (accessed September 16, 2016).
47. Barry and Gan (2011), p. 237.
48. Serreze and Stroeve (2015).
49. Pistone et al. (2014).
50. Arctic Monitoring and Assessment Programme (2017).
51. National Snow & Ice Data Center, "Arctic Sea Ice 2017: Tapping the Brakes in September," October 5, 2017, http://nsidc.org/arcticseaicenews/2017/10/arctic-sea-ice-2017-tapping-the-brakes-in-september/ (accessed October 6, 2017).
52. Arctic Monitoring and Assessment Programme (2017); Stroeve et al. (2014).
53. National Snow & Ice Data Center, "Arctic Sea Ice Maximum at Record Low for Third Straight Year," March 22, 2017, https://nsidc.org/news/newsroom/arctic-sea-ice-maximum-record-low-third-straight-year/ (accessed May 25, 2017).
54. National Snow & Ice Data Center, "2018 Winter Arctic Sea Ice: Bering Down," April 4, 2018, http://nsidc.org/arcticseaicenews/2018/04/2018-winter-arctic-sea-ice-bering-down/ (accessed April 24, 2018); National Snow & Ice Data Center, "Arctic Sea Ice Maximum at Second Lowest in the Satellite Record," March 23, 2018, https://nsidc.org/news/newsroom/arctic-sea-ice-maximum-second-lowest-satellite-record (accessed April 24, 2018). The Arctic winter sea ice maximum extent was 14.5 million square kilometers (5.6 million square miles) on March 17, 2018, as compared with the record lowest maximum of 14.4 square kilometers (5.6 million square miles) set on March 7, 2017.
55. Serreze and Stroeve (2015).
56. Perovich et al. (2016).

57. Arctic Monitoring and Assessment Programme (2017).
58. Perovich et al. (2014).
59. Stroeve et al. (2014).
60. Pithan and Mauritsen (2014).
61. Stroeve et al. (2014).
62. Serreze and Stroeve (2015).
63. Swart (2017).
64. Arctic Monitoring and Assessment Programme (2017); Overland and Wang (2013).
65. Masters (2014).
66. Cohen et al. (2014); Masters (2014); Francis and Vavrus (2012).
67. Screen and Simmonds (2014); Cohen et al. (2014).
68. Screen (2017).
69. Bintanja et al. (2013).
70. Depoorter et al. (2013); Rignot (2013); Pritchard et al. (2012).
71. Holland and Kwok (2012).
72. National Snow & Ice Data Center, "Antarctic Maximum Extent," in "Arctic Sea Ice 2017: Tapping the Brakes in September" (accessed October 6, 2017); "Arctic Sea Ice Maximum at Second Lowest in the Satellite Record" https://nsidc.org/news/newsroom/arctic-sea-ice-maximum-second-lowest-satellite-record (accessed April 24, 2018).

3. IMPERMANENT PERMAFROST

1. Péwé (2014).
2. Kramer (2008); Larmer (2013).
3. Barry and Gan (2011), p. 167; Osterkamp and Burn (2002).
4. Péwé (2014).
5. Péwé (2014); see glossary.
6. Global Terrestrial Network for Permafrost (GTN-P), https://gtnp.arcticportal.org/ (accessed October 10, 2017); Romanovsky et al. (2010). Ground temperatures are measured at the depth of zero annual amplitude.
7. Lachenbruch and Marshall (1986).
8. Exploratorium, Ice Stories, "Tundra and Permafrost," http://icestories.exploratorium.edu/dispatches/big-ideas/tundra-and-permafrost.
9. The reason why water, unlike most liquids, expands upon freezing is explained in chapter 1.
10. As in congelation sea ice (see chap. 2), ice grows vertically with its c-axis oriented horizontally (Corte [1969]). Maximum growth occurs in the downward direction perpendicular to c (i.e., along the a-axis).
11. For a fuller discussion on ice lens formation and frost heaving, see, e.g., Davis (2001), pp. 67–95, and Rempel (2010).

12. Davis (2001), p. 110.
13. Geologists call this type of igneous intrusion a *laccolith*.
14. Davis (2001), pp. 121–131.
15. Karst topography, a landscape in limestone or dolomite riddled by subsurface caves, disappearing streams, sinkholes, and underground drainage, was named after a region between Slovenia and Italy.
16. Grosse et al. (2013), section 8.21.3.
17. Grosse et al. (2013), section 8.21.6. Oriented lakes may also form by alignment along faults or joints in rocks, or along contacts between different rock or soil types.
18. U.S. Army Corps of Engineers, Engineer Research and Development Center, "Permafrost Tunnel Research Facility," https://www.erdc.usace.army.mil/CRREL/Permafrost-Tunnel-Research-Facility/.
19. Quoted in Walker (2007).
20. Rosen, "Speedily Eroding North Slope River Bluff Offers Window Into Ice Age Past." *Anchorage Daily News*, August 29, 2016, https://www.adn.com/arctic/article/ice-rich-bluff-along-north-slope-river-eroding-fast-clip/2015/11/27/.
21. Romanovsky et al. (2010); Arctic Monitoring and Assessment Programme (2017); Intergovernmental Panel on Climate Change (2013), section 4.7.2.1. Permafrost temperature is measured at the depth of zero amplitude (around 20 meters below the surface).
22. Intergovernmental Panel on Climate Change (2013), section 4.7.2.2.
23. Intergovernmental Panel on Climate Change (2013), section 4.7.2.2; Arctic Monitoring and Assessment Programme (2011), section 5.3.2.1.2.
24. Smith et al. (2005).
25. Walker (2007).
26. Avis et al. (2011). See also appendix A.
27. Arctic Foundations, "Building on Thick Ice," *Alaska Business Monthly*, February 2007, http://www.arcticfoundations.com/index.php/news/73-building-on-thick-ice.
28. G. H. Johnston, "CBD-64. Permafrost and Foundations. National Research Council Canada," April 1965, http://web.mit.edu/parmstr/Public/NRCan/CanBldgDigests/cbd064_e.html (accessed November 12, 2018).
29. Wikipedia, "Construction of the Trans-Alaska Pipeline System," https://en.wikipedia.org/wiki/Construction_of_the_Trans-Alaska_Pipeline_System (last modified September 21, 2018).
30. Bird (2008), p. 202.
31. Arctic Monitoring and Assessment Programme, SWIPA (2011), chap. 5, section 5.3.2.5.
32. Are et al. (2008)
33. Barnhart et al. (2014); Lantuit et al. (2012); Overeem et al. (2011).
34. Are et al. (2008).
35. Barnhart et al. (2014).
36. Jones et al. (2009); Lantuit et al. (2012); Barnhart et al. (2014).
37. Jones et al. (2009).

38. Gornitz (2013). See chapter 9 of this book for further discussion of the impacts on Arctic communities.
39. Schuur et al. (2015).
40. Mascarelli (2009).
41. Whiteman et al. (2013).
42. Brewer (2014); Shakhova et al. (2014).
43. Brewer (2014); Shakhova et al. (2014); Anthony et al. (2012); Vonk et al. (2012).
44. Anthony et al. (2012); Anthony et al. (2014); Anthony et al. (2016).
45. Sobek (2014); Anthony et al. (2014).
46. Anthony et al. (2014).
47. Smith et al. (2005).
48. Avis et al. (2011).
49. Vonk et al. (2012).
50. Olefeldt et al. (2016).
51. Anthony et al. (2012); Anthony et al. (2016).
52. Treat and Frolking (2013).
53. Schuur (2016).
54. Schuur (2016); Schuur et al. (2015).
55. Herndon (2018).
56. Gao et al. (2013). However, they did not consider greenhouse gas emissions from thermokarst or from offshore gas hydrates.
57. Thornton and Crill (2015).
58. Shakhova et al. (2010).
59. Ruppel and Kessler (2017).
60. The base of the methane hydrate stability zone appears on seismic reflection profiles as a "bottom simulating reflector," or BSR, which parallels the seafloor bathymetry, not the lithology or stratigraphy.
61. Maslin et al. (2010). Around 8,200 years ago, a massive underwater avalanche of loose sediment charged with methane gas and water cascaded abruptly down the flanks of the continental shelf and slope off the coast of Norway. The Storegga slide, one of the largest known submarine landslides, unleashed some 2,500–3,500 cubic kilometers of debris, spread over 9,500 square kilometers. It also triggered a devastating tsunami that inundated the coasts of northeastern Scotland and western Norway and dumped marine deposits up to 20 meters above sea level in the Shetland Islands. The submarine landslide and associated release of confined methane may have been triggered by icequakes resulting from glacial rebound after the melting of the Greenland Ice Sheet (see chap. 5).
62. Talling et al. (2014).
63. Shakhova et al. (2010); Shakova et al. (2014); Kort et al. (2012).
64. Thornton and Crill (2015).
65. Ruppel and Kessler (2017).
66. Ananthaswamy (2015).

4. DARKENING MOUNTAINS—DISAPPEARING GLACIERS

1. Muir, J. (1902), as quoted by Molnia et al. (2008), p. K134.
2. Molnia (2007).
3. National Oceanic and Atmospheric Administration, "Sea Level Trends," 2017, http://tidesandcurrents.noaa.gov/sltrends/sltrends.html.
4. Motyka et al. (2007).
5. Due to local factors, for example, warm moist air from the Pacific Ocean, local microclimates, and lower mountain elevations than farther north.
6. Gardner et al. (2013).
7. Dyurgerov and Meier, in Williams and Ferrigno (2012).
8. Radic et al. (2014; table 1); Grinsted (2013).
9. Benn and Evans (2010), p. 80; Barry and Gan (2011), p. 92.
10. Jacka (2007), p. 507; Barry and Gan (2011), p. 104. Glide planes represent planes of weakness within ice crystals that slip more readily under pressure. They lie perpendicular to the c-axes, which tend also to be the vertical direction in a glacier. Alignment of ice crystals with this orientation imparts a characteristic crystal fabric. See also figure 1.2 and box 2.1 regarding the crystal structure of ice.
11. Centre for Ice and Climate, Niels Bohr Institute, "The Firn Zone: Transforming Snow to Ice," http://www.iceandclimate.nbi.ku.dk/research/drill_analysing/cutting_and_analysing_ice_cores/analysing_gasses/firn_zone.
12. Bakke and Nesje (2014).
13. Jewelers' rouge used to polish metals is finely powdered hematite, or iron oxide, Fe_2O_3. Its hardness (5–6 on the Mohs scale, where talc = 1 and diamond = 10) is somewhat less than that of most typical rock-forming minerals (Mohs hardness 6–7).
14. Benn and Evans (2010), pp. 62–64.
15. Benn and Evans (2010), p. 66.
16. Benn and Evans (2010), pp. 65–66.
17. Benn and Evans (2010), pp. 68–74; Chu (2014), pp. 33–37.
18. Benn and Evans (2010), pp. 77–78.
19. Jiskoot (2011a), p. 250.
20. Johnson (2013).
21. Jiskoot (2011a), p. 248.
22. See note 9 above.
23. Benn and Evans (2010), p. 116.
24. Jiskoot (2011a), pp. 248–250.
25. Jiskoot (2011a), pp. 248–250.
26. Karpilo Jr. (2009), pp. 153–156.
27. Jiskoot (2011b), p. 416; Barry and Gan (2011), p. 107.
28. European Space Agency, "Chasing Ice," August 21, 2015, http://www.esa.int/Our_Activities/Observing_the_Earth/Copernicus/Sentinel-1/Chasing_ice; Chelsea

Harvey, "One of the World's Fastest Melting Glaciers May Have Just Lost Its Biggest Chunk of Ice on Record, *The Washington Post*, August 19, 2015, https://www.washingtonpost.com/news/energy-environment/wp/2015/08/19/one-of-the-worlds-fastest-melting-glaciers-may-have-just-lost-its-biggest-chunk-of-ice-ever/?utm_term=.4afdc3c25606.

29. Benn and Evans (2010), pp. 169–186; Warren (2014), pp. 105–106.
30. O'Leary and Christoffersen (2013).
31. Molnia (2007); Joughin et al. (2014); Nick et al. (2009). See also chapter 2.
32. Scambos et al. (2009). Ice shelves, as noted in chapter 2, are floating extensions of glaciers or ice sheets.
33. Benn and Evans (2010), pp. 38–39; Karpilo Jr. (2009). Ice sheet cores also preserve important archives of past climates, including information on regional temperature, precipitation, atmospheric circulation patterns, and global atmospheric composition (see chap. 7).
34. Grinsted (2013).
35. Barry and Gan (2011), p. 99.
36. Appendix B briefly describes some important satellite instruments used to monitor the cryosphere. An in-depth review of principles of remote sensing used in glacier and cryosphere studies is given by Tedesco, ed. (2015) and Pellikka and Rees (2010).
37. Raup et al. (2015), p. 133.
38. Raup and Kargel (2012), p. A249.
39. Appendix B; Pellikka and Rees (2010), pp. 16–18.
40. Wikipedia, "Gravity Recovery and Climate Experiment," http://en.wikipedia.org/wiki/Gravity_Recovery_and_Climate_Experiment (last modified February 3, 2019).
41. Boening et al. (2012). But the upward sea level trend resumed soon thereafter.
42. Jacob et al. (2012).
43. Gardner et al. (2013).
44. Raup et al. (2015), p. 141, briefly summarizes how this is accomplished.
45. Mouginot et al. (2014); Rignot et al. (2014).
46. Williams and Ferrigno (2012). Individual chapters cover different glacier regions; e.g., for Alaska, chapter K, see Molnia et al. (2008).
47. World Glacier Monitoring Service, "Latest Glacier Mass Balance Data," http://wgms.ch/latest-glacier-mass-balance-data/ (last updated May 16, 2018); Haeberli (2011).
48. Raup and Kargel (2012), pp. 247–260.
49. E.g., Molnia et al. (2008) for Alaska.
50. Pfeffer et al. (2014). Grinsted (2013) is an example of a study that derives global glacier volume from data in RGI, WGI, and GLIMS.
51. Johann Siegen (1921), "Glacier Tales." Gletscherwelten/World of Glaciers. https://www.jungfraualetsch.ch/wp-content/uploads/unesco_regionalbroschuere_gletscherwelten_web.pdf.
52. Montagnes magique, "Cordée d'alpinistes," by Henry George Willink (1892), https://fresques.ina.fr/montagnes/fiche-media/Montag01006/cordee-d-alpinistes.html (accessed November 12, 2018). *Wilderwurm* translated literally means "wild worm."

53. Nussbaumer et al. (2007).
54. Nussbaumer et al. (2007).
55. Zemp et al. (2008). However, natural climate cycles, such as the decades-long Atlantic Multidecadal Oscillation (AMO), which affects climate patterns over much of the Northern Hemisphere, cause significant variations in glacier mass balance. Enhanced Alpine glacier melting, especially since the 1980s, is associated with an increasing AMO index (Huss et al., 2010).
56. Thompson et al. (2009).
57. Thompson et al. (2011).
58. Mouginot and Rignot (2015); Davies and Glasser (2012).
59. Edward Wong, "Chinese Glacier's Retreat Signals Trouble for Asian Water Supply," *New York Times*, December 9, 2015.
60. See also Immerzeel et al. (2010).
61. Brun et al. (2017).
62. Brun et al. (2017); Maurer et al. (2016); Bolch et al. (2012); Kääb et al. (2012).
63. Brun et al. (2017).
64. Wong, "Chinese Glacier's Retreat."
65. Zemp et al. (2015).
66. Kraaijenbrink et al. (2017).
67. Slangen et al. (2017), their table 1, including peripheral glaciers; see also appendix A.
68. Radic et al. (2014).
69. Arctic amplification is discussed in chapter 2.
70. Wikipedia, "Karakoram," http://en.wikipedia.org/wiki/Karakoram (last modified October 2, 2018; accessed March 11, 2019).
71. Brun et al. (2017); Farinotti (2017).
72. Kapnick et al. (2014); Bolch et al. (2012).
73. Kraaijenbrink et al. (2017).
74. Radic et al. (2014).

5. THE GREENLAND ICE SHEET

1. These paleoclimate proxies include tiny marine organisms called foraminifera, sedimentary evidence from marine cores, and oxygen isotope data from Greenland ice cores.
2. Folger (2017).
3. Folger (2017).
4. Wikipedia, "Greenland," http://en.wikipedia.org/wiki/Greenland (last modified October 5, 2018).
5. McGrath et al. (2013). "Ablation" and "equilibrium line altitude" are defined in chapter 4 and the glossary.
6. Tedesco et al. (2017).
7. Viñas (2012).

8. Nettles and Ekström (2010).
9. Quote from Huntford (1997).
10. Bamber et al. (2013); Morlighem et al. (2014, 2017).
11. Pelto (2015b).
12. Assuming all melted ice is spread out evenly over the ocean. See table 1.1, footnote 2 for conversion of ice mass to sea level rise.
13. Kjeldsen et al. (2015); Khan et al. (2014); Kjaer et al. (2012).
14. Velicogna et al. (2014); Moon et al. (2012).
15. Kjaer et al. (2012); box 5.1.
16. Smith (2012); Bjørk (2012).
17. Helm et al. (2014), table 4; Csatho et al. (2014).
18. Csatho et al. (2014); Moon et al. (2012).
19. Khan et al. (2014).
20. Jiang et al. (2010).
21. Viñas (2012).
22. Howat et al. (2013).
23. Leeson et al. (2014).
24. Poinar et al. (2015).
25. Moon et al. (2012).
26. An et al. (2017).
27. Joughin, Smith, Sheam et al. (2014).
28. A potential instability in a marine-terminating glacier or ice stream where the grounding line rests below sea level and the submarine topography slopes landward (see glossary).
29. An et al. (2017); Joughin, Smith, Sheam et al. (2014).
30. Khan et al. (2014).
31. Morlighem et al. (2014).
32. Morlighem et al. (2014).
33. The North Atlantic Current is the northern extension of the Gulf Stream. Joughin et al. (2012); Straneo and Heimbach (2013).
34. Holland et al. (2008).
35. Joughin et al. (2012); Straneo and Heimbach (2013).
36. Morlighem et al. (2017); Rignot et al. (2015).
37. Krinner and Durand (2012).
38. Nettles and Ekstrøm (2010).
39. Murray et al. (2015); Nettles and Ekström (2010).
40. Tedesco et al. (2017). National Oceanic and Atmospheric Administration, Arctic Program, "Arctic Report Card: Update for 2017," http://www.arctic.noaa.gov/Report-Card/Report-Card-2017/ArtMID/.
41. Straneo and Heimbach (2013).
42. Straneo and Heimbach (2013).
43. Hu et al. (2011).
44. Harper et al. (2012); Harper (2014); Forster et al. (2014).

45. Forster et al. (2014).
46. Poinar et al. (2017).
47. Michael Casey, "The Case of Greenland's Disappearing Lakes," CBS News, January 21, 2015, https://www.cbs.news.com/news/the-case-of-greenlands-disappearing-lakes/; Howat et al. (2015).
48. Willis et al. (2015).
49. Rhonda Zurn and Lacey Nygard, "Atmospheric Warming Heats the Bottom of Ice Sheets, as Well as the Top," January 21, 2015, https://cse.umn.edu/college/news/atmospheric-warming-heats-bottom-ice-sheets-well-top.
50. Stevens et al. (2015).
51. Stevens et al. (2015); Poinar et al. (2015).
52. Smith et al. (2015).
53. Meg Sullivan, "UCLA-Led Study Shows How Rivers of Meltwater on Greenland's Ice Sheet Contribute to Rising Sea Levels," January 12, 2015, http://newsroom.ucla.edu/releases/ucla-study-shows-rivers-meltwater-on-greenlands-ice-sheet-contribute-rising-sea-levels.
54. Lüthi (2010).
55. Doyle et al. (2015).
56. Sundal et al. (2011).
57. Lüthi (2013).
58. Nienow (2014).
59. Bell et al. (2014).
60. Bell et al. (2014).
61. Bamber et al. (2013). Steve Cole, George Hale, and Hannah Johnson, "NASA Data Reveals Mega-Canyon Under Greenland Ice Sheet," NASA, August 29, 2013, https://www.nasa.gov/press/2013/august/nasa-data-reveals-mega-canyon-under-greenland-ice-sheet/#.W7kW2_ZRfIU (accessed July 27, 2015).
62. Morlighem et al. (2014).
63. Morlighem et al. (2014).
64. Morlighem et al. (2017).
65. Cooper et al. (2016).
66. Rogozhina et al. (2016). Iceland is located on the mid-Atlantic Ridge which separates the North American plate from the Eurasian plate. According to plate tectonic theory, the midocean ridge system, which encircles the globe, is where magma ascends from the upper mantle, which helps drive the plates apart. Furthermore, Iceland sits atop a long-lived "hot spot" of anomalously warm magma that has built up the island over time. Tens of millions of years ago Greenland lay over the Icelandic hot spot, but has subsequently drifted northwestward as the plates have spread apart.
67. As cited by Eric Holtaus, "*Slate* Exclusive: Why Greenland's 'Dark Snow' Should Worry You," Future Tense (column), Slate.com, September 16, 2014, http://www.slate.com/blogs/future_tense/2014/09/16/jason_box_s_research_into_greenland_s_dark_snow_raises_more_concerns_about.html (accessed August 5, 2015).

68. E.g., Stibal et al. (2017); Tedesco et al. (2016); Howat et al. (2013).
69. Leeson et al. (2014).
70. Tedesco et al. (2016).
71. Wientjes et al. (2011).

6. ANTARCTICA: THE GIANT ICE LOCKER

1. Secretariat of the Antarctic Treaty, "The Antarctic Treaty," http://www.ats.aq/e/ats.htm.
2. This portion of the treaty is up for review in 2048. A new race to the South Pole may then unfold, as nations compete for increasingly scarce critical mineral resources and as technological advances and improving climate conditions make resource exploitation there more feasible (more in chap. 9). S. Romero, "Array of Players Joining Race for Space at Bottom of the World," *New York Times*, December 30, 2015.
3. Ford (2015).
4. IceCube, South Pole Neutrino Laboratory, "Antarctic Weather," https://icecube.wisc.edu/pole/weather/.
5. Ford (2015).
6. Benn and Evans (2010), p. 210.
7. Rignot et al. (2013).
8. Barry and Gan (2011), pp. 278–279.
9. NASA, Earth Observatory, "Pine Island Glacier Births New 'Berg," September 28, 2017, https://earthobservatory.nasa.gov/images/91066/pine-island-glacier-births-new-berg (accessed May 3, 2018).
10. NASA, Earth Observatory, "Drifting with Ice Island B31," November 18, 2013, https://earthobservatory.nasa.gov/IOTD/view.php?id=83519 (accessed May 10, 2018).
11. Jeong et al. (2016).
12. National Science Foundation Press Release 11-247, "Antarctica's Gamburtsev Subglacial Mountains Mystery Solved," November 16, 2011, https://www.nsf.gov/news/news_summ.jsp?cntn_id=122290 (accessed September 13, 2013).
13. Fretwell et al. (2013).
14. Mount Erebus and Mount Terror were named after the ships on Captain Franklin's ill-fated Arctic expedition (see chaps. 1 and 2). Wikipedia, "Mount Erebus," https://en.wikipedia.org/wiki/Mount_Erebus (last modified September 20, 2018).
15. Van Wyk de Vries et al. (2017).
16. Lough et al. (2013).
17. Schroeder et al. (2014).
18. Some dispute this early date for the first cyanobacteria, placing their origin closer to the "Great Oxidation Event" around 2.5 billion years ago, when atmospheric oxygen levels began to rise (e.g., Rasmussen et al., 2008).
19. Wikipedia, "Blood Falls," https://en.wikipedia.org/wiki/Blood_Falls (last modified September 9, 2018).

20. Ehlmann and Edwards (2014). The discovery of perchlorate in the soils of the Dry Valleys also reinforces the resemblance to Mars. Perchlorate has been found near the South Pole of Mars and by the Curiosity rover now exploring the Gale Crater (Tamparri et al., 2010; King (2015).
21. Corrigan (2011); Hobart M. King, "Hunting Meteorites in Antarctica," Geology.com, http://geology.com/stories/13/antarctica-meteorites/ (accessed January 21, 2016).
22. Bell et al. (2011).
23. Shtarkman et al. (2013).
24. Wright and Siegert (2012).
25. Rignot et al. (2011).
26. Bell (2008a); Bell (2008b).
27. Wright and Siegert (2012).
28. Wingham et al. (2006).
29. Bell (2008b).
30. Tulaczyk and Hossainzadeh (2011); Bell et al. (2011).
31. Zwally et al. (2015).
32. Eric Mack, "Is Antarctica Gaining or Losing Ice? It's Both," *Forbes*, November 4, 2015, https://www.forbes.com/sites/ericmack/2015/11/04/yes-antarctica-is-both-gaining-and-losing-ice-but-really-losing-it/#1b092e309301/ (accessed January 22, 2016).
33. Altimetry: McMillan et al. (2014); Zwally et al. (2015). GRACE gravimetry: Sasgen et al. (2013); Harig and Simons (2015); table 6.1; also chap. 4, appendix B.
34. See also chapter 2; Joughin and Alley (2011); Joughin et al. (2012); Alley et al. (2015).
35. Mouginot et al. (2014).
36. Sutterley et al (2014).
37. Thomas Summer, "West Antarctic Ice Sheet Is Collapsing," *Science News*, May 12, 2014, https://www.sciencemag.org/news/2014/05/west-antarctic-ice-sheet-collapsing (accessed November 13, 2018).
38. Rignot et al. (2014).
39. Joughin et al. (2014).
40. Favier et al. (2014); Joughin, Smith, and Medley (2014).
41. Konrad et al. (2018).
42. Christie et al. (2016). The grounding line of the ice stream feeding the Venable Ice Shelf has held fast, in spite of considerable ice shelf thinning since the 1990s. This shelf may be stuck on underwater "pinning points" that act as brakes on ice motion.
43. Feldmann and Levermann (2015). The marine-based portion of the WAIS is that grounded below present sea level.
44. Joughin, Smith, and Medley (2014).
45. Alley et al. (2015).
46. Whereby a rapidly eroding stream cuts back across the drainage divide and captures the flow of another stream.
47. Feldmann and Levermann (2015).
48. Millan et al. (2017).
49. Paolo et al. (2015); Rignot et al. (2013, 2014); Depoorter et al. (2013); Pritchard et al. (2012).

50. Rignot et al. (2013); Depoorter et al. (2013).
51. See glossary; chap. 2, "Undermining Ice Shelves."
52. Schmidtko et al. (2014).
53. Pollard et al. (2015).
54. MacGregor et al. (2012); Pollard et al. (2015).
55. Alley et al. (2016).
56. Bell et al. (2017); Kingslake et al. (2017).
57. Pritchard et al. (2012).
58. Thinning ice shelves: Pritchard et al. (2012); Rignot et al. (2013); Depoorter et al. (2013); Paolo et al. (2015). Grounding line retreat and ice discharge: Konrad et al. (2018); Mouginot et al. (2014); Rignot et al. (2014); Joughin, Smith, and Medley (2014); Favier et al. (2014); Sutterley et al. (2014).
59. Fürst et al. (2016).
60. Jamieson et al. (2012).
61. Robel et al. (2016).
62. Gomez et al. (2015).
63. Gomez et al. (2015).
64. Barry and Gan (2011), p. 2. The total mass of ice on the WAIS is equivalent to a ~5-meter SLR. The oft-cited statement that complete melting of the WAIS would raise sea level by ~3.3 meters refers to the ice mass subject to MISI. Around 1.8 meters SLR equivalent would survive any likely warming event.
65. Rignot et al. (2013); Pritchard et al. (2012); Paolo et al. (2015).
66. Qiu (2017).
67. Fretwell (2015); Greenbaum et al. (2015).
68. Rintoul et al. (2016).
69. Greenbaum et al. (2015).
70. Li et al. (2015).
71. Mengel and Levermann (2014).

7. FROM GREENHOUSE TO ICEHOUSE

1. Eberle and Greenwood (2012).
2. DeConto et al. (2008); Galeotti et al. (2016).
3. DeConto et al. (2008); Galeotti et al. (2016).
4. Miller et al. (2011).
5. Levy et al. (2016).
6. Salzmann et al. (2016).
7. Bo et al. (2009).
8. Araucariaceae are a family of tall evergreen coniferous trees with spirally arranged needles or broad, flat leaves. They generally occur in Southern Hemisphere tropical or semitropical forests, although some species are also found in drier scrubland. More familiar examples include the monkey puzzle tree and Norfolk Island pine. The colorful

200-million-year-old petrified logs of Petrified Forest National Park in northern Arizona were members of a related, now-extinct species.

9. Naish et al. (2009).
10. Rovere et al. (2014); Rowley et al. (2013); Miller et al. (2012). (Estimates range between 10 and 30 meters higher than present sea levels.)
11. See glossary; also discussion in chapter 6, under "Maybe Not So Fast!"
12. Tibet stays cool in summer because of its high elevation. Yet it remains the warmest spot on Earth at that altitude. This warm spot sets up a continent-scale circulation akin to land and sea breezes. During summer, the zone of highest mean surface temperatures shifts north. Moisture from the Indian Ocean replaces the hot, dry air over the Tibetan plateau. As the moist air approaches the Himalayas, it rises, cools, and falls as rain over India and along the southern slopes of the Himalayas. By the time the monsoon passes over this topographic barrier, the air has dried out, leaving Tibet a near-desert. The uplift of the Himalayas has intensified this circulation pattern.
13. Raymo and Ruddiman (1992).
14. Kent and Muttoni (2013).
15. Livermore et al. (2007).
16. Stickley et al. (2004); Lyle et al. (2007).
17. Haug et al. (2004); O'Dea et al. (2016).
18. Lear and Lunt (2016).
19. Dutton et al. (2015); Foster and Rohling (2013); Grant et al. (2012).
20. DeConto et al. (2008); Galeotti et al. (2016).
21. Triparti et al. (2009).
22. Schneider and Schneider (2010).
23. See note 9.
24. DeConto et al. (2008).
25. Bierman et al. (2014); "2.7-Million-Year-Old Forested Landscape Discovered Under Greenland Ice Sheet," *Sci News*, April 17, 2014, http://www.sci-news.com/geology/science-forested-landscape-greenland-ice-sheet.
26. The half-life of ^{10}Be is 1.4 million years, the time needed for half a given quantity of the isotope to decay. (The presumed age of the oldest ice is around two half-lives of ^{10}Be.)
27. Lamont-Doherty Earth Observatory, "Most of Greenland Ice Melted to Bedrock in Recent Geologic Past, Study Says," December 7, 2016, https://www.ldeo.columbia.edu/news-events/most-greenland-ice-melted-bedrock-recent-geologic-past-study-says.
28. Steig and Wolfe (2008).
29. Dutton et al. (2015); Grant et al. (2012); Rohling et al. (2009).
30. Kerr (2013).
31. Maslin and Brierley (2015).
32. Brook (2008).
33. Galeotti et al. (2016).
34. Naish et al. (2009).
35. Cook et al. (2013).
36. Dutton et al. (2015).

37. Raymo and Mitrovica (2012).
38. Reyes et al. (2014).
39. The volume of the Greenland Ice Sheet is ~7 meters (23 feet) of sea level equivalent. Melting of the entire WAIS would raise sea level by ~5 meters (16 feet). The portion of WAIS ice grounded below sea level and subject to the marine ice sheet instability would yield only ~3.3 (11 feet) meters.
40. Dutton et al. (2015); Kopp et al. (2009), after correcting for glacial isostatic, gravitational, and rotational effects.
41. O'Leary et al. (2013).
42. Rohling et al. (2008).
43. Barlow et al. (2018).
44. Colville et al. (2011).
45. Quoted in Schiermeier (2013).
46. MacGregor et al. (2015).
47. NEEM (North Greenland Eemian Ice Core Drilling) (2013); Colville et al. (2011); Kopp et al. (2009).
48. Carlson and Clark (2012).
49. Mackintosh et al. (2011).
50. Weber et al. (2014).
51. Hillenbrand et al. (2013).
52. Evans (2015).
53. Carlson and Clark (2012); Deschamps et al. (2012).
54. Liu et al. (2016).
55. Carlson and Clark (2012); Li et al. (2012).

8. RETURN TO THE GREENHOUSE

1. The percentage of freshwater may be closer to 77 percent (Barry and Gan, 2011, p. 3).
2. Walker (2013).
3. Archer and Brovkin (2008).
4. National Snow & Ice Data Center, "2017 Ushers In Record Low Extent," February 7, 2017, http://nsidc.org/arcticseaicenews/2017/02/2017-ushers-in-record-low-extent/ (accessed February 17, 2017); Andrea Thompson, "Sea Ice Hits Record Lows at Both Poles," *Scientific American*, February 13, 2017, https://www.scientificamerican.com/article/sea-ice-hits-record-lows-at-both-poles/ (accessed February 20, 2017); Scott Sutherland, "North Pole Temps Spiked by Nearly 30 Celsius Last Week," The Weather Network, February 13, 2017, https://www.theweathernetwork.com/news/articles/arctic-storms-bring-another-winter-heatwave-to-north-pole/79190 (accessed February 20, 2017).
5. Hogg and Gudmundsson (2017); Viñas (2017).
6. Archer (2009).
7. DeConto and Pollard (2016); Pollard et al. (2015).

8. Ritz et al. (2015).
9. Vermeer and Rahmstorf (2009); Rahmstorf (2007).
10. Joughin, Smith, and Medley (2014).
11. Golledge et al. (2015); Sutter et al. (2016).
12. Feldmann and Levermann (2015).
13. Khan et al. (2014).
14. Gomez et al. (2015).
15. Robinson et al. (2012).
16. Golledge et al. (2015).
17. DeConto and Pollard (2016).
18. Wikipedia, "Svante Arrhenius," https://en.wikipedia.org/wiki/Svante_Arrhenius (last modified October 11, 2018; accessed February 14, 2017).
19. Intergovernmental Panel on Climate Change (2013a), chap. 12, box 12.2, p. 1110.
20. NASA, Global Climate Change, "Carbon Dioxide," https://climate.nasa.gov/vital-signs/carbon-dioxide/; CO2.Earth, http://co2.earth. See also fig. 1.7.
21. NASA, Goddard Institute for Space Studies, "NASA, NOAA Data Show 2016 Warmest Year on Record Globally," January 18, 2017, https://www.giss.nasa.gov/research/news/20170118/ (accessed February 15, 2017). While not record-setting, 2018 was the fourth warmest in the global historic record. G. A. Schmidt and D. Arndt, "NOAA/NASA Annual Global Analysis for 2018," https://www.giss.nasa.gov/research/news/20190206/201902briefing.pdf (accessed February 7, 2019).
22. On February 8, 2017, New York City basked in unseasonable, record-shattering, mild 60°–65°F weather; by early the next morning, temperatures had plunged to the upper 20s Fahrenheit and 8–12 inches of snow covered the city.
23. Clark et al. (2016).
24. UNFCCC, "The Paris Agreement," http://unfccc.int/paris_agreement/items/9485.php (accessed February 16, 2017).
25. Hansen et al. (2008). The 2°C limit has been urged by many climate scientists, based on anticipated impacts, and has been adopted by the UNFCCC as a target in the current Paris Agreement.
26. For starters, see chapter 10 in Gornitz (2013).
27. Such a lengthy safe storage period—almost the duration of the Holocene, the epoch following the last ice age—is needed to cool down the radioactive waste products.
28. See, for example, Rosenzweig et al. (2018) and case studies therein.

9. THE IMPORTANCE OF ICE

1. NASA, Goddard Institute for Space Studies, "NASA, NOAA Data Show 2016 Warmest Year on Record Globally," January 18, 2017, https://www.giss.nasa.gov/research/news/20170118/ (accessed May 30, 2017).

2. National Snow & Ice Data Center, "2017 Ushers In Record Low Extent," February 7, 2017, http://nsidc.org/arcticseaicenews/2017/02/2017-ushers-in-record-low-extent/ (accessed February 17, 2017); Scott Sutherland, "North Pole Temps Spiked by Nearly 30 Celsius Last Week," The Weather Network, February 13, 2017, https://www.theweathernetwork.com/news/articles/arctic-storms-bring-another-winter-heatwave-to-north-pole/79190 (accessed February 20, 2017).
3. National Oceanic and Atmospheric Administration, "Unprecedented Arctic Weather Has Scientists on Edge," February 17, 2017, http://www.noaa.gov/news/unprecedented-arctic-weather-has-scientists-on-edge (accessed February 23, 2017).
4. Derocher (2008).
5. Struzik (2009).
6. Mele and Victor (2016); Goode (2016).
7. Pilkey et al. (2016).
8. Wikipedia, "Polar Bear," https://en.wikipedia.org/wiki/Polar_bear (last modified September 19, 2018; accessed May 31, 2017).
9. See glossary.
10. National Snow & Ice Data Center, "Arctic Sea Ice Maximum at Second Lowest in the Satellite Record," March 23, 2018, http://nsidc.org/arcticseaicenews/2018/03/arctic-sea-ice-maximum-second-lowest/ (accessed April 24, 2018); National Snow & Ice Data Center, "Arctic Sea Ice Maximum at Record Low for Third Straight Year," March 22, 2017, https://nsidc.org/news/newsroom/arctic-sea-ice-maximum-record-low-third-straight-year/ (accessed May 25, 2017); National Snow & Ice Data Center, "2016 Ties with 2007 for Second Lowest Arctic Sea Ice Minimum," September 16, 2016, http://nsidc.org/arcticseaicenews/2016/09/2016-ties-with-2007-for-second-lowest-arctic-sea-ice-minimum/ (accessed September 16, 2016).
11. "Arctic Sea Ice Maximum at Second Lowest in the Satellite Record."
12. Theroux (2016).
13. E.g., Pongracz et al. (2017).
14. Morello (2010).
15. Quoted in Rosen (2017).
16. United Nations Environment Programme (2007).
17. Ehrlich (2010), pp. 30, 69.
18. Ehrlich (2010), pp. 216, 241.
19. Boelman (2011).
20. Struzic (2011).
21. Miller and Ruiz (2014).
22. Quoted in Rosen (2017).
23. Cohen et al. (2014); Masters (2014).
24. Overland et al. (2016).
25. Gillis and Fountain (2016); Vaidyanathan and Patterson (2015).
26. Gautier et al. (2009).

27. Wikipedia, "Klondike Gold Rush," https://en.wikipedia.org/wiki/Klondike_Gold_Rush (last modified October 4, 2018; accessed June 13, 2017).
28. Shigley et al. (2016).
29. Shigley et al. (2016).
30. "A New Perfection Found in Diamonds Created by an Asteroid in Siberian Crater 35 Million Years Ago," *Siberian Times*, September 16, 2013, http://siberiantimes.com/science/casestudy/features/a-new-perfection-found-in-diamonds-created-by-an-asteroid-in-siberian-crater-35-million-years-ago/ (accessed June 15, 2017).
31. Gautier et al. (2009).
32. Wikipedia, "Natural Resources of the Arctic," https://en.wikipedia.org/wiki/Natural_resources_of_the_Arctic (last modified May 14, 2018; accessed June 20, 2017).
33. Glasby and Voytekhovshy (2009).
34. Jennifer Kingsley, "When Coal Leaves Center Stage in Longyearbyen," Arctic Deeply (archive), News Deeply, March 3, 2016, https://www.newsdeeply.com/arctic/articles/2016/03/03/when-coal-leaves-center-stage-in-longyearbyen (accessed June 21, 2017).
35. Kingsley, "When Coal Leaves Center Stage in Longyearbyen." See also Wikipedia, "Natural Resources of the Arctic."
36. Wikipedia, "Red Dog Mine," https://en.wikipedia.org/wiki/Red_Dog_mine (last modified September 4, 2018; accessed June 21, 2017).
37. Wikipedia, "Mary River Mine," https://en.wikipedia.org/wiki/Mary_River_Mine (last modified September 28, 2018); Baffinland, "Who We Are," http://www.baffinland.com/about-us/who-we-are/?lang=en; Nick Murray, "Baffinland Iron Mines' Phase 2 Plan Gets Sent Back to Nunavut Planning Commission," CBC News, December 20, 2016, http://www.cbc.ca/news/canada/north/nirb-baffinland-phase-2-planning-commission-1.3904189 (accessed June 21, 2017).
38. Wikipedia, "Mary River Mine." https://en.wikipedia.org/wiki/Mary_River_Mine (last updated November 15, 2018; accessed November 16, 2018); Nunatsiaq News, "Nunavut Board to Finish with Mary River Railway Proposal by June 2019," *Nunatsiaq News*, October 17, 2018. http://nunatsiaq.com/stories/article/65674nunavut_board_to_finish_with_mary_river_railway_proposal_by_june_2019/.
39. Gray (2016)
40. Haecker (2016).
41. Av Mieke Coppes, "No More Crystal Serenity in the Northwest Passage," *High North News*, December 13, 2017, http://www.highnorthnews.com/no-more-crystal-serenity-in-the-northwest-passage/ (accessed March 5, 2018); Jane George, "Norwegian Cruise Company Plans Northwest Passage Transit with Hybrid Vessel," *Nunatsiaq News*, December 11, 2017, http://nunatsiaq.com/stories/article/65674norwegian_cruise_company_plans_nw_passage_transit_with_hybrid_vessel/ (accessed March 5, 2018).
42. Arctic Monitoring and Assessment Programme (2017); Overland and Wang (2013).
43. Lulu (2017).

44. Miller and Ruiz (2014).
45. Schiffman (2016).
46. Clearly, it's much more complex than that. Chapter 4 discusses the glacial retreat more fully.
47. Harriman (2013).
48. Saavedra et al. (2018).
49. Navab (2011); Bradley et al. (2006).
50. Pritchard (2017).
51. Laghari (2013).
52. Bolch (2017); Pritchard (2017).
53. Immerzeel et al. (2010).
54. Huss and Hock (2018); Kraaijenbrink et al. (2017); Immerzeel et al (2010).
55. Huss and Hock (2018); Kraaijenbrink et al., 2017.
56. Huss and Hock (2018); appendix A.
57. Huss and Hock (2018); appendix A.
58. Fyfe et al. (2017).
59. Fyfe et al. (2017).
60. Sierra Nevada Conservatory, "California's Primary Watershed," 2011, http://www.sierranevada.ca.gov/our-region/ca-primary-watershed; California-Nevada Climate Applications Program (CNAP), "Sierra Nevada Snowpack," March 2016, https://www.swcasc.arizona.edu/sites/default/files/Snowpack.pdf/ (both accessed December 3, 2018).
61. Natural Resources Defense Council (NRDC), "California Snowpack and the Drought," fact sheet, April 2014, https://www.nrdc.org/sites/default/files/ca-snowpack-and drought-FS.pdf (accessed August 18, 2017).
62. Larson (2015).
63. Between 1880 and 2015. Swiss Academy of Sciences, "Welcome to the Glacier Monitoring in Switzerland (GLAMOS)," http://glaciology.ethz.ch/messnetz/ (accessed August 10, 2017).
64. Fischer et al. (2014).
65. Tagliabue (2013).
66. Huggel et al. (2012); Tagliabue (2013).
67. Keane (2017).
68. Clague et al. (2012).
69. McGuire (2012), p. 245.
70. Huggel et al. (2012); Tagliabue (2013).
71. Horstmann (2004).
72. Veh et al. (2018).
73. McGuire (2012), p. 264. See also Nettles and Ekström (2010), cited in chap. 5.
74. Van Wyk de Vries et al. (2017); Schroeder et al. (2014).
75. Intergovernmental Panel on Climate Change (2013a), chap. 13, table 13.5, p. 1182 (RCP8.5, 2081–2100). See appendix A for explanation of RCP scenarios.

76. DeConto and Pollard (2016). See also Pollard et al. (2015).
77. Carbognin et al. (2009).
78. Sweet et al. (2014).
79. Strauss et al. (2016).
80. Blake et al. (2011).
81. Orton et al. (2016).
82. Special Initiative for Rebuilding and Resiliency (2013).
83. National Oceanic and Atmospheric Administration (2017).
84. Neumann et al. (2015).
85. Hauer et al. (2016).
86. Gornitz (2018).
87. Morton et al. (2004).
88. Gornitz (2013), chap. 8, pp. 188–190.
89. Gornitz (2013), chap. 9.
90. E.g., Aerts et al. (2009).
91. Rotterdam Climate Initiative (2010).
92. Stalenberg (2012).
93. When Canal Street lived up to its name. Originally a "sluggish stream," it became a ditch to drain local swamps. It soon turned into a stinking sewer, was covered, and by 1820, Canal Street was built over it. Wikipedia, "Canal Street (Manhattan)," https://en.wikipedia.org/wiki/Canal_Street_(Manhattan) (last modified July 27, 2018; accessed July 27, 2017).
94. Nordenson et al. (2010).
95. Lafarge Holcim Foundation, "The Dryline: Urban Flood Protection Infrastructure," http://lafargeholcim-foundation.org/projects/the-dryline (accessed June 7, 2016).
96. Mele and Victor (2016).
97. Goode (2016).
98. De Melker and Saltzman (2016).
99. Green (2016); Hino et al. (2017).
100. Hauer (2017).
101. Hino et al. (2017).
102. Polar night affects all areas north of the polar circle (66.5°N) to some extent. The duration of polar night increases with latitude, reaching 6 months of the year at the North Pole. But conversely, the pole experiences six months of sunlight during the second half of the year. Farther south, there may be several hours of dim twilight, even when the sun lies below the horizon, because of the way the atmosphere bends the rays of light.

APPENDIX A. ANTICIPATING FUTURE CLIMATE CHANGE

1. Intergovernmental Panel on Climate Change (2013a).
2. A negative surface mass balance (SMB) implies that ice losses (surface melting, runoff, and sublimation [ice to water vapor]) exceed accumulation (snowfall). See glossary.

APPENDIX B. EYES IN THE SKY—MONITORING THE CRYOSPHERE FROM ABOVE

1. NASA, "The Landsat Program," https://landsat.gsfc.nasa.gov/; USGS, "Landsat Missions," https://landsat.usgs.gov/ (accessed January 18, 2018).
2. NASA, "Landsat 8," https://landsat.gsfc.nasa.gov/landsat-data-continuity-mission/; USGS, "Landsat Missions: Imaging the Earth Since 1972," https://landsat.usgs.gov/landsat-missions-timeline (accessed January 18, 2018).
3. Satellite Imaging Corp., "Satellite Sensors (0.31m–2m)," https://www.satimagingcorp.com/satellite-sensors/.
4. USGS, Land Processes Distributed Active Archive Center, "ASTER Overview," https://lpdaac.usgs.gov/dataset_discovery/aster/(accessed January 9, 2017); https://lpdaac.user_resources/data_in_action/global_land_ice_measurements_space_observing_glaciers (accessed January 30, 2018).
5. Raup and Kargel (2012), p. A249.
6. NASA, "MODIS Web, Design Concept," https://modis.gsfc.nasa.gov/about/design.php/; NASA, Terra, "MODIS," https://terra.nasa.gov/about/terra-instruments/modis/ (accessed January 9, 2018).
7. NASA, "ICESat-2: Instrument," http://icesat.gsfc/nasa/gov/icesat2/instrument.php; NASA, ICESat-2, "Science," https://icesat-2.gsfc.nasa.gov/science/ (accessed January 9, 2017).
8. NASA, "IceBridge," https://www.nasa.gov/mission_pages/icebridge/index.html (accessed January 9, 2018).
9. Wikipedia, "Special Sensor Microwave/Imager," https://en.wikipedia.org/wiki/Special_sensor_microwave/imager (last modified May 27, 2018; accessed February 12, 2019).
10. For more mission details, see http://www.jpl.nasa.gov, https://sealevel.jpl.nasa.gov/missions/jason3/, and http://www.cnes.fr.
11. NASA, "Understanding Sea Level; Key Indicators; Global Mean Sea Level," https://sealevel.nasa.gov/understanding-sea-level/key-indicators/global-mean-sea-level (covers January 1993 to October 9, 2017; accessed January 5, 2018); CNES/Aviso+, "Mean Sea Level Rise," https://www.aviso.altimetry.fr/en/data/products/ocean-indicators-products/mean-sea-level.html (sea level rise plot covers January 5, 1993, to November 17, 2017; accessed February 5, 2018).
12. European Space Agency, "Overview," November 6, 2013, https://earth.esa.int/web/guest/missions/esa-eo-missions/cryosat/mission-summary (accessed January 31, 2018).
13. SIRAL Altimeter. https://www.aviso.altimetry.fr/en/missions/current-missions/cryosat/instruments/siral.html/ (accessed November 27, 2018).
14. European Space Agency, Copernicus, "Synthetic Aperture Radar Missions," http://www.esa.int/Our_Activities/Observing_the_Earth/Copernicus/SAR_missions/ (accessed January 31, 2018).
15. Geoimage, "TerraSAR-X," https://www.geoimage.com.au/satellite/TerraSar/ (accessed January 31, 2018).

16. Canadian Space Agency, "RADARSAT-2," December 14, 2017, http://www.asc.gc.ca/eng/satellites/radarsat2/default.asp/; http://www.asc-csa.gc.ca/eng/satellites/radarsat/radarsat-tableau.asp (both accessed Nov. 27, 2018).
17. Wikipedia, "Gravity Recovery and Climate Experiment," http://en.wikipedia.org/wiki/Gravity_Recovery_and_Climate_Experiment (last modified February 3, 2019; accessed February 12, 2019).

BIBLIOGRAPHY

Achilli, A. et al. 2013. "Reconciling Migration Models to the Americas with the Variation of North American Native Mitogenomes." *Proceedings of the National Academy of Sciences* 110(35):14308–14313. doi: 10.1073/pnas.1306290110.

Aerts, J., Major, D. C., Bowman, M. J., Dircke, P., and Marfai, M. A. 2009. *Connecting Delta Cities: Coastal Cities, Flood Risk Management and Adaptation to Climate Change*. Amsterdam: Vrije Universiteit Press.

Alley, K. E., Scambos, T. A., Siegfried, M. R., and Fricker, H. A. 2016. "Impacts of Warm Water on Antarctic Ice Shelf Stability Through Basal Channel Formation." *Nature Geoscience* 9:290–294.

Alley, R. B. 2000. *The Two-Mile Time Machine*. Princeton, NJ: Princeton University Press.

Alley, R. B. 2007. "'C'ing Arctic Climate with Black Ice." *Science* 317:1333–1334.

Alley, R. B., Anandakrishnan, S., Christianson, K., Horgan, H. J., Muto, A., Parizek, B. R., Pollard, D., and Walker, R. T. 2015. "Oceanic Forcing of Ice-Sheet Retreat: West Antarctica and More." *Annual Reviews of Earth and Planetary Sciences* 43:207–231.

An, L., and 8 others. 2017. "Bed Elevation of Jakobshavn Isbrae, West Greenland, from High-Resolution Airborne Gravity and Other Data." *Geophysical Research Letters* 44:3728–3736. doi:10.1002/2017GL073245.

Ananthaswamy, A. 2015. "The Methane Apocalypse." *New Scientist*, 23 May, 38–41.

Anthony, K. M. W., Anthony, P., Grosse, G., and Chanton, J. 2012. "Geologic Methane Seeps Along Boundaries of Arctic Permafrost Thaw and Melting Glaciers." *Nature Geoscience* 5:419–426.

Anthony, K. M. W., Zimov, S. A., Grosse, G., Jones, M. C., Anthony, P. M., Chapin, F. S. III, Finlay, J. C., Mack, M. C., Davydov, S., Frenzel, P., and Frolking, S. 2014. "A Shift of Thermokarst Lakes from Carbon Sources to Sinks During the Holocene Epoch." *Nature* 511:452–456.

Anthony, K. W., Daanen, R., Anthony, P., Schneider von Deimling, T., Ping, C. L., Chanton, J. P., and Grosse, G. 2016. "Methane Emissions Proportional to Permafrost Carbon Thawed in Arctic Lakes Since the 1950s." *Nature Geoscience* 9:679–682.

Archer, D. 2009. *The Long Thaw: How Humans Are Changing the Next 100,000 Years of Earth's Climate*. Princeton, NJ: Princeton University Press.

Archer, D., and Brovkin, V. 2008. "The Millennial Atmospheric Lifetime of Anthropogenic CO_2." *Climatic Change* 90:283–297.

Arctic Monitoring and Assessment Programme (AMAP). 2011. SWIPA (Snow, Water, Ice and Permafrost in the Arctic). Chap. 5, "Permafrost." http://amap.no/swipa/SWIPA2100 Executive_SummaryVI.pdf.

Arctic Monitoring and Assessment Programme (AMAP). 2017a. "Snow, Water, Ice and Permafrost in the Arctic: Summary for Policy-Makers." https://www.amap.no/documents/doc/Snow-Water-Ice-and-Permafrost-Summary-for-Policy-makers/1532.

Arctic Monitoring and Assessment Programme (AMAP). 2017b. "Snow, Water, Ice and Permafrost in the Arctic (SWIPA): 2017." https://www.amap.no/documents/doc/Snow-Water-Ice-and-Permafrost-in-the-Arctic-SWIPA-2017/1610.

Are, F., Reimnitz, E., Grigoriev, M., Hobberston, H.-W. M., and Rachold, V. 2008. "The Influence of Cryogenic Processes on the Erosional Arctic Shoreface." *Journal of Coastal Research* 24(1):110–121.

Avis, C. A., Weaver, A. J., and Meissner, K. J. 2011. "Reduction in Areal Extent of High-Latitude Wetlands in Response to Permafrost Thaw." *Nature Geoscience* 4:444–448. doi:10/1038/NGEO1160.

Bakke, J., and Nesje, A. 2014. "Equilibrium Line Altitude." In *Encyclopedia of Snow, Ice and Glaciers*, ed. V. J. Singh, P. Singh, and U. K. Haritashya, 268–277. Dordrecht: Springer.

Bamber, J. L., Riva, R. E. M., Vermeersen, B. L. A., and LeBrocq, A. M. 2009. "Assessment of the Potential Sea-Level Rise from a Collapse of the West Antarctic Ice Sheet." *Science* 324:901–903.

Bamber, J. L., Siegert, M. J., Griggs, J. A., Marshall, S. J., and Spada, G. 2013. "Paleofluvial Mega-Canyon Beneath the Central Greenland Ice Sheet." *Science* 341:997–999.

Bard, E., Hamelin, B., and Delanghe-Sabatier, D. 2010. "Deglacial Meltwater Pulse 1B and Younger Dryas Sea Levels Revisited with Boreholes at Tahiti." *Science* 327:1235–1237.

Bard, E., Hamelin, B., and Fairbanks, R. G. 1990. "U-Th Ages Obtained by Mass Spectrometry in Corals from Barbados: Sea Level During the Past 130,000 Years." *Nature* 346:456–458.

Barlow, N. L. M. and 15 others. 2018. "Lack of Evidence for a Substantial Sea-Level Fluctuation Within the Last Interglacial." *Nature Geoscience* 11:627–634.

Barnhart, K. R., Overeen, I., and Anderson, R. S. 2014. "The Effect of Changing Sea Ice on the Vulnerability of Arctic Coasts." *The Cryosphere Discussion* 8:2277–2329.

Barry, R., and Gan, T. Y. 2011. Chap. 3, "Glaciers and Ice Caps"; Chap. 4, "Ice Sheets"; and Chap. 5, "Frozen Ground and Permafrost." *The Global Cryosphere: Past, Present, and Future*. Cambridge, UK: Cambridge University Press.

Bell, R. E. 2008a. "The Unquiet Ice." *Scientific American*, February, 60–67.

Bell, R. E. 2008b. "The Role of Subglacial Water in Ice-Sheet Mass Balance." *Nature Geoscience* 1:297–304.

Bell, R. E., Ferraccioli, F., Creyts, T. T., and 9 others. 2011. "Widespread Persistent Thickening of the East Antarctic Ice Sheet by Freezing from the Base." *Science* 331:1592–1595.

Bell, R., Studinger, M., Tikku, A. A., Clarke, G. K. C., Gutner, M. M., and Meertens, C. 2002. "Origin and Fate of Lake Vostok Water Frozen to the Base of the East Antarctic Ice Sheet." *Nature* 416:307–310.

Bell, R. E., Tinto, K., Das, I., Wolovick, M., Chu, W., Creyts, T. T., Frearson, N., Abdi, A., and Paden, J. D. 2014. "Deformation, Warming, and Softening of Greenland's Ice by Refreezing Meltwater." *Nature Geoscience* 7:497–502.

Bell, R. E., and 9 others. 2017. "Antarctic Ice Shelf Potentially Stabilized by Export of Meltwater in Surface Rivers." *Nature* 544:344–348.

Belmecheri, S., Babst, F., Wahl, E. R., Stahle, D. W., and Trouet, V. 2016. "Multi-Century Evaluation of Sierra Nevada Snowpack." *Nature Climate Change* 6:3. https://www.nature.com/articles/nclimate2809.

Benn, D. I., and Evans, D. J. A. 2010. *Glaciers and Glaciation*. London: Hodder Education, Hachette UK, 802p.

Bierman, P. R., Corbett, L. B., Graly, J. A., Neumann, T. A., Lini, A., Crosby, B. T., and Rood, D. H. 2014. "Preservation of a Preglacial Landscape Under the Center of the Greenland Ice Sheet." *Science* 344 (6182):402–405.

Bingham, R. G., Ferraccioli, F., King, E. C., Larter, R. D., Pritchard, H. D., Smith, A. M., and Vaughan, D. C. 2012. "Inland Thinning of West Antarctic Ice Sheet Steered Along Subglacial Rifts." *Nature* 487:468–471.

Bintanja, R., Van Oldenborgh, G. J., Drijfhout, S. S., Wouter, B., and Katsman, C. A. 2013. "Important Role for Ocean Warming and Increased Ice-Shelf Melt in Antarctic Sea-Ice Expansion." *Nature Geoscience* 6:376–379.

Bird, E. 2008. *Coastal Geomorphology: An Introduction*. Chichester, UK: Wiley.

Bjørk, A. A. 2012. "An Aerial View of 80 Years of Climate-Related Glacier Fluctuations in Southeast Greenland." *Nature Geoscience* 5:427–432.

Blake, E. S., Landsea, C. W., and Gibney, E. J. 2011. "The Deadliest, Costliest, and Most Intense United States Tropical Cyclones from 1851 to 2010 (and Other Frequently Requested Hurricane Facts)." NOAA Technical Memorandum NWS NHC-6. National Weather Service, National Hurricane Center, Miami, FL.

Bo, S., and 8 others. 2009. "Gamburtsev Mountains and the Origin and Early Evolution of the Antarctic Ice Sheet." *Nature* 459:690–693.

Boelman, N. 2011. "The Dangers of an Early Spring." *New York Times*, June 7. https://scientistatwork.blogs.nytimes.com/2011/06/07/the-dangers-of-an-early-spring.

Boening, C., Willis, J., Landerer, F., Nerem, R. S., and Fasullo, J. 2012. "The 2011 La Niña: So Strong, the Oceans Fell." *Geophysical Research Letters* 39(19). doi:10.1029/2012GL053055.

Bolch, T. 2017. "Asian Glaciers Are a Reliable Water Source." *Nature* 545:161–162.

Bolch, T., and 11 others. 2012. "The State and Fate of Himalayan Glaciers." *Science* 336:310–314.

Bradley, R. S., Vuille, M., Diaz, H. F., and Vergara, W. 2006. "Threats to Water Supplies in the Tropical Andes." *Science* 312:1755–1756.

Bresson, D. 2012. "The Science Behind the Iceberg That Sank the Titanic." *History of Geology* (blog), *Scientific American*, April 14. https://blogs.scientificamerican.com/history-of-geology/the-science-behind-the-iceberg-that-sank-titanic/.

Brewer, P. 2014. "Arctic Shelf Methane Sounds Alarm." *Nature Geoscience* 7:6–7.

Broecker, W. 2010. "The Great Ocean Conveyor: Discovering the Trigger for Abrupt Climate Change." Princeton, NJ: Princeton University Press.

Bromwich, D. H., Nicolas, J. P., Monaghan, A. J., Lazzara, M. A., Keller, L. M., Weidner, G. A., and Wilson, A. B. 2014. "Central West Antarctica Among the Most Rapidly Warming Regions on Earth. Corrigendum." *Nature Geoscience* 7:76.

Brook, E. 2008. "Windows on the Greenhouse." *Nature* 453:291–292.

Brun, F., Berthier, E., Wagnon, P., Kääb, A., and Treichler, D. 2017. "A Spatially Resolved Estimate of High Mountain Asia Glacier Mass Balances from 2000 to 2016." *Nature Geoscience* 10:668–674.

Bruneau, S. E. 2004. *Icebergs of Newfoundland and Labrador*. St. John's, NL, Canada: Flanker Press, 64p.

Callaghan, T. V., and 20 others. 2011. "The Changing Face of Arctic Snow Cover: A Synthesis of Observed and Projected Changes." *Ambio* 40:17–31.

Carbognin, L., Teatini, P., Tomasin, A., and Tosi, L. 2009. "Global Change and Relative Sea Level Rise at Venice: What Impact in Terms of Flooding." *Climate Dynamics*, 35:1039–1047. doi:10.1007/s00382-009-0617-5.

Carlson, A. E., and Clark, P. U. 2012. "Ice Sheet Sources of Sea Level Rise and Freshwater Discharge During the Last Deglaciation." *Reviews of Geophysics* 50, RG4007/2012, 1–72.

Christie, F. D. W., Bingham, R. G., Gourmelen, N., Tett, S. F. B., and Muto, A. 2016. "Four-Decade Record of Pervasive Grounding Line Retreat Along the Bellingshausen Margin of West Antarctica." *Geophysical Research Letters* 43:5741–5749.

Chu, V. 2014. "Greenland Ice Sheet Hydrology: A Review." *Progress in Physical Geography* 38(1):19–54.

Clague, J. J., Huggel, C., Korup, O., and McGuire, W. 2012. "Climate Change and Hazardous Processes in High Mountains." *Revista de la Asociación Geológica Argentina* 69:328–338.

Clark, P. U., McCabe, A. M., Mix, A. C., and Weaver, A. J. 2004. "Rapid Rise of Sea Level 19,000 Years Ago and Its Global Implications." *Science* 304:1141–1144.

Clark, P. U., and 21 others. 2016. "Consequences of Twenty-First Century Policy for Multi-Millennial Climate and Sea Level Change." *Nature Climate Change* 6:360–369.

Coghlan, A. 2016. "From Lush to Slush in 40 Million Years." *New Scientist*, May 14, 13.

Cohen, J., Screen, J. A., Furtado, J. C., Barlow, M., Whittleson, D., Coumou, D., Francis, J., Dethlo, K., Entekhabi, D., Overland, J., and Jones, J. 2014. "Recent Arctic Amplification and Extreme Mid-Latitude Weather." *Nature Geoscience* 7:627–630.

Colville, E. J., Carlson, A. E., Beard, B. L., Hatfield, R. G., Stoner, J. S., Reyes, A. V., and Ullman, D. J. 2011. "Sr-Nd-Pb Isotope Evidence for Ice-Sheet Presence on Southern Greenland During the Last Interglacial." *Science* 333:620–623.

Comiso, J. C. 2012. "Large Decadal Decline of the Arctic Multiyear Ice Cover." *Journal of Climate* 25(4):1176–1193.

Cook, A. J., and Vaughan, D. G. 2010. "Overview of Areal Changes of the Ice Shelves on the Antarctic Peninsula Over the Past 50 Years." *The Cryosphere* 4:77–98.

Cook, C. P., and IODP Expedition Scientists. 2013. "Dynamic Behaviour of the East Antarctic Ice Sheet During Pliocene Warmth." *Nature Geoscience* 6:765–769.

Cooper, M. A., Michaelides, K., and Bamber, J. L. 2016. "Paleofluvial Landscape Inheritance for Jakobshavn Isbrae Catchment, Greenland." *Geophysical Research Letters* 43:6350–6357. doi:10.1002/2016GL069458.

Corrigan, C. 2011. "Antarctica: The Best Place on Earth to Collect Meteorites." *Elements* 7 (5):296.

Corte, A. E. 1969. "Geocryology and Engineering." In *Reviews in Engineering Geology*, vol. 2, ed. D. A. Varnes and G. Kiersch, 119–186. Boulder, CO: Geological Society of America.

Cronin, T. M., Vogt, P. R., Willard, D. A., Thunell, R., Halka, J., and Berke, M. "Rapid Sea Level Rise and Ice Sheet Response to 8,200-Year Climate Event." 2007. *Geophysical Research Letters* 34: L20603.

Csatho, B. M., and 9 others. 2014. "Laser Altimetry Reveals Complex Pattern of Greenland Ice Sheet Dynamics." *Proceedings of the National Academy of Sciences* 111(52):18478–18483.

Davies, B. 2017. "Ice Shelf Collapse." AntarcticGlaciers.org. http://www.antarcticglaciers.org/glaciers-and-climate/shrinking-ice-shelves/ice-shelves/ (last updated June 19, 2017).

Davies, B. J., and Glasser, N. F. 2012. "Accelerating Shrinkage of Patagonian Glaciers from the 'Little Ice Age' (~AD 1870) to 2011." *Journal of Glaciology* 58(212):1063–1084.

Davis, N. 2001. *Permafrost: A Guide to Frozen Ground in Transition*. Fairbanks: University of Alaska Press, 351p.

DeConto, R. M., and Pollard, D. 2016. "Contribution of Antarctica to Past and Future Sea-Level Rise." *Nature* 531:591–597.

DeConto, R. M., Pollard, D., Wilson, P. A., Pälike, H., Lear, C. H., and Pagani, M. 2008. "Thresholds for Cenozoic Bipolar Glaciation." *Nature* 455:652–656.

De Melker, S., and Saltzman, M. 2016. "Native Community in Louisiana Relocates as Land Washes Away." *PBS NewsHour*, July 30. https://www.pbs.org/newshour/show/native-community-louisiana-relocates-land-washes-away.

Depoorter, M. A., Bamber, J. L., Griggs, J. A., Lenaerts, J. T. M., Ligtenberg, S. R. M., Van den Broeke, M. R., and Moholdt, G. 2013. "Calving Fluxes and Basal Melt Rates of Antarctic Ice Shelves." *Nature* 502:89–92.

Derocher, A. E., 2008. "Polar Bears in a Warming Arctic." In *Sudden and Disruptive Climate Change: Exploring the Risks and How We Can Avoid Them*, ed. M. A. McCracken, F. Moore, and J. C. Topping Jr., 193–204. London: Earthscan.

Deschamps, P., and 8 others. 2012. "Ice-Sheet Collapse and Sea-Level Rise at the Bølling Warming 14,600 Years Ago." *Nature* 483:559–564.

Diekmann, B. 2007. "Sedimentary Patterns in the Late Quaternary Southern Ocean." *Deep-Sea Research* II, 54[21–22]:2350–2366.

Diemand, D. 2001, 2008. "Icebergs." In *Encyclopedia of Ocean Science*, 2nd ed., ed. J. H. Steele, K. K. Turekian, and S. A. Thorpe, 1255–1264. Amsterdam: Elsevier.

Dieng, H. B., Cazenave, A., Meyssignac, B., and Ablain, M. 2017. "New Estimate of the Current Rate of Sea Level Rise from a Sea Level Budget Approach." *Geophysical Research Letters* 44:1–8. doi:10.1002/2017GL073308.

Doyle, S. H., and 15 others. 2015. "Amplified Melt and Flow of the Greenland Ice Sheet Driven by Late-Summer Cyclonic Rainfall." *Nature Geoscience* 8:647–656.

Dudek, D. 2012. "Chasing Ice Pursues Chilling Evidence of Climate Change" (review). *Journal Sentinel*. http://archive.jsonline.com/entertainment/featured/chasing-ice-pursues-chilling-evidence-of-climate-change-5j7v692-184339691.html/.

Dutton, A., Carlson, A. E., Long, A. J., Milne, G. A., Clark, P. U., DeConto, R., Horton, B. P., Rahmstorf, S., and Raymo, M. E. 2015. "Sea Level Rise Due to Polar Ice-Sheet Mass Loss During Past Warm Periods." *Science* 349(6244). doi:10.1126/science.aaa4019.

Dyurgerov, M. B., and Meier, M. F. 2012. "Glacier Mass Changes and Their Effect on the Earth System." In *State of the Earth's Cryosphere at the Beginning of the 21st century—Glaciers, Global Snow Cover, Floating Ice, and Permafrost and Periglacial Environments*, ed. R. S. Williams Jr. and J. G. Ferrigno, A192–246. U.S. Geological Survey Professional Paper 1386–A.

Eberle, J. J., and Greenwood, D. R. 2012. "Life at the Top of the Greenhouse Eocene World—A Review of the Eocene Flora and Vertebrate Fauna from Canada's High Arctic." *GSA Bulletin* 124(1/2):3–23.

Ehlmann, B. L., and Edwards, C. S. 2014. "Mineralogy of the Martian Surface." *Annual Review of Earth and Planetary Sciences* 42:291–315.

Ehrlich, G. 2010. *In the Empire of Ice: Encounters in a Changing Landscape*. Washington, D.C.: National Geographic Society.

Enderlin, E. M., Howat, I. M., Jeong, S., Noh, M.-J., Van Angelen, J. H., and Van den Broeke, M. R. 2014. "An Improved Mass Budget for the Greenland Ice Sheet." *Geophysical Research Letters* 41:866–872. doi:10.1002/2013GL059010.

Evans, J. 2015. "Antarctic Ice Growth and Retreat." *Nature Geoscience* 8:585.

Fairbanks, R. G. 1989. "17,000-Year Glacio-eustatic Sea Level Record: Influence of Glacial Melting Rates on the Younger Dryas Event and Deep-Ocean Circulation." *Nature* 342:637–642.

Farinotti, D. 2017. "Asia's Glacier Changes." *Nature Geoscience* 10:621–622.

Favier, L., Durand, G., Cornford, S. L., Gudmundsson, G. H., Gagliardini, O., Gillet-Chaulet, F., Zwinger, T., Payne, A. J., and Le Brocq, A. M. 2014. "Retreat of Pine Island Glacier Controlled by Marine Ice-Sheet Instability." *Nature Climate Change* 4:117–121.

Feldmann, J. and Levermann, A. 2015. "Collapse of the West Antarctic Ice Sheet After Local Mobilization of the Amundsen Basin." *Proceedings of the National Academy of Sciences* 112(46):14191–14196.

Fettweis, X., and 6 others. 2013. "Estimating the Greenland Ice Sheet Surface Mass Balance Contribution to Future Sea Level Rise Using the Regional Atmospheric Climate Model MAR." *The Cryosphere* 7:369–489.

Fischer, M., Huss, M., Barboux, C., and Hoelzle, M. 2014. "The New Swiss Glacier Inventory SGI2010: Relevance of Using High-Resolution Source Data in Areas Dominated by Very Small Glaciers." *Arctic, Antarctic, and Alpine Research* 46(4): 933–945.

Folger, T. 2017. "Darkness at the Edge of the World." *Smithsonian*, March: 28–39, 82.

Ford, A. B. 2015. "Antarctica." *Encyclopaedia Britannica Online*, http://www.britannica.com/place/Antarctica.

Forsberg, R., Sørensen, L., and Simonsen, S., 2017. "Greenland and Antarctica Ice Sheet Mass Changes and Effects on Global Sea Level." *Surveys in Geophysics* 38:89–104.

Forster, R. R., and 12 others. 2014. "Greenland Liquid Meltwater Storage in Firn Within the Greenland Ice Sheet." *Nature Geoscience* 7:95–98.

Foster, G. L., and Rohling, E. J. 2013. "Relationship Between Sea Level and Climate Forcing by CO_2 on Geologic Timescales." *Proceedings of the National Academy of Sciences* 110:1209–1214.

Fox, D. 2012. "Witness to an Antarctic Meltdown." *Scientific American*, July, 54–61.

Francis, J. A., and Vavrus, S. J. 2012. "Evidence Linking Arctic Extreme Amplification to Extreme Weather in Mid-Latitudes." *Geophysical Research Letters* 39, L06801. doi:10.1029/2012GL051000.

Fretwell, P. 2015. "Entry Beneath Ice." *Nature Geoscience* 8:253–254.

Fretwell, P., and 59 others. 2013. "Bedmap2: Improved Ice Bed, Surface and Thickness Datasets for Antarctica." *The Cryosphere* 7:375–393.

Fürst, J. J., Duand, G., Gillet-Chaulet, F., Tavard, L., Rankl, M., Braun, M., and Gagliardini, O. 2016. "The Safety Band of Antarctic Ice Shelves." *Nature Climate Change* 6:479–482.

Fürst, J. J., Goelzer, H., and Huybrechts, P. 2015. "Ice-Dynamic Projections of the Greenland Ice Sheet in Response to Atmospheric and Oceanic Warming." *The Cryosphere* 9:1039–1062.

Fyfe, J. C., and 11 others. 2017. "Large Near-Term Projected Snowpack Loss over the Western United States." *Nature Communications* 8:14996. doi:10.1038/ncomms14996.

Galeotti, S., and 12 others. 2016. "Antarctic Ice Sheet Variability Across the Eocene-Oligocene Boundary Climate Transition." *Science* 352(6281):76–80.

Gao, X., Schlosser, C. A., Sokolov, A., Anthony, K. W., Zhuang, Q., and Kicklighter, D. 2013. "Permafrost Degradation and Methane: Low Risk of Biogeochemical Climate-Warming Feedback." *Environmental Research Letters* 8(3):035014. doi:10.1088/1748-9326/8/3/035014.

Gardner, A. S., and 15 others. 2013. "A Reconciled Estimate of Glacier Contributions to Sea Level Rise: 2003–2009." *Science* 340:852–857.

Gautier, D. L., et al. 2009. "Assessment of Undiscovered Oil and Gas in the Arctic." *Science* 324:1175–1179.

Gillis, J., and Fountain, H. 2016. "Global Warming Cited as Wildfires Increase in Fragile Boreal Forest." *New York Times*, May 10. https://www.nytimes.com/2016/05/11/science/global-warming-cited-as-wildfires-increase-in-fragile-boreal-forest.html.

Glasby, G. P., and Voytekhovsky, Y. L. 2009. "Arctic Russia: Minerals and Mineral Resources." Geochemical Society. https://www.geochemsoc.org/publications/geochemicalnews/gn140jul09/arcticrussianmineralsandmin/ (accessed June 13, 2017; no longer available online).

Golledge, N. R., Kowalewski, D. E., Naish, T. R., Levy, R. H., Fogwill, C. J., and Gasson, E. G. W. 2015. "The Multi-Millennial Antarctic Commitment to Future Sea-Level Rise." *Nature* 526:421–425.

Gomez, N., Pollard, D., and Holland, D. 2015. "Sea-Level Feedback Lower Projections of Future Antarctic Ice-Sheet Mass Loss." *Nature Communications*. doi:10.1038/ncomms9798/.

Goode, E. 2016. "2016 Climate Change Pushes Towns in Alaska to Wrenching Choice." *New York Times*, November 29, A1, A11–13.

Gornitz, V. 2013. *Rising Seas: Past, Present, Future*. New York: Columbia University Press.

Gornitz, V. 2018. Case Study 9.8, "Coastal Hazard and Action Plans in Miami, Florida, USA"; Chapter 9, "Coastal Zones in Urban Areas" (R. J. Dawson, M. Shah Alam Khan, and V. Gornitz, coordinating lead authors). In *Climate Change and Cities: Second Assessment Report of the Urban Climate Change Research Network*, ed. C. Rosenzweig, W. D. Solecki, P. Romero-Lankao, S. Mehrotra, S. Dhakal, and S. A. Ibrahim. New York: Cambridge University Press.

Grant, K. M., Rohling, E. J., Bar-Matthews, M., Ayalon, A., Medina-Elizalde, M., Ramsey, C. B., Satow, C., and Roberts, A. P. 2012. "Rapid Coupling Between Ice Volume and Polar Temperature Over the Past 150,000 Years." *Nature* 491:744–747.

Gray, B. 2016. "As Greenland Ramps Up Mining, Who Will Benefit?" *Arctic Deeply: News Deeply*, March 17. https://www.newsdeeply.com/arctic/articles/2016/03/17/as-greenland-ramps-up-mining-who-will-benefit/.

Green, M. 2016. "Contested Territory." *Nature Climate Change* 6:817–818.

Greenbaum, J. S., and 10 others. 2015. "Ocean Access to a Cavity Beneath Totten Glacier in East Antarctica." *Nature Geoscience* 8:294–298.

Grinsted, A. 2013. "An Estimate of Global Glacier Volume." *The Cryosphere* 7:121–151.

Grosse, G., Jones, B., and Arp, C. 2013. "Thermokarst Lakes, Drainage, and Drained Basins." In *Treatise on Geomorphology*, vol. 8, *Glacial and Periglacial Geomorphology*, ed. J. F. Shroder, R. Giardino, and J. Harbor, 325–353. San Diego: Academic Press.

Hadley, O. L., and Kirchstetter, T. W. 2012. "Black-Carbon Reduction of Snow Albedo." *Nature Climate Change* 2:437–440.

Haeberli, W. 2011. "Glacier Mass Balance." In *Encyclopedia of Snow, Ice and Glaciers*, ed. V. J. Singh, P. Singh, and U. K. Haritashya, 399–408. Dordrecht: Springer.

Haecker, D. 2016. "Luxury Ship Crystal Serenity to Arrive in Nome on Sunday." *The Nome Nugget*, August 8. http://www.nomenugget.net/news/luxury-cruise-ship-crystal-serenity-arrive-nome-sunday.

Hanebuth, T., Stattegger, K., and Grootes, P. M. 2000. "Rapid flooding of the Sunda Shelf: A Late-Glacial Sea-Level Record." *Science* 288:1033–1035.

Hansen, J., Sato, M., Kharecha, P., Beerling, D., Masson-Delmotte, V., Pagani, M., Raymo, M., Royer, D., and Zachos, J. C. 2008. "Target Atmospheric CO_2: Where Should Humanity Aim?" *Open Atmospheric Science Journal* 2:217–231.

Hansen, J., Sato, M., Russell, G., and Kharecha, P. 2013. "Climate Sensitivity, Sea Level, and Atmospheric CO_2." *Philosophical Transactions of the Royal Society A* 371. doi:10.1098/rsta.2012.0294.

Hanson, S., Nicholls, R., Ranger, N., Hallegate, S., Corfee-Morlot, J., Herweijer, C., and Chateau, J. 2011. "A Global Ranking of Port Cities with High Exposure to Climate Extremes." *Climatic Change* 104:89–111.

Harig, C., and Simons, F. J. 2015. "Accelerated West Antarctic Ice Mass Loss Continues to Outpace East Antarctic Gains." *Earth and Planetary Science Letters* 415:134–141.

Harper, J. 2014. "Greenland's Lurking Aquifer." *Nature Geoscience* 7:86–87.

Harper, J., Humphrey, N., Pfeffer, W. T., Brown, J., and Fettweis, X. 2012. "Greenland Ice-Sheet Contribution to Sea-Level Rise Buffered by Meltwater Storage in Firn." *Nature* 491:240–243.

Harriman, L. 2013. "Where Will the Water Go? Impacts of Accelerated Glacier Melt in the Tropical Andes." UNEP Global Environmental Alert Service (GEAS), September 2013. https://na.unep.net/geas/getUNEPPageWithArticleIDScript.php?article_id=104.

Hauer, M. E. 2017. "Migration Induced by Sea-Level Rise Could Reshape the US Population Landscape." *Nature Climate Change* 7:321–325.

Hauer, M. E., Evans, J. M., and Mishra, D. R. 2016. "Millions Projected to be at Risk from Sea-Level Rise in the Continental United States." *Nature Climate Change* 6:691–695.

Haug, G. H., Tiedemann, R., and Keigwin, L. D. 2004. "How the Isthmus of Panama Put Ice in the Arctic." *Oceanus* 42(2):1–4. http://www.whoi.edu/oceanus/feature/how-the-isthmus-of-panama-put-ice-in-the-arctic.

Hay, C. C., Morrow, E., Kopp, R. E., and Mitrovica, J. X. 2015. "Probabilistic Reanalysis of Twentieth-Century Sea-Level Rise." *Nature* 517:481–484.

Haykin, S., Lewis, E. O., Rayney, R. K., and Rossiter, J. R. 1994. *Remote Sensing of Sea Ice and Icebergs*. New York: Wiley.

Helm, V., Humbert, A., and Miller, H. 2014. "Elevation and Elevation Change of Greenland and Antarctica Derived from CryoSat-2." *The Cryosphere* 8:1539–1559.

Henderson, B. 2009. "Who Discovered the North Pole?" *Smithsonian Magazine*, April. http://www.smithsonianmag.com/history-archaeology/Cook-vs-Peary.html.

Herndon, E. M. 2018. "Permafrost Slowly Exhales Methane." *Nature Climate Change* 8:273–274.

Higgins, J. A., and 8 others. 2015. "Atmospheric Composition 1 Million Years Ago from Blue Ice in the Allan Hills, Antarctica." *Proceedings of the National Academy of Sciences* 112:6887–6891.

Higginson, M. J. 2009. "Geochemical Proxies (Non-Isotopic)." In *Encyclopedia of Paleoclimatology and Ancient Environments*, ed. V. Gornitz, 341–354. Dordrecht: Springer.

Hillenbrand, C.-D., and 10 others. 2013. "Grounding Line Retreat of the West Antarctic Ice Sheet from Inner Pine Island Bay." *Geology* 41(1):35–38.

Hino, M., Field, C. B., and Mach, K. J. 2017. "Managed Retreat as a Response to Natural Hazard Risk." *Nature Climate Change* 7:364–370.

Hogg, A. E., and Gudmundsson, G. H. 2017. "Impacts of the Larsen-C Ice Shelf Calving Event." *Nature Climate Change* 7:540–542.

Holland, D. M., Thomas, R. H., De Young, B., Ribergaard, M. H., and Lyberth, B. 2008. "Acceleration of Jakobshavn Isbrae Triggered by Warm Subsurface Ocean Waters." *Nature Geoscience* 11:659–664.

Holland, P. R., and 8 others. 2015. "Oceanic and Atmospheric Forcing of Larsen C Ice-Shelf Thinning." *The Cryosphere* 9:1005–1024.

Holland, P. R., and Kwok, R. 2012. "Wind-Driven Trends in Antarctic Sea-Ice Drift." *Nature Geoscience* 5:872–875.

Horstmann, B. 2004. "Glacial Lake Outburst Floods in Nepal and Switzerland: New Threats Due to Climate Change." *Germanwatch*. https://germanwatch.org/en/2753.

Howat, I. M., et al. 2013. "Expansion of Meltwater Lakes on the Greenland Ice Sheet." *The Cryosphere* 7:201–204.

Howat, I. M., Porter, C., Noh, M. J., Smith, B. E., and Jeong, S. 2015. "Brief Communication: Sudden Drainage of a Subglacial Lake Beneath the Greenland Ice Sheet." *The Cryosphere* 9:103–108.

Hu, A., Meehl, G. A., Han, W., and Yin, J. 2011. "Effect of the Potential Melting of the Greenland Ice Sheet in the Meridional Overturning Circulation and Global Climate in the Future." *Deep-Sea Research Part II* 58:1914–1926.

Huggel, C., Clague, J. J., and Korup, O. 2012. "Is Climate Change Responsible for Changing Landslide Activity in High Mountains?" *Earth Surface Processes and Landforms* 37:77–91.

Huntford, R. 1997. *Nansen*. London: Little, Brown Book Group.

Huss, M., and Hock, R. 2018. "Global-Scale Hydrological Response to Future Glacier Mass Loss." *Nature Climate Change* 8:135–140.

Immerzeel, W. W., Van Beek, L. P. H., and Bierkens, M. F. P. 2010. "Climate Change Will Affect the Asian Water Towers." *Science* 328:1382–1385.

Ingstad, H., and Ingstad, A. S. 2001. *The Viking Discovery of America: The Excavation of a Norse Settlement in L'Anse aux Meadows, Newfoundland*. New York: Checkmark.

Intergovernmental Panel on Climate Change. 2013a. Chap. 4, "The Cryosphere," Section 4.4.2.3, "Antarctica." *Climate Change 2013: The Physical Science Basis. Contribution of Working Group I to the Fifth Assessment Report of the Intergovernmental Panel on Climate Change.* Ed. T. Stocker, Q. Dahe, G.-K. Plattner, M. M. B. Tignor, S. K. Allen, J. Boschung, A. Nauels, Y. Xia, V. Bex, and P. M. Midgley. Cambridge, UK: Cambridge University Press.

Intergovernmental Panel on Climate Change. 2013b. *Summary for Policymakers. Climatic Change: The Physical Basis. Contribution of Working Group I to the Fifth Assessment Report of the Intergovernmental Panel on Climate Change.* Ed. L. Alexander, S. Allen, L. Bindoff, J. Church, and others. Cambridge, UK: Cambridge University Press.

Jacka, T. H. 2007. "Ice Crystal Size and Orientation." In *Encyclopedia of the Antarctic,* vol. 1, ed. B. Riffenburgh. New York: Routledge, Taylor & Francis Group.

Jacob, T., Wahr, J., Pfeffer, W. T., and Swenson, S. 2012. "Recent Contributions of Glaciers and Ice Caps to Sea Level Rise." *Nature* 482:514–518.

Jamieson, S. S. R., Vieli, A., Livingstone, S. J., Cofaigh, C. O., Stokes, C., Hillenbrand, C. D., and Dowdeswell, J. A. 2012. "Ice-Stream Stability on a Reverse Bed Slope." *Nature Geoscience* 5:799–802. doi:10.1038/NGEO1600.

Jeong, S., Howat, I. M., and Bassis, J. N. 2016. "Accelerated Ice Shelf Rifting and Retreat at Pine Island Glacier, West Antarctica." *Geophysical Research Letters* 43:1–6. doi:10.1002/2016GL071360.

Jevrejeva, S., Matthews, A., and Slangen, A. 2017. "The Twentieth-Century Sea Level Budget: Recent Progress and Challenges." *Surveys in Geophysics* 38:295–307.

Jiang, Y., Dixon, T. H., and Wdowinski, S. 2010. "Accelerating Uplift in the North Atlantic as an Indicator of Ice Loss." *Nature Geoscience* 3:404–407.

Jiskoot, H. 2011a. "Dynamics of Glaciers." In *Encyclopedia of Snow, Ice and Glaciers,* ed. V. J. Singh, P. Singh, and U. K. Haritashya, 245–256. Dordrecht: Springer.

Jiskoot, H. 2011b. "Glacier Surging." In *Encyclopedia of Snow, Ice and Glaciers,* ed. V. J. Singh, P. Singh, and U. K. Haritashya, 415–428. Dordrecht: Springer.

Johnson, K. 2013. "Alaska Looks for Answers in Glacier's Summer Surges." *New York Times,* July 22. https://www.nytimes.com/2013/07/23/us/alaska-looks-for-answers-in-glaciers-summer-flood-surges.html.

Jones, B. M., Arp, C. D., Jorgenson, M. T., Hinkel, K. M., and Schmutz, J. A. 2009. "Increase in the Rate and Uniformity of Coastline Erosion in Arctic Alaska." *Geophysical Research Letters* 36:1–5, L03503. doi:10.1029/2008GL036205, 5p.

Jones, N. 2012. "Arctic Ice Turns to the Dark Side." *Nature Climate Change* 2:479.

Jones, N. 2016. "Abrupt Sea Level Rise Looms as Increasingly Realistic Threat." Yale Envronment 360: Reporting, Analysis, Opinion & Debate, May 5. https://e360.yale.edu/features/abrupt_sea_level_rise_realistic_greenland_antarctica.

Joughin, I., and Alley, R. B. 2011. "Stability of the West Antarctic Ice Sheet in a Warming World." *Nature Geoscience* 4:506–513.

Joughin, I., Alley, R. B., and Holland, D. M. 2012. "Ice Sheet Response to Oceanic Forcing." *Science* 338:1172–1176.

Joughin, I., Smith, B. E., and Medley, B. 2014. "Marine Ice Sheet Collapse Potentially Under Way for the Thwaites Glacier Basin, West Antarctica." *Science*, May 16. http://science.sciencemag.org/content/344/6185/735.

Joughin, I., Smith, B. E., Sheam, D. E., and Floricioiu, D. 2014. "Brief Communication: Further Summer Speedup of Jakobshavn Isbrae." *The Cryosphere* 8:209–214.

Kääb, A., Bethier, E., Nuth, C., Gardelle, J., and Arnaud, Y., 2012. "Contrasting Patterns of Early Twenty-First-Century Glacier Mass Change in the Himalayas." *Nature* 488:495–498.

Kapnick, S. B., Delworth, T. L., Ashfaq, M., Malyshev, S., and Milly, P. C. D. 2014. "Snowfall Less Sensitive to Warming in Karakoram Than in Himalayas Due to a Unique Seasonal Cycle." *Nature Geoscience* 7:834–840.

Karpilo, R. D., Jr. 2009. "Glacier Monitoring Techniques." In *Geological Monitoring*, ed. R. Young and R. Norby, 141–162. Boulder, CO: Geological Society of America.

Keane, J. T. 2017. "Catastrophic Glacier Collapse." *Nature Geoscience* 11:84–87.

Kent, D. V., and Muttoni, G. 2013. "Modulation of Late Cretaceous and Cenozoic Climate by Variable Drawdown of Atmospheric pCO_2 from Weathering of Basaltic Provinces on Continental Drifting Through the Equatorial Humid Belt." *Climate of the Past* 9:525–546.

Kerr, R. A. 2013. "How to Make a Great Ice Age, Again and Again and Again." *Science* 341:599.

Khan, S. A., and 12 others. 2014. "Sustained Mass Loss of the Northeast Greenland Ice Sheet Triggered by Regional Warming." *Nature Climate Change* 4:292–299.

Khazendar, A., Borstad, C. P., Scheuchl, B., Rignot, E., and Seroussi, H. 2015. "The Evolving Instability of the Remnant Larsen B Ice Shelf and Its Tributary Glaciers." *Earth and Planetary Science Letters* 419:199–210.

King, H. 2015. "Hunting Meteorites in Antarctica." Geology.com. http://geology.com/stories/13/antarctica-meteorites/.

Kingslake, J., Ely, J. C., Das, I., and Bell, R. E. 2017. "Widespread Movement of Meltwater onto and Across Antarctic Ice Shelves." *Nature* 544:349–352.

Kjaer, K. H., and 13 others. 2012. "Aerial Photographs Reveal Late 20th Century Dynamic Ice Loss in Northwestern Greenland." *Science* 337:569–573.

Kjeldsen, K. K., and 15 others. 2015. "Spatial and Temporal Distribution of Mass Loss from Greenland Ice Sheet Since AD 1900." *Nature* 528:396–400.

Konrad, H., Shepherd, A., Gilbert, L., Hogg, A. E., McMillan, M., Muir, A., and Slater, T. 2018. "Net Retreat of Antarctic Glacier Grounding Lines." *Nature Geoscience*, 11:258–262. doi:10.1038/s41561-018-0082-z.

Kopp, R. E., Horton, R. E., Little, C. M., Mitrovica, J. X., Oppenheimer, M., Rasmussen, D. J., Strauss, B. H., and Tebaldi, C. 2014. "Probabilistic 21st and 22nd Century Sea-Level Projections at a Global Network of Tide Gauge Sites." *Earth's Future* 2:383–406. doi:10.1002/2014EF000239.

Kopp, R. E., Simons, F. J., Mitrovica, J. X., Maloof, A. C., and Oppenheimer, M. 2009. "Probabilistic Assessment of Sea Level During the Last Interglacial Stage." *Nature* 462:863–868.

Kort, E. A., and 11 others. 2012. "Atmospheric Observations of Arctic Ocean Methane Emissions up to 82° North." *Nature Geoscience* 5:318–321.

Kraaijenbrink, P. D. A., Bierkens, M. F. P., Lutz, A. F., and Immerzeel, W. W. 2017. "Impact of a Global Temperature Rise of 1.5 Degrees Celsius on Asia's Glaciers." *Nature* 549:257–260.

Krajick, K. 2001. *Barren Lands: An Epic Search for Diamonds in the North American Arctic.* New York: Henry Holt and Co.

Kramer, A. E. 2008. "Trade in Mammoth Ivory, Helped by Global Thaw, Flourishes in Russia." *New York Times*, March 25. https://www.nytimes.com/2008/03/25/world/europe/25iht-mammoth.4.11415717.html.

Krinner, G., and Durand, G. 2012. "Future of the Greenland Ice Sheet." *Nature Climate Change* 2:396–397.

Lachenbruch, A. H., and Marshall, B. V. 1986. "Changing Climate: Geothermal Evidence from Permafrost in the Alaskan Arctic." *Science* 234:689–696.

Laghari, J. R. 2013. "Melting Glaciers Bring Energy Uncertainty." *Nature* 502:617–618.

Lantuit, H., and 23 others. 2012. "The Arctic Coastal Dynamics Database: A New Classification Scheme and Statistics on Arctic Permafrost Conditions." *Estuaries and Coasts* 35:383–400.

Larmer, B. 2013. "Of Mammoths and Man." *National Geographic Magazine*, April. https://www.nationalgeographic.com/magazine/2013/04/tracking-mammothsl/.

Larson, N. 2015. "Blankets Cover Swiss Glacier in Vain Effort to Halt Icemelt." https://phys.org/news/2015-09-blankets-swiss-glacier-vain-effort.html (posted September 15; accessed August 9, 2017).

Lear, C., and Lunt, D. J. 2016. "How Antarctica Got Its Ice." *Science*. 352(6281):34–35. 10.1126/science.aad6284.

Leclercq, P. Q., Oerlemans, J., and Cogley, J. G. 2011. "Estimating the Glacier Contribution to Sea-Level Rise for the Period 1800–2005." *Surveys of Geophysics* 32:519–535.

Leeson, A. A., Shepherd A., Briggs, K., Howat, I., Fettweis, X., Morlighem, M., and Rignot, E. 2014. "Supraglacial Lakes on the Greenland Ice Sheet Advance Inland Under Warming Climate." *Nature Climate Change* 5:51–55.

Levy, R., Harwood, D., and SMS Science Team. 2016. "Antarctic Sensitivity to Atmospheric CO_2 Variations in the Early to Mid-Miocene." *Proceedings of the National Academy of Science* 113(13):3453–3458.

Li, X., Rignot, E., Morlighem, M., Mouginot, J., and Scheuchl, B. 2015. "Grounding Line Retreat of Totten Glacier, East Antarctica, 1996 to 2013." *Geophysical Research Letters* 42:8049–8056. doi:10.1002/2015GL065701.

Li, Y.-X., Törnqvist, T. E., Nevitt, J. M., and Kohl, B. 2012. "Synchronizing a Sea-Level Jump, Final Lake Agassiz Drainage, and Abrupt Cooling 8,200 Years Ago." *Earth and Planetary Sciences* 315–316:41–50.

Liu, J., Milne, G. A., Kopp, R. E., Clark, P. U., and Shennan, I. 2016. "Sea-Level Constraints on the Amplitude and Source Distribution of Meltwater Pulse 1A." *Nature Geoscience* 9:130–134.

Livermore, R., et al. 2007. "Drake Passage and Cenozoic Climate: An Open and Shut Case?" *Geochemistry, Geophysics, Geosystems* 8(1):1–11. doi:10.1029/2005GC001224.

Lough, A. C., and 9 others. 2013. "Seismic Detection of an Active Subglacial Magmatic Complex in Marie Byrd Land, Antarctica." *Nature Geoscience* 6:1031–1035.

Lulu, J. 2017. "China, Greenland and Competition for the Arctic." *Asia Dialogue—The Online Magazine of the University of Nottingham Asia Research Institute.* http://cpianalysis.org/2017/01/02/china-greenland-and-competition-for-the-arctic/ (posted January 2; accessed June 22, 2017).

Lüthi, M. P. 2010. "Greenland's Glacial Basics." *Nature* 468:776–777.

Lüthi, M. 2013. "Gauging Greenland's Subglacial Water." *Science* 341:721–722.

Lyle, M., Gibbs, S., Moore, T. C., and Rea, D. K. 2007. "Late Oligocene Initiation of the Antarctic Circumpolar Current: Evidence from the South Pacific." *Geology* 35:691–694.

MacGregor, J. A., Catanica, G. A., Markowski, M. S., and Andrews, A. G. 2012. "Widespread Rifting and Retreat of Ice-Shelf Margins in the Eastern Amundsen Sea Embayment Between 1972 and 2011." *Journal of Glaciology* 58(209):458–466. doi:10.3189/2012JoG11J262.

MacGregor, J. A., Fahnstock, M. A., Catania, G. A., Paden, J. D., Gogineni, S. P., Young, S. K., Rybarski, S. C., Mabrey, A. N., Wagman, B. A., and Morlighem, M. 2015. "Radiostatigraphy and Age Structure of the Greenland Ice Sheet." *Journal of Geophysical Research: Earth Surface* 120:212–241. doi:10.1002/2014JF003215.

Mackintosh, A., and 11 others. 2011. "Retreat of the East Antarctic Ice Sheet During the Last Glacial Termination." *Nature Geoscience* 4:195–202.

Mann, M. E. 2012. *The Hockey Stick and the Climate Wars: Dispatches from the Front Lines*. New York: Columbia University Press.

Marko, J. R., Fissel, D. B., Alvarez, M. M. S., Ross, E., and Kerr, R. 2014. "Iceberg Severity off the East Coast of North America in Relation to Upstream Sea Ice Variability: An Update." Ocean-St. John's Conference, September 14–19. doi:1109/OCEANS.2014.7003128.

Marzeion, B., Champollion, N., Haeberli, W., Langley, K., Leclercq, P., and Paul, F. 2017. "Observation-Based Estimates of Global Glacier Mass Change and Its Contribution to Sea-Level Change." *Surveys in Geophysics* 38:105–130.

Mascarelli, A. L. 2009. "A Sleeping Giant?" *Nature Reports Climate Change* 3:46–49. doi:10.1038/climate.2009.24.

Maslin, M., Owen, M., Betts, R., Day, S., Dunkley Jones, T., and Ridgwell, A. 2010. "Gas Hydrates: Past and Future Geohazard?" *Philosophical Transactions of the Royal Society A*, 368(1919):2369–2393.

Maslin, M. A., and Brierley, C. M. 2015. "The Role of Orbital Forcing in the Early Middle Pleistocene Transition." *Quaternary International* 389:47–55.

Masters, D., et al. 2012. "Comparison of Global Mean Sea Level Time Series from TOPEX/Poseidon, Jason-1, and Jason-2." *Marine Geodesy* 35:20–41.

Masters, J. 2014. "The Jet Stream Is Getting Weird." *Scientific American*, December, 69–75.

Matilsky, B., and Duwadi, J. 2008. "In Search of Iceberg: Tracing the 1859 Expedition of the Painter Frederic Edwin Church to Newfoundland and Labrador." Vanishing Ice: Artists on the Frontline of Global Climate Change. http://www.vanishing-ice.org/archive.is/wJ77h.

Maurer, J. M., Rupper, S. B., and Schaefer, J. M. 2016. "Quantifying Ice Loss in the Eastern Himalayas Since 1974 Using Declassified Spy Satellite Imagery." *The Cryosphere* 10:2203–2215.

McDougall, A. H., Avis, C. A., and Weaver, A. J. 2012. "Significant Contribution to Climate Warming from the Permafrost Carbon Feedback." *Nature Geoscience* 5:719–721.

McGrath, D., Colgan, W., Bayou, N., Muto, A., and Steffen, K. 2013. "Recent Warming at Summit, Greenland: Global Context and Implications." *Geophysical Research Letters* 40(1–6). doi:10.1002/grl.50456.

McGuire, W. 2012. *Waking the Giant: How Changing Climate Triggers Earthquakes, Tsunamis, and Volcanoes*. New York: Oxford University Press.

McMillan, M., Shepherd, A., Sundal, A., Briggs, K., Muir, A., Ridout, A., Hogg, A., and Wingham, D. 2014. "Increased Ice Losses from Antarctica Detected by CryoSat-2." *Geophysical Research Letters* 41:3899–3905. doi:10.1002/2014GL060111.

Meehl, G. A. M., and 9 others. 2012. "Relative Outcomes of Climate Change Mitigation Related to Global Temperature Versus Sea-Level Rise." *Nature Climate Change* 2:576–580.

Mele, C., and Victor, D. 2016. "Alaskan Village Votes to Relocate to the Mainland." *New York Times*, August 20, A14.

Mengel, M., and Levermann, A. 2014. "Ice Plug Prevents Irreversible Discharge from East Antarctica." *Nature Climate Change* 4:451–455.

Mengel, M., Levermann, A., Frieler, K., Robinson, A., Marzeion, B., and Winkelmann, R. 2016. "Future Sea Level Rise Constrained by Observations and Long-Term Commitment." *Proceedings of the National Academy of Sciences* 113(10):2597–2602.

Mercer, J. H. 1978. "West Antarctic Ice Sheet and CO_2 Greenhouse Effect—Threat of Disaster." *Nature* 271:321–325.

Millan, R., Righnot, E., Bernier, V., Morlighem, M., and Dutrieux, P. 2017. "Bathymetry of the Amundsen Sea Embayment Sector of West Antarctica from Operation IceBridge Gravity and Other Data." *Geophysical Research Letters* 44:1360–1368.

Miller, A. W., and Ruiz, G. M. 2014. "Arctic Shipping and Marine Invaders." *Nature Climate Change* 4:413–416.

Miller, K., Wright, J. D., Browning, J. V., and 7 others. 2012. "High Tide of the Warm Pliocene: Implications of Global Sea Level for Antarctic Deglaciation." *Geology* 40(5):407–410.

Miller, K. G., Mountain, G. S., Wright, J. D., and Browning, J. V. 2011. "A 180-Million-Year Record of Sea Level and Ice Volume Variations from Continental Margin and Deep-Sea Isotopic Records." *Oceanography* 24(2):40–53.

Molnia, B. F. 2007. "Late Nineteenth to Early Twenty-First Century Behavior of Alaskan Glaciers as Indicators of Changing Regional Climate." *Global and Planetary Change* 56:23–56.

Molnia, B. F., with contributions by Krimmel, R. M., Trabant, D. C., Marsh, R. S., and Manly, W. F. 2008. "Glaciers of North America—Glaciers of Alaska." In *Satellite Image Atlas of the World*, ed. R. S. Williams and J. G. Ferrigno. United States Geological Survey Professional Paper 1386-K. http://pubs.usgs.gov/pp/p1386k.

Moon, T., Smith, B., and Howat, I. 2012. "21st-Century Evolution of Greenland Outlet Glacier Velocities." *Science* 336:576–578.

Morello, L. 2010. "Another Symbol of the Arctic's Complex Ecosystem Finds Itself on Thin Ice." *New York Times*, August 10. http://www.nytimes.com/cwire/2010/08/10/10climatewire-another-symbol-of-the-arctics-complex-ecosyst-8466.html.

Morlighem, M., Rignot, E., Mouginot, J., Seroussi, H., and Larour, E. 2014. "Deeply Incised Submarine Glacial Valleys Beneath the Greenland Ice Sheet." *Nature Geoscience* 7:418–422.

Morlighem, M., and 31 others. 2017. "BedMachine v3: Complete Bed Topography and Ocean Bathymetry Mapping of Greenland from Multibeam Echo Sounding Combined with Mass Conservation." *Geophysical Research Letters* 44:11051–11061. doi:10.1002/2017/2017GL074954.

Morton, R. A., Miller, T. L., and Moore, L. J. 2004. *National Assessment of Shoreline Change: Part 1, Historic Shoreline Changes and Associated Coastal Land Loss Along the U.S. Gulf of Mexico*. U.S. Geological Survey Open-File Report 2004–1043.

Motyka, R. J., Larsen, C. F., Freymueller, J. T., and Echelmeyer, K. A. 2007. "Post Little Ice Age Rebound in the Glacier Bay Region." In *Proceedings of the Fourth Glacier Bay Science Symposium, Oct. 26–28, 2004*, ed. J. F. Piatt and S. M. Gende, 57–59. United States Geological Survey Scientific Investigations Report 2007–5047.

Mouginot, J., and Rignot, W. 2015. "Ice Motion of the Patagonian Icefields of South America 1984–2014." *Geophysical Research Letters* 42:1441–1449.

Mouginot, J., Rignot, E., and Scheuchl, B. 2014. "Sustained Increase in Ice Discharge from the Amundsen Sea Embayment, West Antarctica, from 1973 to 2013." *Geophysical Research Letters* 41:1576–1584. doi:10.1002/2013GL059069.

Murray, T., and 11 others. 2015. "Reverse Glacier Motion During Iceberg Calving and the Cause of Glacial Earthquakes." *Science* 349(6245):305–308.

Naish, T., and 55 others. 2009. "Obliquity-Paced Pliocene West Antarctic Ice Sheet Oscillations." *Nature* 458:322–328.

National Aeronautics and Space Administration. 2013. "NASA Finds 2012 Sustained Long-Term Climate Warming Trend." http://www.giss.nasa.gov/research/news/20130115/.

National Oceanic and Atmospheric Administration. 2017a. "Greenland Ice Sheet." Arctic Report Card: Update for 2017. http://www.arctic.noaa.gov/Report-Card/Report-Card-2017/ArtMID/.

National Oceanic and Atmospheric Administration. 2017b. "Tides and Currents Map." Tides & Currents. http://www.tidesandcurrents.noaa.gov.

National Research Council (NRC), Polar Research Board. 2012. *Lessons and Legacies of the International Polar Year 2007–2008*. Washington, DC: National Academies Press, 152p.

Navab, V. 2011. "Andes Under Stress: Qoyllur Rit'i and the Retreat of the Glaciers." *American Indian Magazine*, Summer, 35–37.

NEEM. 2013. "Eemian Interglacial Constructed from a Greenland Folded Ice Core." *Nature* 493:489–494.

Nerem, R. S., Chambers, D., Choe, C., and Mitchum, G. T. 2010. "Estimating Mean Sea Level Change from the TOPEX and Jason Altimeter Missions." *Marine Geodesy* 33(1), Supp. 1:435.

Nettles, M., and Ekström, G. 2010. "Glacial Earthquakes in Greenland and Antarctica." *Annual Reviews of Earth & Planetary Sciences* 38:467–491.

Neumann, B., Vafeidis, A. T., Zimmermann, J., and Nicholls, R. J. 2015. "Future Coastal Population Growth and Exposure to Sea-Level Rise and Coastal Flooding—A Global Assessment." *PLoS ONE* 10(3): e0118571,1–34. doi:10.1371.journal.pone.0118571.

Nick, F. M., Vieli, A., Howat, I. M., and Joughin, I. 2009. "Large-Scale Changes in Greenland Outlet Glacier Dynamics Triggered at the Terminus." *Nature Geoscience* 2:110–114.

Nienow, P. 2014. "The Plumbing of Greenland's Ice." *Nature* 514:38–39.

Nordenson, G., Seavitt, C., and Yarinsky, A. 2010. *On the Water/Palisade Bay*. Ostfindern, Germany: Hatje Kantz Verlag.

Nussbaumer, S. U., Zumbühl, H. J., and Steiner, D. 2007. "Fluctuations of the 'Mer de Glace' (Mont Blanc Area, France) AD 1500–2050: An Interdisciplinary Approach Using New Historical Data and Neural Network Simulations, Part I: The History of the Mer de Glace AD 1570–2003 According to Pictorial and Written Documents." *Zeitschrift für Gletscherkunde und Glazialgeologie* 40:3–183. (2005/2006).

O'Dea, A., and 34 others. 2016. "Formation of the Isthmus of Panama." *Science Advances* 2016.2: e1600883.

O'Leary, M., and Christoffersen, P. 2013. "Calving on Tidewater Glaciers Amplified by Submarine Frontal Melting." *The Cryosphere* 7:119–128.

O'Leary, M. J., Hearty, P. J., Thompson, W. G., Raymo, M. E., Mitrovica, J. X., and Webster, J. M. 2013. "Ice Sheet Collapse Following a Prolonged Stable Sea Level During the Last Interglacial." *Nature Geoscience* 6:796–800.

Olefeldt, D., and 10 others. 2016. "Circumpolar Distribution and Carbon Storage of Thermokarst Landscapes." *Nature Communications* 7:13043. doi:10.1038/ncomms13043.

Orton, P. M., Hall, T. M., Talke, S. A., Blumberg, A. F., Georgas, N., and Vinogradov, S. 2016. "A Validated Tropical-Extratropical Flood Hazard Assessment for New York Harbor." *Journal of Geophysical Research: Oceans* 121, 26p. doi:10.1002/2016JC011679.

Osterkamp, T. E., and Burn, C. R. 2002. "Permafrost." In *Encyclopedia of Atmospheric Sciences*, ed. J. R. Holton, J. A. Curry, and J. A. Payle, 1717–1729. Amsterdam: Elsevier Sciences.

Overduin, P. P., and Solomon, S. M., lead authors. 2011. "2.1 Physical State of the Circum-Arctic Coast." In *State of the Arctic Coast 2010—Scientific Review and Outlook*, ed. Forbes, D. L. International Arctic Science Committee, Land-Ocean Interactions in the Coastal

Zone, Arctic Monitoring and Assessment Programme, International Permafrost Association: Helmholtz-Zentrum, Geesthacht, Germany. http://arcticcoasts.org.

Overeem, I., Anderson, R. S., Wobus, C. W., Clow, G. D., Urban, F. E., and Matell, N. 2011. "Sea Ice Loss Enhances Wave Action at the Arctic Coast." *Geophysical Research Letters* 38, L17503, 6p. doi:10.1029/2011GL048681.

Overland, J. E., and Wang, M. 2013. "When Will the Summer Arctic Be Nearly Sea Ice Free?" *Geophysical Research Letters* 40:2097–2101. doi:10.1002/grl.50316.

Overland, J. E., and 8 others. 2016. "Nonlinear Response of Mid-Latitude Weather to the Changing Arctic." *Nature Climate Change* 6:992–999.

Paolo, F. S., Fricker, H. A., and Padman, L. 2015. "Volume Loss from Antarctic Ice Shelves is Accelerating." *Science* 348:327–331.

Pellikka, P., and Rees, W. G. 2010. *Remote Sensing of Glaciers: Techniques for Topographic, Spatial and Thematic Mapping of Glaciers*. Boca Raton, FL: CRC Press.

Pelto, M. 2015a. "Developing Instability of Verdi Ice Shelf, Antarctica: Extensive Rift Formation." Blogosphere, American Geophysical Union. http://blogs.agu.org/fromaglaciersperspective/2015/05/17/developing-instability-of-verdi-ice-shelf-antarctica-extensive-rift-formation/ (posted May 17).

Pelto, M. 2015b. "Greenland Glacier Change Index." Blogosphere, American Geophysical Union. http://blogs.agu.org/fromaglaciersperspective/2015/04/02/greenland-glacier (posted April 2).

Perovich, D., and 7 others. 2016. "Sea Ice." NOAA, Arctic Program, Arctic Report Card: Update for 2016. https://www.arctic.noaa.gov/Report-Card/Report-Card-2016/ArtMID/5022/ArticleID/286/Sea-Ice.

Perovich, D., Richter-Menge, J., Polashenski, C., Elder, B., Arbetter, T., and Brennick, O. 2014. "Sea Ice Mass Balance Observations from the North Pole Environmental Observatory." *Geophysical Research Letters* 41:2019–2025. doi:10.1002/2014GL059356.

Péwé, T. L. 2014. "Permafrost." In *Encyclopaedia Britannica Online Academic Edition*, Encyclopaedia Britannica Inc. http://www.britannica.com/EBchecked/topic/452187/permafrost/.

Pfeffer, W. T., and 19 others. 2014. "The Randolph Glacier Inventory: A Globally Complete Inventory of Glaciers." *Journal of Glaciology* 60(221): 537–551.

Pilkey, O. H., Pilkey-Jarvis, L., and Pilkey, K. C. 2016. *Retreat from a Rising Sea: Hard Choices in an Age of Climate Change*. New York: Columbia University Press, 214p.

Pistone, K., Eisenman, I., and Ramanathan, V. 2014. "Observational Determination of Albedo Decrease by Vanishing Arctic Sea Ice." *Proceedings of the National Academy of Sciences* 111(9):3322–3326.

Pithan, F., and Mauritsen, T. 2014. "Arctic Amplification Dominated by Temperature Feedbacks in Contemporary Climate Models." *Nature Geoscience* 7:181–184.

Poinar, K., Joughin, I., Das, S. B., Behn, M. D., Lenaerts, J. T. M., and Van den Broeke, M. R. 2015. "Limits to Future Expansion of Surface-Melt-Enhanced Ice Flow into the Interior of Western Greenland." *Geophysical Research Letters* 42:1800–1807. doi:10.1002/2015GL063192.

Poinar, K., Joughin, I., Lilien, D., Brucker, L., Kehri, L., and Nowicki, S. 2017. "Drainage of Southeast Greenland Firn Aquifer Water Through Crevasses to the Bed." *Frontiers in Earth Science* 5:5. doi:10.3389/feart.2017.00005.

Pollack, H. 2009, 2010. *A World Without Ice*. New York: Avery, Penguin Group, 290p.

Pollard, D., DeConto, R. M., and Alley, R. B. 2015. "Potential Antarctic Ice Sheet Retreat Driven by Hydrofracturing and Ice Cliff Failure." *Earth and Planetary Science* 412:112–121.

Pongracz, J. D., Paetkau, D., Branigan, M., and Richardson, E. 2017. "Recent Hybridization Between a Polar Bear and Grizzly Bears in the Canadian Arctic." *Arctic* 70(2):151–160.

Pritchard, H. D. 2017. "Asia's Glaciers Are a Regionally Important Buffer Against Drought." *Nature* 545:169–174.

Pritchard, H. D., Ligtenberg, S. R. M., Fricker, H. A., Vaughan, D. G., Van den Broeke, M. R., and Padman, L. 2012. "Antarctic Ice-Sheet Loss Driven by Basal Melting of Ice Shelves." *Nature* 484:502–505.

Qiu, J. 2013. "Chain Reaction Shattered Huge Antarctica Ice Shelf." *Nature*, August 9. https://www.nature.com/news/chain-reaction-shattered-huge-antarctica-ice-shelf-1.13540. doi:10.1038/nature.2013.13540.

Qiu, J. 2017. "Antarctica's Sleeping Ice Giant Could Wake Soon." *Nature*, April 12. https://www.nature.com/news/antarctica-s-sleeping-ice-giant-could-wake-soon-1.21808.

Quinn, P. K. 2007. "Arctic Haze: Current Trends and Knowledge Gaps." *Tellus* 59B:99–114.

Radic, V., Bliss, A., Beedlow, A. C., Hock, R., Miles, E., and Cogley, J. C. 2014. "Regional and Global Projections of Twenty-First Century Glacier Mass Changes in Response to Climate Scenarios from Global Climate Models." *Climatc Dynamics* 42:37–58. doi:10.1007/s00382-013-1719-7.

Rahmstorf, S. 2007. "A Semi-Empirical Approach to Projecting Future Sea-Level Rise." *Science* 315:368–370.

Rahmstorf, S., Foster, G., and Cazenave, A. 2012. "Comparing Climate Projections to Observations up to 2011." *Environmental Research Letters* 7:044035, 5p. doi:10.1088/1748-9326/7/4/044035.

Rahmstorf, S., Perrette, M., and Vermeer, M. 2012. "Testing the Robustness of Semi-Empirical Sea Level Projections." *Climate Dynamics* 39:861–875.

Rasmussen et al. 2008. "Reassessing the First Appearance of Eukaryotes and Cyanobacteria." *Nature* 455:1101–1105.

Raup, B. H., Andreassen, L. A., Bolch, T., and Bevan, S. 2015. "Remote Sensing of Glaciers." In *Remote Sensing of the Cryosphere*, ed. M. Tedesco, 123–156. New York: Wiley-Blackwell.

Raup, B. H., and Kargel, J. S. 2012. "Global Land Ice Measurements from Space (GLIMS)." In "State of the Earth's Cryosphere at the Beginning of the 21st Century—Glaciers, Global Snow Cover, Floating Ice, and Permafrost and Periglacial Environments," ed. R. S. Williams Jr. and J. G. Ferrigno, A247–260. U.S. Geological Survey Professional Paper 1386–A.

Raymo, M. E., and Mitrovica, J. X. 2012. "Collapse of Polar Ice Sheet During the Stage 11 Interglacial." *Nature* 483:453–456.

Raymo, M. E., and Ruddiman, W. F. 1992. "Tectonic Forcing of Late Cenozoic Climate." *Nature* 359:117–122.

Reager, J. T., Gardner, A. S., Famiglietti, J. S., Wiese, D. N., Eiker, A., and Lo, M.-H. 2016. "A Decade of Sea-Level Rise Slowed by Climate-Driven Hydrology." *Science* 351:699–702.

Rebesco, M., and 11 others. 2014. "Boundary Condition of Grounding Lines Prior to Collapse, Larsen-B Ice Shelf, Antarctica." *Science* 345:1354–1358.

Rempel, A. W. 2010. "Frost Heave." *Journal of Glaciology* 56(2000):1122–1128.

Revkin, A. C. 2008. "A Farewell to Ice." *New York Times*, March 18. https://dotearth.blogs.nytimes.com/2008/03/18/a-farewell-to-ice/.

Reyes, A. V., Carlson, A. E., Beard, B. L., Hatfield, R. G., Stoner, J. S., Winsor, K., Welke, B., and Ullman, D. J. 2014. "South Greenland Ice-Sheet Collapse During Marine Isotope Stage 11." *Nature* 510:525–528.

Rietbroek, R., Brunnabend, S.-E., Kusche, J., Schöter, J., and Dahle, C. 2016. "Revisiting the Contemporary Sea-Level Budget on Global and Regional Scales." *Proceedings of the National Academy of Sciences* 113(6):1504–1509.

Rignot, E., Casassa, G., Gogineni, P., Krabill, W., Rivera, A., and Thomas, R. 2004. "Accelerated Ice Discharge from the Antarctic Peninsula Following the Collapse of Larsen B Ice Shelf." *Geophysical Research Letters* 31:1–4, L18401. doi:10.1029/2004GL020697.

Rignot, E., Fenty, I., Xu, Y., Cai, Y., and Kemp, C. 2015. "Undercutting of Marine-Terminating Glaciers in West Greenland." *Geophysical Research Letters* 42: 5909–5917.

Rignot, E., Jacobs, S., Mouginot, J., and Scheuchl, B. 2013. "Ice-Shelf Melting Around Antarctica." *Science* 341:266–270.

Rignot, E., Mouginot, J., Morlighem, M., Seroussi, H., and Scheuchi, B. 2014. "Widespread, Rapid Grounding Line Retreat of Pine Island, Thwaites, Smith, and Kohler Glaciers, West Antarctica, from 1992 to 2011." *Geophysical Research Letters* 41:3502–3509. doi:10.1002/20014GL060140.

Rignot, E., Mouginot, J., and Scheuchl, B. 2011. "Ice Flow of the Antarctic Ice Sheet." *Science* 333:1427–1430.

Rintoul, S. R., Silvano, A., Pena-Molino, B., Van Wijk, E., Rosenberg, M., Greenbaum, J. S., and Blankenship D. D. 2016. "Ocean Heat Drives Rapid Basal Melt of the Totten Ice Shelf." *Science Advances* 2: e1601610, 5p.

Ritz, C., Edwards, T. L., Durand, G., Payne, A. J., Peyaud, V., and Hindmarsh, R. C. A. 2015. "Potential Sea-Level Rise from Antarctic Ice-Sheet Instability Constrained by Observations." *Nature* 528:115–118.

Robel, A. A., Schoof, C., and Tziperman, E. 2016. "Persistence and Variability of Ice Stream Grounding Lines on Retrograde Bed Slopes." *The Cryosphere* 10:1883–1896. doi:105194/tc-2016-18.

Robinson, A., Calov, R., and Ganopolski, A. 2012. "Multistability and Critical Thresholds of the Greenland Ice Sheet." *Nature Climate Change* 2:429–432.

Robinson, K. S. 2017. *New York 2017*. New York: Orbit, Hachette Book Group.

Rogozhina, I., and 9 others. 2016. "Melting at the Base of the Greenland Ice Sheet Explained by Iceland Hotspot History." *Nature Geoscience* 9:366–369.

Rohling, E. J., Grant, K., Bolshaw, M., Roberts, A. P., Siddall, M., Hemleben, C., and Kucera, M. 2009. "Antarctic Temperature and Global Sea Level Closely Coupled Over the Past Five Glacial Cycles." *Nature Geoscience* 2:500–504.

Rohling, E. J., Grant, K., Hemleben, C., Siddall, M., Hoogakker, B. A. A., Bolshaw, M., and Kucera, M. 2008. "High Rates of Sea-Level Rise During the Last Interglacial Period." *Nature Geoscience* 1:38–42.

Romanovsky, V. E., Smith, S. L., and Christiansen, H. H. 2010. "Permafrost Thermal State in the Polar Northern Hemisphere During the International Polar Year 2007–2009: A Synthesis." *Permafrost and Periglacial Processes* 21:106–116.

Rosen, J. 2017. "Arctic 2.0: What Happens After All the Ice Goes?" *Nature* 542:152–154.

Rosenzweig, C., Solecki, W., Romero-Lankao, P., Mehrotra, S., Dhakal, S., and Ali Ibrahim, S., eds. 2018. *Climate Change and Cities: Second Assessment Report of the Urban Climate Change Research Network*. Cambridge, UK: Cambridge University Press.

Rotterdam Climate Initiative-Climate Proof 2010. *Rotterdam Climate Adaptation Program 2010: The Rotterdam Challenge on Water and Climate Adaptation*. http://www.rotterdamclimateinitiative.nl/documents/2015-en-ouder/RCP/English/RCP_ENG_def.pdf.

Rovere, A., Raymo, M. E., Mitrovica, J. X., Hearty, P. J., O'Leary, M. J., and Inglis, J. D. 2014. "The Mid-Pliocene Sea-Level Conundrum: Glacial Isostasy, Eustasy, and Dynamic Topography." *Earth and Planetary Science Letters* 387:27–33.

Rowley, D. B., Forte, A. M., Moucha, R., Mitrovica, J. X., Simmons, N. A., and Grand, S. P. 2013. "Dynamic Topography Change of the Eastern United States Since 3 Million Years Ago." *Science* 340:1560–1563.

Ruppel, C. D., and Kessler, J. D. 2017. "The Interaction of Climate Change and Methane Hydrates." *Reviews of Geophysics* 55:126–168. doi:10.1002/2016RG000534.

Saavedra, F. A., Kampf, S. K., Fassnacht, S. R., and Sibold, J. S. 2018. "Changes in Andes Snow Cover from MODIS Data, 2000–2016." *The Cryosphere* 12:1027–1046.

Sallenger, A. H., Jr., Doran, K. S., and Howd, P. A. 2012. "Hotspot of Accelerated Sea-Level Rise on the Atlantic Coast of North America." *Nature Climate Change* 2(12):884–888.

Salzmann, U., Strother, S., Sangiorgi, F., Bijl, P., Pross, J., Woodward, J., Escutia, C., and Brinkhuis, H. 2016. "Oligocene to Miocene Terrestrial Climate Change and the Demise of Forests on Wilkes Land, East Antarctica." *Geophysical Research Abstracts* 18, EGU2016-2717.

Sasgen, I., Konrad, H., Ivins, E. R., Van den Broeke, M. R., Bamber, J. L., Martinec, Z., and Klemann, V. 2013. "Antarctic Ice-Mass Balance 2003 to 2012: Regional Reanalysis of GRACE Satellite Gravimetry Measurements with Improved Estimate of Glacial-Isostatic Adjustment Based on GPS Uplift Rates." *The Cryosphere* 7:1499–1512.

Scambos, T. A., Bohlander, J. A., Shuman, C. A., and Skvarca, P. 2004. "Glacier Acceleration and Thinning After Ice Shelf Collapse in the Larsen B Embayment, Antarctica." *Geophysical Research Letters* 31:1–4, L18402. doi:10.1029/2004GL020670.

Scambos, T., and 7 others. 2009. "Ice Shelf Disintegration by Plate Bending and Hydro-Fracturing: Satellite Observations and Model Results of the 2008 Wilkins Ice Shelf Break-Ups." *Earth & Planetary Science Letters* 280:51–60.

Schaefer, J., and 8 others. 2016. "Greenland Was Nearly Ice-Free for Extended Periods During the Pleistocene." *Nature* 540:252–255.

Schiermeier, Q. 2013. "Greenland Defied Ancient Warming." *Nature* 493:459–460.

Schiffman, R. 2016. "Why Are Gray Whales Moving to the Ocean Next Door?" *Discover Magazine*, April. http://discovermagazine.com/2016/april/13-why-are-gray-whales-moving-to-the-ocean-next-door (posted February 25; accessed June 28, 2017).

Schmidtko, S., Heywood, K. J., Thompson, A. F., and Aoki, S. 2014. "Multidecadal Warming of Antarctic Waters." *Science* 346:1227–1231.

Schneider, B., and Schneider, R. 2010. "Global Warmth with Little Extra CO_2." *Nature Geoscience* 3:6–7.

Schroeder, D. M., Blankenship, D. D., Young, D. A., and Quartini, E. 2014. "Evidence for Elevated and Spatially Variable Geothermal Flux Beneath the West Antarctic Ice Sheet." *Proceedings of the National Academy of Sciences* 111(25):9070–9072.

Screen, J. A. 2017. "Far-Flung Effects of Arctic Warming." *Nature Geoscience* 10:253–254.

Screen, J. A., and Simmonds, I. 2014. "Amplified Mid-Latitude Planetary Waves Favour Particular Regional Weather Extremes." *Nature Climate Change* 4:704–709.

Schuur, T. 2016. "The Permafrost Prediction." *Scientific American*, December, 57–61.

Schuur, E. A. G., and 16 others. 2015. "Climate Change and the Permafrost Carbon Feedback." *Nature* 520:171–179.

Serreze, M., and Stroeve, J. 2015. "Arctic Sea Ice Trends, Variability, and Implications for Seasonal Ice Forecasting." *Philosophical Transactions of the Royal Society A* 373:20140159.

Shakhova, N., and 10 others. 2014. "Ebullition and Storm-Induced Methane Release from the East Siberian Arctic Shelf." *Nature Geoscience* 7:64–70.

Shakhova, N., Semiletov, I., Salyuk, A., Yusupov, V., Kosmach, D., and Gustafsson, O. 2010. "Extensive Methane Venting to the Atmosphere from Sediments of the East Siberian Arctic Shelf." *Science* 327:1246–1249.

Shepherd, A., and 46 others. 2012. "A Reconciled Estimate of Ice-Sheet Mass Balance." *Science* 338:1183–1189.

Shigley, J. E., Shor, R., Padua, P., Breeding, C. M., Shirey, S. B., and Ashbury, D. 2016. "Mining Diamonds in the Canadian Arctic: The Diavik Mine." *Gems & Gemology* 52(2):104–131.

Shtarkman, Y. M., Kocer, Z. A., Edgar, R., Veerapaneni, R. S., D'Elia, T., Morris, P. F., and Rogers, S. O. 2013. "Subglacial Lake Vostok (Antarctica) Accretion Ice Contains a Diverse Set of Sequences from Aquatic, Marine, and Sediment-Inhabiting Bacteria and Eukarya." *PLoS ONE* 8(7): e67221. doi:10.1371/journal.pone.0067221/.

Slangen, A. B. A., Adloff, F., Jevrejeva, S., LeClercq, P. W., Marzeion, B., Wada, Y., and Winkelmann, R. 2017. "A Review of Recent Updates of Sea-Level Projections at Global and Regional Scales." *Surveys of Geophysics* 38:385–406.

Smith, B. E. 2012. "Repeat Warming in Greenland." *Nature Geoscience* 5:369–370.

Smith, L. C., Sheng, Y., MacDonald, G. M., and Hinzman, L. D. 2005. "Disappearing Arctic Lakes." *Science* 308:1429.

Smith, L. C., and 15 others. 2015. "Efficient Meltwater Drainage Through Supraglacial Streams and Rivers on the Southwest Greenland Ice Sheet." *Proceedings of the National Academy of Sciences* 112(4):1001–1006.

Sobek, S. 2014. "Cold Carbon Storage." *Nature* 511:415–417.

Special Initiative for Rebuilding and Resiliency (SIRR). 2013. "A Stronger, More Resilient New York." PlaNYC, the City of New York. http://www.nyc.gov/html/sirr/html/report/report.shtml/.

Stalenberg, B. 2012. "Innovative Flood Defenses in Highly Urbanized Water Cities." In *Climate Adaptation and Flood Risk in Coastal Cities*, ed. J. Aerts, W. Botzen, M. J. Bowman, P. J. Ward, and P. Dircke, 145–164. London: Earthscan.

Stanford, J. D., Hemingway, R., Rohling, E. J., Challenor, P. G., Medina-Elizalde, M., and Lester, A. J. 2011. "Sea-Level Probability for the Last Deglaciation: A Statistical Analysis of Far-Field Record." *Global and Planetary Change* 79:193–203.

Steig, E. J. 2016. "Cooling in the Antarctic." *Nature* 535:358–359.

Steig, E. J., Ding, Q., Battisti, D. S., and Jenkins, A. 2012. "Tropical Forcing of Circumpolar Deep Water Inflow and Outlet Glacier Thinning in the Amundsen Sea Embayment, West Antarctica." *Annals of Glaciology* 56(60):19–28.

Steig, E. J., and Wolfe, A. P. 2008. "Sprucing Up Greenland." *Science* 320:1595.

Stevens, L. A., Behn, M. D., McGuire, J. J., Das, S. B., Joughin, I., Herring, T., Sheean, D. E., and King, M. A. 2015. "Greenland Supraglacial Lake Drainages Triggered by Hydrologically Induced Basal Slip." *Nature* 522:73–76.

Stibal, M., and 9 others. 2017. "Algae Drive Enhanced Darkening of Bare Ice on the Greenland Ice Sheet." *Geophysical Research Letters* 44:11463–11471.

Stickley, C. E., et al. 2004. "Timing and Nature of Deepening of the Tasmanian Gateway." *Paleoceanography* 19 (PA4027):1–18. doi:10.1029/2004PA001022.

Straneo, F., and Heimbach, P. 2013. "North Atlantic Warming and the Retreat of Greenland's Outlet Glaciers." *Nature* 504:36–43.

Strauss, B. H., Kopp, R. E., Sweet, W. V., and Bittermann, K. 2016. "Unnatural Coastal Floods: Sea Level Rise and the Human Fingerprint on U.S. Floods Since 1950 (1–16)." Climate Central Research Report.

Stroeve, J. C., Markus, T., Boisvert, L., Miller, J., and Barrett, A., 2014. "Changes in Arctic Melt Season and Implications for Sea Ice Loss." *Geophysical Research Letters* 41:1216–1225. doi:10.1002/2013GL058951.

Stroeve, J., Serreze, M. C., Holland, M. M., Kay, J. E., Maslanik, J., and Barrett, A. P. 2012. "The Arctic's Rapidly Shrinking Sea Ice Cover: A Research Synthesis." *Climatic Change* 110:1005–1027.

Struzik, E. 2009. *The Big Thaw: Travels in the Melting North*. Mississauga, ON: Wiley Canada, Ltd.

Struzik, E. 2011. "Arctic Roamers: The Move of Southern Species into Far North." Yale Environment 360, Yale School of Forestry & Environmental Studies. https://e360.yale.edu/features/arctic_roamers_the_move_of_southern_species_into_far_north (posted February 14; accessed June 6, 2017).

Summerhayes, C. P. 2008. "International Collaboration in Antarctica: The International Polar Years, the International Geophysical Year, and the Scientific Committee in Antarctic Research." *Polar Record* 44(231):321–334.

Sundal, A. V., Shepherd, A., Nienow, P., Hanna, E., Palmer, S., and Huybrechts, P. 2011. "Melt-Induced Speed-Up of Greenland Ice Sheet Offset by Efficient Subglacial Drainage." *Nature* 469:521–524.

Sutter, J., Gierz, P., Grosfeld, K., Thoma, M., and Lohmann, G. 2016. "Ocean Temperature Thresholds for Last Interglacial West Antarctic Ice Sheet Collapse." *Geophysical Research Letters* 43:2675–2682.

Sutterley, T. C., Velicogna, I., Rignot, E., Mouginot, J., Flament, T., Van den Broeke, M. R., Van Wessem, J. M., and Reijmer, C. H. 2014. "Mass Loss of the Amundsen Sea Embankment of West Antarctica from Four Independent Techniques." *Geophysical Research Letters* 41:8421–8428.

Swart, N. 2017. "Natural Causes of Arctic Sea-Ice Loss." *Nature Climate Change* 7:239–241.

Sweet, W., Park, J., Marra, J., Zervas, C., and Gill, S. 2014. "Sea Level Rise and Nuisance Flood Frequency Changes Around the United States." NOAA Technical Report NOS CO-OPS 073.

Tagliabue, J. 2013. "As Glaciers Melt, Alpine Mountains Lose Their Glue, Threatening Swiss Village." *New York Times*, May 29. https://www.nytimes.com/2013/05/30/world/europe/in-swiss-alps-glacial-melting-unglues-mountains.html.

Talling, P. J., Clare, M., Urlaub, M., Pope, E., Hunt, J. E., and Watt, S. F. L. 2014. "Large Submarine Landslides on Continental Slopes: Geohazards, Methane Release, and Climate Change." *Oceanography* 27(2):32–45.

Tamparri, L. K., and 11 others. 2010. "McMurdo Dry Valleys, Antarctica—A Mars Phoenix Mission Analog." Lunar and Planetary Institute, *41st Lunar and Planetary Science Conference*, The Woodlands, Texas, March 1–5, 2010. https://www.lpi.usra.edu/meetings/lpsc2010/pdf/2464.pdf.

Tedesco, M., ed. 2015. *Remote Sensing of the Cryosphere*. New York: Wiley-Blackwell.

Tedesco, M., Box, J. E., Van As, D., Van de Wal, R. S. W., and Velicogna, I. 2017. "Greenland Ice Sheet." Arctic Report Card: Update for 2017, NOAA. http://www.arctic.noaa.gov/Report-Card/Report-Card-2017/ArtMID/.

Tedesco, M., Doherty, S., Fettweis, X., Alexander, P., Jeyanranatnam, J., and Stoeve, J. 2016. "The Darkening of Greenland Ice Sheet: Trends, Drivers, and Projections (1981–2100)." *The Cryosphere* 10:477–496.

Theroux, M. 2016. "Polar Bears: Close Encounters of the Furred Kind in Canada." Lonely Planet. https://www.lonelyplanet.com/canada/manitoba/churchill/travel-tips-and-articles/polar-bears-close-encounters-of-the-furred-kind.

Thompson, L., Brecher, H. H., Mosley-Thompson, E., Hardy, D. R., and Mark, B. G. 2009. "Glacier Loss on Kilimanjaro Continues Unabated." *Proceedings of the National Academy of Sciences* 107(47):19770–19775.

Thompson, L., Mosley-Thompson, E., Davis, M. E., and Brecher, H. H. 2011. "Tropical Glaciers, Recorders and Indicators of Climate Change, Are Disappearing Globally." *Annals of Glaciology* 52(59):22–33.

Thornton, B. F., and Crill, P. 2015. "Microbial Lid on Subsea Methane." *Nature Climate Change* 5:723–724.

Treat, C. C., and Frolking, S. 2013. "A Permafrost Carbon Bomb?" *Nature Climate Change* 3:865–866.

Triparti, A. K., Roberts, C. D., and Eagle, R. A. 2009. "Coupling of CO_2 and Ice Sheet Stability over Major Climate Transitions of the Last 20 Million Years." *Science* 326:1394–1397.

Tulaczyk, S., and Hossainzadeh, S. 2011. "Antarctica's Deep Frozen 'Lakes.'" *Science* 331:1524–1525.

United Nations Environment Programme (UNEP). 2007. *Global Outlook for Ice & Snow*. Nairobi: UNEP, 235p.

United Nations Framework Convention on Climate Change (UNFCCC). 2010. "Report of the Conference of the Parties on Its Sixteenth Session Held in Cancun from 29 November to 10 December 2010." http://unfccc.int/resource/docs/2010/cop16/eng/07a01.pdf.

Vaidyanathan, G., and Patterson, B. 2015. "Fires Rapidly Consume More Forests and Peat in the Arctic." *Scientific American*, November 20. https://www.scientificamerican.com/article/fires-rapidly-consume-more-forests-and-peat-in-the-arctic/.

Van den Broeke, M. R., et al. 2016. "On the Recent Contribution of the Greenland Ice Sheet to Sea Level Change." *The Cryosphere* 10:1933–1946.

Van Wyk de Vries, M., Bingham, R. G., and Hein, A. S. 2017. "A New Volcanic Province: An Inventory of Subglacial Volcanoes in West Antarctica." In *Exploration of Subsurface Antarctica: Uncovering Past Changes and Modern Processes*, ed. M. J. Siegert, S. S. R. Jamieson, and D. A. White. London: Geological Society of London.

Veh, G., Korup, O., Roessner, S., and Walz, A. 2018. "Detecting Himalayan Glacial Lake Outburst Floods from Landsat Time Series." *Remote Sensing of the Environment* 207:84–97.

Velicogna, I., Sutterley, T. C., and Van den Broeke, M. R. 2014. "Regional Acceleration in Ice Mass Loss from Greenland and Antarctica Using GRACE Time-Variable Gravity Data." *Journal of Geophysical Research Space Physics* 41(22):8130–8137. doi:10.1002/2014GL061052.

Vermeer, M., and Rahmstorf, S. 2009. "Global Sea Level Linked to Global Temperature." *Proceedings of the National Academy of Sciences* 106(51):21527–21532.

Viñas, M.-J. 2012. "Satellites See Unprecedented Greenland Ice Sheet Surface Melt." NASA's Earth Science News Team, Goddard Space Flight Center, Greenbelt, MD. http://www.nasa.gov/topics/earth/features/greenland-melt.html (last updated July 24, 2012).

Viñas, M.-J. 2015. "Study Shows Global Sea Ice Diminishing, Despite Antarctic Gains." NASA. https://climate.nasa.gov/news/2237/study-shows-global-sea-ice-diminishing-despite-antarctic-gains/ (posted February 11).

Viñas, M.-J. 2017. "Massive Iceberg Breaks Off from Antarctica." NASA. https://climate.nasa.gov/news/2606/massive-iceberg-breaks-off-from-antarctica/ (posted July 12; accessed July 13, 2017).

Vonk, J. E., and 12 others. 2012. "Activation of Old Carbon by Erosion of Coastal and Subsea Permafrost in Arctic Siberia." *Nature* 489:137–140.

Wadhams, P. 2003. "How Does Arctic Sea Ice Form and Decay?" Pacific Marine Environmental Laboratory (PMEL) Arctic Zone, NOAA. https://www.pmel.noaa.gov/arctic-zone/essay_wadhams.html.

Wadhams, P. 2013. "Iceberg: Ice Formation." *Encyclopaedia Britannica Online*. http://www.britannica.com/EBchecked/topic/281212/iceberg/.

Waldman, S. 2017. "Forget That Big Iceberg—A Smaller One in the Arctic Is More Troubling." *Scientific American*, July 31. https://www.scientificamerican.com/article/forget-that-big-iceberg-a-smaller-one-in-the-arctic-is-more-troubling.

Walker, G. 2007. "A World Melting from the Top Down." *Nature* 446:718–721.

Walker, G. 2013. *Antarctica: An Intimate Portrait of a Mysterious Continent*. Boston: Houghton Mifflin Harcourt.

Ward, P. D. 2010. *The Flooded Earth: Our Future in a World Without Ice Caps*. New York: Basic Books.

Warren, C. R. 2014. "Calving Glaciers." In *Encyclopedia of Snow, Ice and Glaciers*, ed. V. J. Singh, P. Singh, and U. K. Haritashya, 105–106. Dordrecht: Springer.

Weber, M. E., Clark, P. U., and 10 others. 2014. "Millennial-Scale Variability in Antarctic Ice-Sheet Discharge During the Last Deglaciation." *Nature* 510:134–138.

Whiteman, G., Hope, C., and Wadhams, P. 2013. "Vast Costs of Arctic Change." *Nature* 499:401–403.

Wientjes, I. G. M., Van de Wal, R. S. W., Reichart, G. J., Sluijs, A., and Oerlemans, J. 2011. "Dust from the Dark Region in the Western Ablation Zone of the Greenland Ice Sheet." *The Cryosphere* 5:589–601.

Williams, R. S., Jr., and Ferrigno, J. G., eds. 2012. "State of the Earth's Cryosphere at the Beginning of the 21st Century: Glaciers, Global Snow Cover, Floating Ice, and Permafrost and Periglacial Environments." U.S. Geological Survey Professional Paper 1386–A, 546 p.

Willis, M. J., et al. 2015. "Recharge of a Subglacial Lake by Surface Meltwater in Northeast Greenland." *Nature* 518:223–228.

Wingham, D. J., Siegert, M. J., Shepherd, A., and Muir, A. S. 2006. "Rapid Discharge Connects Antarctic Subglacial Lakes." *Nature* 440:1033–1036.

Wong, E. 2015. "Chinese Glacier's Retreat Signals Trouble for Asian Water Supply." *New York Times*, December 8. https://www.nytimes.com/2015/12/09/world/asia/chinese-glaciers-retreat-signals-trouble-for-asian-water-supply.html.

Wright, A., and Siegert, M. J. 2012. "A Fourth Inventory of Subglacial Antarctic Lakes." *Antarctic Science* 24(6):659–664.

Yin, J., Griffies, S. M., and Stouffer, R. J. 2010. "Spatial Variability of Sea Level Rise in Twenty-First Century Projections." *Journal of Climate* 23:4585–4608.

Yokoyama, Y., Lambeck, K., De Deckker, P., Johnston, P. and Fifield, L. K. 2000. "Timing of the Last Glacial Maximum from Observed Sea-Level Minima. *Nature*, 406:713–716.

Yu, S.-Y., Berglund, B.E., Sandgren, P., and Lambeck, K. 2007. "Evidence for a Rapid Sea-Level Rise 7,600 Years Ago." *Geology* 35:891–894.

Zemp, M., Paul, F., Hoelzle, M., and Haeberli, W. 2008. "Glacier Fluctuations in the European Alps, 1850–2000." In *Darkening Peaks: Glacier Retreat, Science, and Society*, ed. B. Orlove, E. Wiegandt, and B. H. Luckman, 152–167. Berkeley: University of California Press.

Zemp, M., and 38 others. 2015. "Historically Unprecedented Global Glacier Decline in the Early 21st Century." *Journal of Glaciology* 61(228). doi:10.3189/2015JoG15J017.

Zwally, H. J., Li, J., Robbins, J. W., Saba, J. L., Yi, D., and Brenner, A. C. 2015. "Mass Gains of the Antarctic Ice Sheet Exceed Losses." *Journal of Glaciology* 61(230):1019–1032.

INDEX

Page numbers in *italics* represent figures or tables.